Techniques of Classical Mechanics

From Lagrangian to Newtonian mechanics

Techniques of Classical Mechanics

From Lagrangian to Newtonian mechanics

Samya Zain
Department of Physics, Susquehanna University, USA

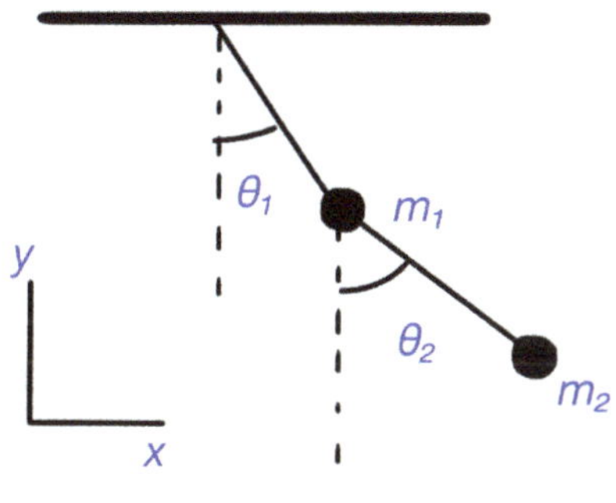

IOP Publishing, Bristol, UK

ISBN 978-0-7503-2076-4 (ebook)
ISBN 978-0-7503-2074-0 (print)
ISBN 978-0-7503-2075-7 (mobi)

DOI 10.1088/2053-2563/aae1b9

Version: 20190101

IOP Expanding Physics
ISSN 2053-2563 (online)
ISSN 2054-7315 (print)

British Library Cataloguing-in-Publication Data: A catalogue record for this book is available from thc British Library.

Published by IOP Publishing, wholly owned by The Institute of Physics, London

IOP Publishing, Temple Circus, Temple Way, Bristol, BS1 6HG, UK

US Office: IOP Publishing, Inc., 190 North Independence Mall West, Suite 601, Philadelphia, PA 19106, USA

To my husband, who is my rock, and
to my family, who have been putting up with me since either my birth or theirs!

Contents

Appendices

Preface

Techniques of Classical Mechanics: From Newtonian to Lagrangian Mechanics is, as the name implies, intended to be a gentle transition from introductory freshman physics to more advanced topics in mechanics. This book is the outcome of an undergraduate physics course given at Susquehanna University to Junior and Senior physics and math majors. There is sufficient material for a one-semester four-credit hour long course. The underlying philosophy of this book is that physics is a cohesive set of ideas, knowledge and experience that needs to be understood in its entirety and should not be compartmentalized.

Mechanics is one of the earliest physics topics to be developed and historically forms the basis of many fields in physics and engineering. So a deep understanding and appreciation of the field is essential for any serious scholar in the field. I will introduce Lagrangian and Hamiltonian mechanics quite early in the text and then address physics problems utilizing as many techniques as I can. My hope is that this text will be the bridge that helps students transition from introductory physics to graduate level physics.

I recognize that mathematics is fundamental and indispensable to the study of physics. A fair amount of familiarity with calculus up to multivariate calculus and deferential equations is assumed. Generally, I will assume that students will recognize Newton's force law as a second order differential equation and will be able to recognize its solutions as sinusoidal and exponential functions. However, where possible I will avoid deriving proofs in favor of problem solving methods. For details of mathematical proofs of concepts please check other sources.

Main ideas for understanding

In this text I have put physics concepts before the historic development of physics topics. Where necessary I have prioritized concepts according to their fundamental significance rather than historic development.

When I started writing this book I pictured sitting and talking with a student. I have tried to keep it as conversational and down-to-earth as possible. Students should be able to pick it up and learn on their own just as easily as they would in a classroom environment. Worked examples are included as part of the text in nearly every section. Additional concept questions and exercises are included for further practice and understanding.

Wherever possible I will introduce physics concepts by showing various methods that could be used to solve them. For example, I will introduce a physics problem then show how to solve it using many different techniques, say by solving it by energy conservation, Newtonian mechanics, Lagrangian mechanics, Hamiltonian mechanics, etc. I hope that by doing this students/readers will develop a feel for the subject and when required they will have in their basket of knowledge the correct tools to approach the various kinds of physics problems.

The main ideas that we will tackle in the course of this textbook include the following topics:

Big idea 1: Kinematics
Big idea 2: Conservation laws
Big idea 3: Simple harmonic motion
Big idea 4: Gravitation
Big idea 5: Circular motion
Big idea 6: Fluids
Big idea 7: Resistance due to fluids
Big idea 8: Solids
Big idea 9: Motion of rigid bodies
Big idea 10: System of particles
Big ideal 11: Scattering theory

Mainly for instructors

I realize that you may see some concepts repeated in multiple chapters. This repetition is intentional. I feel that reviews and re-reviews of these important concepts can never hurt. Some questions at the end may go just beyond the material in the chapter. Such content is intended for students to apply the concepts that they have learnt during the course of the chapter and the book overall.

Mainly for students

Before we begin, I just have a few additional things to comment on. Remember, physics must *not* be memorized but rather understood. Know that understanding does not come quickly or easily, it takes patience and resilience. Remember how you learnt to play an instrument? Did you become an expert overnight? Or did it take hours and hours of practice? Physics is just the same, you need to actively learn it, cramming overnight before the exam is not be the ticket to understanding and hence success.

Easy how-tos to succeed in physics:

1. Manage your time.
2. Commit yourself whole-heartedly.
3. Find a calendar-method that works for you. Get organized.
4. Find a study partner; motivate your partner and yourself.
5. Actively seek help when needed.
6. Set a scheduled dedicated time to study physics every day. Thirty minutes a day, every day will get you more than pulling-an-all-nighter the day of the quiz.
7. Prepare before coming to class, be a good listener and take good notes.
8. And the most important of all—remember, someone very wise once said: 'None of the tricks of success will work unless you do'.

Half your battle is—just show up. So read the book, do the examples and exercises, come to class, ask questions, share your knowledge, communicate and most important of all, work hard and enjoy the ride.

Samya Zain
30 August 2018

Acknowledgments

I am indebted to many people who have been invaluable during this whole process. First and foremost is Dr Grosse, who sat through the course and when I first mentioned the idea of writing this book, encouraged me to put words to paper, and Andrew Steely who was responsible for the first read, poor guy!

Special thanks must be extended to my reviewer and editor, Professor Ken Brakke, Department of Mathematics, Susquehanna University. Without your helpful comments, suggestions and tons and tons of corrections this work would not be possible. Thank you so much, I really appreciate it.

I also have to acknowledge Dr M L Klotz, Department of Psychology, for always being encouraging and patient with me, and Robert Everly, for not making 'too' much fun of my crazy ideas. And to the Susquehanna students, Anton Soloviev, Hannah Cooper, Andrew Steely, Rakee Cromwell and Ben Horine for transcribing my notes, and to Hareem Zain for drawing pictures and letting me use his artwork as a part of this endeavor.

My heartfelt gratitude also extends to my family, my dad Dr Iqbal and mom Safia Bano, and my brother Dr Aamir Iqbal, who have always believed in me and have encouraged me to be my best self. To my very patient husband Zain, for putting up with the long hours of me typing away and for 'understanding' without once attempting to actually read my manuscripts! My kids, Hareem, Sami and Eman, for being (nearly always!) willing to postpone the Hershey park trip and not complaining about lots and lots of pizza nights.

Author biography

Samya Zain

Samya Zain grew up in Lahore, Pakistan. In seventh grade she decided that she wanted to learn more physics, since physics explains how the universe works, and additionally allowed her not to go into the medical field like the rest of her family. She completed her undergraduate work at the University of the Punjab, Lahore, Pakistan and was awarded the gold medal for the highest score in the annual final exam. Then with her husband and four-year-old son she went on to graduate work in physics at the university at Albany, State University of New York in 2001. She received her PhD in experimental particle physics in 2006 and continued as a post-doctoral fellow at the Albany High Energy Physics group. In the fall of 2008, she joined Susquehanna University as an assistant professor, where she currently serves as an associate professor and chair of the department of physics.

When not working on physics she can often be found with her nose in a book. She makes time to read every day (even when she has to hide from her family in order to do so). She paints when the mood strikes her, loves visiting used bookstores and collecting books and absolutely refuses to let any of her collection go.

Chapter 1

Foundations

1.1 The nature of science

Science is the observation and precise description of natural phenomena. Math is the language of all sciences, for example, physics, chemistry, biology, geology, astronomy, engineering, etc

Physics serves as a foundation for other sciences. Physics is fundamentally the study of matter and waves, from the smallest particles, such as quarks and electrons, to the largest bodies, such as galaxies. The difference in scale between the smallest and largest objects is of the order of 10^{40}!

Physicists use the language of mathematics to construct physical theories to describe the world around us. The description might not be perfect, but it is remarkable how well we still do with our descriptions. So in general we can think of physics as a *toolbox of ideas* which can be used in building scientific models to describe the world around us.

Before we begin a formal discussion let us define a few important concepts that we will use:

- *Coordinate system.* A coordinate system is a way of using sets of numbers to identify points in space. It is an imaginary grid of lines that fills all space. A coordinate system is used to identify the location of a particular particle or an event.
- *Origin.* An origin is an agreed upon reference point in a particular coordinate system. All measurements are made with respect to the origin in the coordinate system.
- *Systems of coordinates.* The most common coordinates are *rectangular*, *spherical* and *cylindrical* coordinates, the details of each are given in section 1.7.
- *Event.* If an action occurs at a particular point of space at a particular time then in physics we call it an *event*. For example, the first instantaneous contact between colliding cars in an accident is called an event.

  However, for the complete description of the entire collision we need to define many points that occur over multiple units of time. We call this complicated process a *collection* of events. To simplify our systems we often consider mass no matter how small or large to occupy one point in space.

- *Ideal particle.* In classical physics an ideal particle is an approximation and is regarded as a point-like object. In most cases we will regard complicated objects, such as a driver and car where the driver may be drinking coffee, as a single object with a particular mass (m). Classical physics often simplifies systems by considering objects to occupy a single point of space in our coordinate system.
- *Complete description.* In classical physics we need to know the position of a particle at all times. Hence, for a *complete description* of the particle state, we need to know the mass of the object and the position of it in space at any given time. This means that measurements of position, time and mass are fundamentally important to physics and understanding physical systems.
- *Three dimensions of space.* Since we live in a three-dimensional space, it is obvious that we will require three coordinates to locate an object relative to a user-defined origin. In general we use Cartesian coordinates (x, y, z), but we also use spherical and cylindrical coordinates. The specifics and details of each coordinate system are given in section 1.7. In order to measure an event we have to do the following:
    1. Set an origin, called a user-defined origin.
    2. Define a coordinate system with respect to the user-defined origin.
    3. Measure the coordinates of the event with respect to the user-defined origin and coordinate system.

  This is called a *relative measurement* since it is based on the definitions of the origin and coordinate system. If we change the origin or the orientation of the coordinate system it will appear that the event has changed, which cannot happen. To overcome this issue we have to define different coordinate systems called *frames of reference*, which can be transformed to each other such that the event under consideration remains physically constant and unchanged. Measurement of position is always relative and only becomes meaningful once a coordinate system is defined.
- *Reference frames.* A coordinate system and a synchronized time together make up a reference frame. Both of them determine the location of a particular event. When the coordinates of the particle are constant over time, the reference frame is called the *rest frame* of the particle. For example, we are on planet Earth, hence Earth is a rest frame for us and we call it *Earth's rest frame* (abbreviated as ERF). In Earth's rest frame we can ignore the motion of the Earth with respect to the Sun and only concentrate on the motion of the car driving down a highway at 55 miles/hour.

1.2 Units

At the heart of it all, physics is an experimental science and experiments involve measurements. Measurements in physics involve many varied and sometimes complex quantities such as force, energy, momentum, magnetic field, capacitance, etc.

The surprising thing is that all these complex qualities can be expressed using a few fundamental quantities, for example, the measurement of time (T), length (L) and mass (M). Historically, the scientific community decided that these units, e.g. seconds, meters and kilograms, are convenient to use in physical experiments and decided that these will be called *fundamental units*, or SI units, which is abbreviated from the French, Le Système International d'Unités (the International System of Units). However, everyone does not agree on the standard units and the history of these debates is quite interesting.

We should always be able to convert from one system of units to another given the *dictionary* of units.

1.2.1 Different systems of units

There are three systems of units currently in use around the world. These are:

- The *MKS system*, also called the metric system, which uses meter (m), kilogram (kg) and seconds (s) as the base units for measurements. This system is the most commonly used around the world.
- The *CGS system*, also called the Gaussian system, which uses centimeter (cm), grams (gm) and seconds (s) as base units for measurements.
- The *US-engineering system*, also called the British system, is most commonly used in the US. In this system the unit of length is a foot and the unit of time is a second. This system does not have a base unit for mass, instead it uses the unit of the pound-mass, symbolized as (lb-mass).

Units of...	MKS	CGS	US-engineering
Length	meter (m)	centimeter (cm)	foot (ft)
Mass	kilogram (kg)	gram (gm)	pound-mass (lb-mass)
Time	second (s)	second (s)	second (s)
Temperature	kelvin (K)	kelvin (K)	Fahrenheit (°F)

1.3 International system of units (SI)

The International System of Units (SI) specifies a set of base units and derived units. The names of base SI units are written in lowercase, for example, meter (m), kilogram (kg) and time (s).

Derived units are associated with derived quantities such as the newton, which is a unit of weight or force ($\mathrm{kg\,m\,s^{-2}}$), and watt, which is a unit of power ($\mathrm{kg\,m^2\,s^{-3}}$). Interesting fact: the symbols for units named after scientists are often capitalized.

1.3.1 Base SI units

In this section let us discuss the base units of time, mass and length.

1.3.1.1 Time: the second

What is time and what is the arrow of time? This is a deep philosophical question, and that is a discussion for a later time. At this point all we have to remember is that time is measured in SI units: seconds. The dictionary of conversion units of time includes:

1 millisecond	= 1 ms	= 10^{-3} s
1 microsecond	= 1 μs	= 10^{-6} s
1 nanosecond	= 1 ns	= 10^{-9} s

1.3.1.2 Mass: the kilogram

Mass is something we seem to be very familiar with in everyday life, however, the concept of mass has a very deep relationship to both length and time.

Mass is the only base unit that is defined by an artifact, a platinum–iridium cylinder stored at the International Bureau of Weights and Measures, in France. The mass of this cylinder is defined to be exactly one *kilogram*. It is expected that in November 2018 the kilogram will be defined in terms of the quantum mechanical Planck's constant (h), and will be given an exact value based on measurements from around the world. Please note the physical kilogram mass will not be eliminated as they are the most practical method of everyday measurements. But the difference will be that the SI mass unit will be based on a fundamental constant of nature rather than an artifact.

Conversion of the kilogram is given in the table below.

1 kilogram	= 1 kg	= 10^{3} g
1 gram	= 1 gm	= 10^{-3} kg
1 milligram	= 1 mg	= 10^{-6} kg
1 microgram	= 1 μg	= 10^{-9} kg
1 nanogram	= 1 ng	= 10^{-12} kg

1.3.1.3 Length: the meter

Despite the fact that length and time appear independent to us, there is a deep connection between them. We will discuss this briefly when we discuss gravitation, but you will understand it in detail when you study special relativity in an undergraduate course in topics of modern physics. For now the dictionary of conversion of units of length includes:

1 kilometer	= 1 km	= 10^{3} m
1 centimeter	= 1 cm	= 10^{-2} m
1 millimeter	= 1 mm	= 10^{-3} m
1 micrometer	= 1 μm	= 10^{-6} m
1 nanometer	= 1 nm	= 10^{-9} m

Example 1.1

The given speed on a highway is 55 miles per hour. Convert this speed to meters per second. (*Hint*: 1 mile = 1.6 km.)

Solution

$$55\,\frac{\text{miles}}{\text{hour}} = \frac{55}{1}\,\frac{\text{miles}}{\text{hour}} \times \frac{1609}{1}\,\frac{\text{m}}{\text{miles}} \times \frac{1}{3600}\,\frac{\text{hours}}{\text{second}}$$
$$= 24.58\frac{\text{m}}{\text{s}}.$$

1.3.2 Derived units

The International System of Units (SI) specifies a set of seven base units. From these seven base units other SI units of measurement are derived. We call these *derived units*. Each of the derived units can be expressed as a product of one or more base units. For example, the SI derived unit of area is the square meter (m^2) and of density is kilogram per meter cubed ($kg\,m^{-3}$).

Some important facts about derived quantities include:

- The dimension of any derived physical quantity is some product of the base quantities, for example, density is kilogram per meter cubed.
- When the values of the base quantities change, the derived quantities will also change by the same factor, for example, when mass is doubled, the density of the object also doubles.
- When two or more derived quantities of the same units are added, their sum will also have the same units, for example sum of two speeds, given in ($m\,s^{-1}$) will also have units of ($m\,s^{-1}$).
- Products and ratios of derived quantities usually have units that are not the same as the original quantities. For example, force (N) per unit area (m^2) is N m^{-2}, which is a pascal (Pa), a unit of pressure.

Example 1.2

Density is a derived unit given in $kg\,m^{-3}$ in the MKS system. The density of water is 997 $kg\,m^{-3}$, convert this to:

(a) CGS units.
(b) US-engineering or British units.

Solution

(a) In the CGS unit the base units are grams and centimeters, and 1 kg = 1000 g and 1 m = 100 cm. Hence:

$$997\,\frac{\text{kg}}{\text{m}^3} = \frac{997}{1}\,\frac{\text{kg}}{\text{m}^3} \times \frac{1000}{1}\,\frac{\text{g}}{\text{kg}} \times \frac{1}{1000^3}\,\frac{\text{m}^3}{\text{cm}^3}$$
$$= 0.997\,\frac{\text{g}}{\text{cm}^3}.$$

(b) Similarly 0.454 kg = 1 lb-mass and 1 foot = 0.305 m. Hence:

$$997\,\frac{\text{kg}}{\text{m}^3} = \frac{997}{1}\,\frac{\text{kg}}{\text{m}^3} \times \frac{1}{0.454}\,\frac{\text{lb} - \text{mass}}{\text{kg}} \times \frac{0.305^3}{1}\,\frac{\text{m}^3}{\text{ft}^3}$$
$$= 62\frac{\text{lb} - \text{mass}}{\text{ft}^3}.$$

Concept questions 1.1

1. Suppose a really good thief stole the standard kilogram from the town of Sèvres, France, where it is kept. Would this be the end of the metric system?
2. Name some of the ways that ancient people kept time, from the Egyptians to the Romans to the Arabs and the Chinese.
3. Could you take length, mass and density as your three fundamental units of measurement? If so, what are the advantages or disadvantages to this system?
4. Could you think of setting up your own system of fundamental units? Which units do you think you can use if you are not allowed to use the standard length, time and mass?

Exercises 1.1

1. Calculate your height in each of the three systems of units, the MKS, CGS and British units.
2. What is the difference between a nautical and a regular mile? In which cases are each used and does it change the distance you have to travel if you use one over the other?
3. The accepted age of planet Earth is 4.5×10^9 yr. Convert the age of Earth to months, hours and seconds.
4. Most of our body is made up of water. If we assume that the density of a human body is the same as density of water, can you approximate your volume?

1.4 Dimensional analysis

Dimensional analysis, also called the factor-label method or the unit factor method, is a very useful problem-solving technique.

The concept of physical dimension was introduced by the French mathematician Joseph Fourier in 1822 CE. The actual origins of dimensional analysis have been disputed by historians; but most agree that the main credit still goes to Fourier. Dimensional analysis is based on the idea that physical laws should be independent of the units used to measure the physical variables. For example $\vec{F} = m\vec{a}$ is independent of the units employed to measure m and $\vec{a}$. This led to the conclusion that meaningful laws must be independent of the units of measurement. At the most basic level, a collection of basic physical dimension includes dimensions for mass = $[M]$, length = $[L]$ and time = $[T]$.

- *Position* (x) has the dimension [L] (length).
- *Velocity*, which is the derivative of position with respect to time (velocity = dx/dt = m s^{-1}), has dimension [L][T]$^{-1}$ where [L] represents length (dx) with SI unit (m), [T] represents time (dt) with SI unit (s).
- *Acceleration*, the second derivative of position (acceleration = $\frac{dv}{dt} = \frac{d^2x}{dt^2}$), has dimension [$L$][$T$]$^{-2}$.
- *Force* ($F = ma = m\frac{d^2x}{dt^2}$) has dimension [$M$][$L$][$T$]$^{-2}$.

Dimensional analysis is used to understand and describe relationships between the physical quantities involved in a particular phenomenon without being bogged down with the various system of units. Dimensional analysis was first used by Lord Rayleigh in 1872 CE to illustrate why the sky is blue. Rayleigh first published the technique in his 1877 book *The Theory of Sound.*

Example 1.3. Centripetal force

How to set up an equation for force on a mass attached to a string revolving in a circle? What does it depend on? Ignore drag forces in this case.

Solution

The force (F_c) on a mass attached to a string rotating in a circle is called a centripetal force[1]. To find the equation of force, let us first discuss the quantities that a rotating mass must depend on. We know from experience a heavier mass will require more force to keep it moving in a circle, similarly if the movement is faster then more force will be needed. Lastly, we know from observation that the length of the string will also be a factor. This means F_c will depend on the following quantities, attached mass (m), speed of rotation (v) and length of string (r), as

$$F_c \propto [m]^{\alpha}[v]^{\beta}[r]^{\gamma}. \tag{1.1}$$

The left-hand side (lhs) of the above equation (1.1) is force. The unit of force is a newton which is given as

$$\text{newton} = m\vec{a} = m\frac{d\vec{v}}{dt} = \frac{d^2\vec{x}}{dt^2} = [M][L][T]^{-2},$$

and the right-hand side (rhs) of equation (1.1) will be given as

$$[M]^{\alpha}\left[\frac{L}{T}\right]^{\beta}[L]^{\gamma} = [M]^{\alpha}[L]^{\beta-\gamma}[T]^{-\beta}.$$

Equating the above equations we can write

[1] Centripetal force is a force that acts on a body moving in a circular path and is directed toward the center around which the body is moving.

$$[M]^1[L]^1[T]^{-2} = [M]^{\alpha}[L]^{\beta+\gamma}[T]^{-\beta}.$$

Equating powers,

$$\begin{aligned}&\text{mass } [M]\text{: } 1 = \alpha\\ &\text{length } [L]\text{: } 1 = \beta + \gamma\\ &\text{time } [T]\text{: } -2 = -\beta.\end{aligned}$$

Solving the above expressions ($\alpha = 1; \beta = 2; \gamma = -1$) and putting values in equation (1.1), we obtain

$$\begin{aligned}F_c &= [m]^1[v]^2[r]^{-1}\\ F_c &= \frac{mv^2}{r}.\end{aligned} \tag{1.2}$$

Example 1.4. Period of a mass attached to a spring

Use dimensional analysis to find an equation for a period[2] (T) for a mass (m) attached to an idealized massless spring. Assume drag plays no role in this case.

Solution

Let us first find all the parameters that could play a role. We have a mass (m) attached to a spring, we will see in chapter 7 that this system will undergo a simple harmonic motion, and under this motion the factors that are important will be:

- Mass (m).
- Restoring constant (k)—given dimensions of restoring force constant are $[M][T]^{-2}$.

These dependences can be expressed mathematically as

$$\begin{aligned}T &\propto m^{\alpha}k^{\beta}\\ T &\propto [M]^{\alpha}[M]^{\beta}[T]^{-2\beta}.\end{aligned} \tag{1.3}$$

Equating powers,

$$\begin{aligned}&\text{mass } [M]\text{: } 0 = \alpha + \beta\\ &\text{time } [T]\text{: } 1 = -2\beta.\end{aligned}$$

Solving the above expressions ($\alpha = \frac{1}{2}; \beta = -\frac{1}{2}$) and putting values in equation (1.3) we obtain

$$\begin{aligned}T &\propto m^{\frac{1}{2}}k^{-\frac{1}{2}}\\ T &\propto \sqrt{\frac{m}{k}}.\end{aligned}$$

[2] The wave period is the time it takes to complete one cycle. The standard unit of a wave period is in seconds.

Note that in this case dimensional analysis does not give us the constant of proportionality but it does tell us how the period (T) depends on the mass and spring constant. Given a particular value of mass we can easily find the constant of proportionality by measuring the period.

1.4.1 Planck length (l_p)

Planck length is a unit named after Max Planck, who was the first to propose it. Planck length is a unit of length in Planck units and is represented by the formula and given numerically as

$$l_p \equiv \sqrt{\frac{\hbar G}{c^3}} = 1.616 \times 10^{-35}\,\text{m}, \tag{1.4}$$

where $\hbar$ is the reduced Planck constant, mathematically given as $\hbar = \frac{h}{2\pi}$, with a numerical value of (6.63×10^{-34}kg m^2 s^{-1}); G is the universal gravitational constant, with a numerical value of (6.67×10^{-11} m^3 kg^{-1} s^{-2}), as discussed in detail in chapter 8, and c is the speed of light (3×10^8 m s^{-1}).

Note that $l_p \approx 10^{-20}$ times smaller than the diameter of a proton, which is unbelievably small. Planck length is very important for theoretical calculations, however, it is not very important for us. It is just the shortest possible length that can be measured.

- In some forms of quantum gravity, Planck length is the length scale at which the structure of space–time becomes dominated by quantum effects and hence it becomes impossible to find the difference between two locations one Planck length apart.
- Precise effects of quantum gravity are unknown; however, we think that space–time might have a discrete structure at these lengths. What does that even mean? We really do not know.

Example 1.5. Planck length

The Planck length depends on the following fundamental constants:

- G, the universal gravitational constant, with units m^3 kg^{-1} s^{-2}.
- c, the speed of light, with units m s^{-1}.
- $\hbar$, the reduced Planck constant, with units kg m^2 s^{-1}.

We can therefore write the Plank length in terms of these dependences as

$$l_p \propto c^\alpha G^\beta \hbar^\gamma, \tag{1.5}$$

whose dimensional relationship must be

$$[L]^1 \propto [LT^{-1}]^\alpha [L^3T^{-2}M^{-1}]^\beta [ML^2T^{-1}]^\gamma$$
$$[L]^1[T]^0[M]^0 \propto [L]^{\alpha+3\beta+2\gamma}[T]^{-\alpha-2\beta-\gamma}[M]^{-\beta+\gamma}.$$

Equating powers,

$$\text{length } [L]:\ 1 = \alpha + 3\beta + 2\gamma$$
$$\text{time } [T]:\ 0 = -\alpha - 2\beta - \gamma$$
$$\text{mass } [M]:\ 0 = -\beta + \gamma.$$

Solving the above expressions ($\alpha = -\frac{3}{2}; \beta = \frac{1}{2}; \gamma = \frac{1}{2}$) and putting the values in equation (1.5) we obtain

$$l_p \propto c^{-\frac{3}{2}} G^{\frac{1}{2}} \hbar^{\frac{1}{2}}$$
$$l_p \propto \sqrt{\frac{G\hbar}{c^3}}.$$

Use the values to find the numerical value of Planck length:

$$l_p \approx 10^{-35}\text{m}.$$

1.4.2 Planck time (t_p)

Planck time is also named after Max Planck. It is the time required for light to travel in a vacuum a distance of 1 Planck length (which is 1.6×10^{-35} m or about 10^{-20} times the size of a proton). It is the unit of time in the system of natural units called *Planck units*,

$$t_p \equiv \sqrt{\frac{\hbar G}{c^5}} \approx 5.39 \times 10^{-44}\ \text{s}. \tag{1.6}$$

Planck time represents a rough time scale at which quantum gravitational effects are likely to become important. Human and hence scientific effects are billions of billions of billions of Planck times, so these effects are nearly impossible to detect. So far the smallest direct time measurement interval is $\approx 1.2 \times 10^{-17}$s, or 12 as, which is about 2.2×10^{26} Planck times.

The term t_p is called the *quantum of time*, the smallest measurement of time that has any meaning. Within the framework of what we understand today, we can only say that theoretically the Universe came into existence when it already had an age of 10^{-43} s!

1.4.3 Planck's mass (m_p)

Planck's mass is nature's maximum allowed mass for point particles (called *quanta*). In other words, an elementary particle capable of holding a single elementary charge. Unlike the other Planck base units, Planck mass has a scale more or less conceivable to humans. Traditionally Planck mass $\cong$ the mass of a flea egg.

Concept questions 1.2

1. Suppose the diameter of a penny is 0.75 inches. How many pennies will you need to line up end to end to cover 1 mile in length?
2. Suppose we have a quantity $y = y(t)$ having units $[M][L]^{-1}$, given as

$$y(t) = 4e^{-2t},$$

where t has units of time. What are the units associated with the numbers 4 and 2 in the equation?
3. Given the density of water is approximately $\rho_{\text{water}} = 1.0 \text{ g cm}^{-3}$ and $R_{\text{Earth}} = 6.4 \times 10^3$ km, estimate the mass of water on Earth. *Hint*: assume oceans cover 75% of the Earth's surface and assume the average depth of the ocean to be approximately 1 km.

Exercises 1.2

1. Use dimensional analysis to determine the speed of sound in a gas. *Hint*: the speed of sound in gas is proportional to the pressure (p), density (ρ) and the volume (V) of the gas.
2. Find the equation for the time taken to drop a ball from height h using dimensional analysis.
3. What is the Planck mass? Determine using dimensional analysis:
   (a) The formula for the Planck mass.
   (b) The numerical value for the Planck mass.

1.5 A quick review of vectors

1.5.1 Vectors

According to the Merriam-Webster dictionary, a vector is a quantity, for example velocity or acceleration, that has a size and direction. Vector can be formally defined as follows,

> 'A quantity that has a magnitude and a direction relative to a chosen reference frame. Vector quantities are represented pictorially by an arrow whose head represents the direction of the quantity and whose length represents the magnitude of the quantity'.

Familiar examples of a vector include velocity, acceleration and force.

A given vector $\vec{A}$ is specified by defining its magnitude and direction relative to a chosen reference frame. A typical example of a vector in three dimensions in unit vector notation[3] is written as

$$\vec{A} = (a_x, a_y, a_z) = a_x\hat{i} + a_y\hat{j} + a_z\hat{k}. \tag{1.7}$$

If $\vec{A}$ is a velocity vector then $\vec{a}_x$ is the velocity of the particle along the x-direction, $\vec{a}_y$ is the velocity of the particle along the y-direction and $\vec{a}_z$ is the velocity of the particle along the z-direction.

1.5.2 Scalars

A scalar, on the other hand, is a physical quantity that is able to be completely specified by a number. Familiar examples include temperature, density, volume, etc. Mathematically, scalars are treated as ordinary real numbers. They obey the rules of regular algebraic addition, subtraction, multiplication, division, etc, along with the rules for dimensioned quantities.

1.5.3 Formal definitions and rules of vector algebra

Vectors have the following properties:

- *Vector equality*: $\vec{A} = \vec{B}$ iff $a_x = b_x$; $a_y = b_y$; $a_z = b_z$.
- *Vector magnitude*: $|\vec{A}| = \sqrt{a_x^2 + a_y^2 + a_z^2}$.
- *Vector addition*: $\vec{A} + \vec{B} = (a_x + b_x)\hat{i} + (a_y + b_y)\hat{j} + (a_z + b_z)\hat{k}$.
- *Vector scalar multiplication*: $c\vec{A} = ca_x\hat{i} + ca_y\hat{j} + ca_z\hat{k}$.
- *Null vector*: $\vec{0} = (0, 0, 0)$ direction undetermined.
- *Commutative law*: $\vec{A} + \vec{B} = \vec{B} + \vec{A}$.
- *Associate law*: $\vec{A} + (\vec{B} + \vec{C}) = (\vec{A} + \vec{B}) + \vec{C}$.
- *Distributive law*: $c(\vec{A} + \vec{B}) = c\vec{A} + c\vec{B}$.
- *Unit coordinate vectors*: vectors with unit magnitude usually denoted with the caret symbol (^). For Cartesian coordinates:

$$\hat{i} = e_x = (1, 0, 0)$$
$$\hat{j} = e_y = (0, 1, 0)$$
$$\hat{k} = e_z = (0, 0, 1).$$

The unit vectors are also called *basis vectors*.

[3] A unit vector in vector space is a vector of length 1. A general vector can be represented as a sum of multiples of unit vectors. A unit vector is often denoted by an addition of a *hat* on the vector: example, $\hat{i}$ usually represents the x-direction in Cartesian coordinates, whereas $\hat{j}$ represents the y-direction and $\hat{k}$ represents the z-direction.

1.5.4 Scalar product for vectors

Given two vectors $\vec{A}$ and $\vec{B}$, the scalar product $(\vec{A} \cdot \vec{B})$ is defined as

$$\vec{A} \cdot \vec{B} = a_x b_x + a_y b_y + a_z b_z = |A||B|\cos\theta, \tag{1.8}$$

here θ is the angle between the two vectors $\vec{A}$ and $\vec{B}$. The scalar product is also called the *dot product* or *inner product*. Some properties of the scalar product are as follows:

- The scalar product is commutative:

$$\vec{A} \cdot \vec{B} = \vec{B} \cdot \vec{A}.$$

- The scalar product is distributive:

$$\vec{A} \cdot (\vec{B} + \vec{C}) = \vec{A} \cdot \vec{B} + \vec{A} \cdot \vec{C}.$$

- For scalar products of basis vectors:

$$\hat{i} \cdot \hat{i} = \hat{j} \cdot \hat{j} = \hat{k} \cdot \hat{k} = 1$$
$$\text{otherwise } (\hat{i} \cdot \hat{j}) = (\hat{i} \cdot \hat{k}) = (\hat{k} \cdot \hat{j}) = 0$$

1.5.5 Law of cosines

Let $|\vec{A}|$, $|\vec{B}|$ and $|\vec{C}|$ be the magnitudes of three sides of a triangle as shown in figure 1.1. and the opposite angles be $\hat{a}, \hat{b}$ and $\hat{c}$, then the law of cosines can be written as

$$\begin{aligned} A^2 &= B^2 + C^2 + 2BC\cos\hat{a} \\ B^2 &= A^2 + C^2 + 2AC\cos\hat{b} \\ C^2 &= A^2 + B^2 + 2AB\cos\hat{c}. \end{aligned} \tag{1.9}$$

Example 1.6

Find the angle between the vectors $\vec{A} = 2\hat{i} + \hat{j} + \hat{k}$ and $\vec{B} = \hat{i} - \hat{j} + 2\hat{k}$.

Solution

Since $\vec{A} \cdot \vec{B} = AB\cos\theta$, where θ is the angle between the two vectors, then

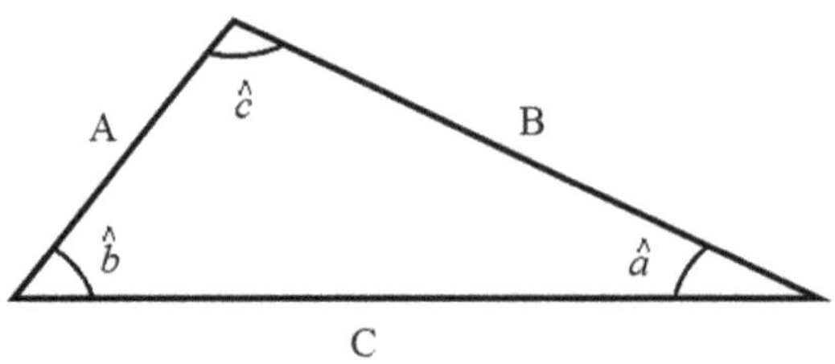

Figure 1.1. Law of cosines.

$$
\begin{aligned}
\cos\theta &= \frac{\vec{A}\cdot\vec{B}}{AB} \\
&= \frac{a_x b_x + a_y b_y + a_z b_z}{\sqrt{a_x^2 + a_y^2 + a_z^2}\sqrt{b_x^2 + b_y^2 + b_z^2}} \\
&= \frac{(2)(1) + (1)(-1) + (1)(2)}{\sqrt{3}\sqrt{6}} \\
&= \frac{3}{4.2} \\
\cos\theta &= 0.71
\end{aligned}
$$

Then,

$$
\begin{aligned}
\theta &= \cos^{-1}(0.71) \\
\theta &= 44^\circ
\end{aligned}
$$

1.5.6 Vector product for vectors

Given two vectors $\vec{A}$ and $\vec{B}$ the vector product ($\vec{A} \times \vec{B}$) is defined as

$$\vec{A} \times \vec{B} = |\vec{A}||\vec{B}| \sin\theta. \tag{1.10}$$

In terms of its components the vector product will be defined as

$$\vec{A} \times \vec{B} = \begin{vmatrix} \hat{\imath} & \hat{\jmath} & \hat{k} \\ a_x & a_y & a_z \\ b_x & b_y & b_z \end{vmatrix}$$

$$\vec{A} \times \vec{B} = (a_y b_z - a_z b_y)\hat{\imath} - (a_x b_z - a_z b_x)\hat{\jmath} + (a_x b_y - a_y b_x)\hat{k}. \tag{1.11}$$

The vector product is also called the *cross product*. Some properties of the vector product are as follows:

- The vector product is anti-commutative: $\vec{A} \times \vec{B} = -(\vec{B} \times \vec{A})$.
- $n(\vec{A} \times \vec{B}) = (n\vec{A}) \times \vec{B} = \vec{A} \times (n\vec{B})$.
- The vector product is distributive: $\vec{A} \times (\vec{B} + \vec{C}) = \vec{A} \times \vec{B} + \vec{A} \times \vec{C}$.
- For vector products of basis vectors:

$$
\begin{aligned}
&\hat{\imath} \times \hat{\imath} = \hat{\jmath} \times \hat{\jmath} = \hat{k} \times \hat{k} = 0 \\
&\hat{\imath} \times \hat{\jmath} = \hat{k};\ \hat{\jmath} \times \hat{k} = \hat{\imath};\ \hat{k} \times \hat{\imath} = \hat{\jmath} \\
&\hat{\imath} \times \hat{k} = -\hat{\jmath};\ \hat{k} \times \hat{\jmath} = -\hat{\imath};\ \hat{\jmath} \times \hat{\imath} = -\hat{k}.
\end{aligned}
$$

Torque ($\vec{\tau}$)[4] is a common example where cross products find a home. Torque is mathematically defined as

[4] Torque is a twisting force that tends to cause rotation in any object that it is applied to.

$$\vec{\tau} = \vec{r} \times \vec{F} = rF \sin\theta\hat{\tau}, \tag{1.12}$$

where r is called the moment arm, $\vec{F}$ is applied force and $\hat{\tau}$ is a unit vector perpendicular to $\vec{r}$ and $\vec{F}$, given by the right-hand rule. I have only defined torque here, for more details see section 1.12.

Example 1.7

Given the two vectors $\vec{A} = 2\hat{\imath} + 2\hat{\jmath} - \hat{k}$ and $\vec{B} = \hat{\imath} - 2\hat{\jmath} + 2\hat{k}$, find $\vec{A} \times \vec{B}$.

Solution

$$\vec{A} \times \vec{B} = \begin{vmatrix} \hat{\imath} & \hat{\jmath} & \hat{k} \\ 2 & 2 & -1 \\ 1 & -2 & 2 \end{vmatrix}$$

$$\vec{A} \times \vec{B} = \hat{\imath}(-4 - 2) - \hat{\jmath}(4 + 1) + \hat{k}(-4 - 2)$$

$$\vec{A} \times \vec{B} = -6\hat{\imath} - 5\hat{\jmath} - 6\hat{k}.$$

Example 1.8

Find a unit vector *normal*[5] to the plane containing the two vectors $\vec{A}$ and $\vec{B}$; $\vec{A} = 2\hat{\imath} + \hat{\jmath} - \hat{k}$ and $\vec{B} = \hat{\imath} - \hat{\jmath} + 2\hat{k}$.

Solution

$$\begin{aligned} \hat{n} &= \frac{\vec{A} \times \vec{B}}{|\vec{A} \times \vec{B}|} \\ &= \frac{\hat{\imath} - 5\hat{\jmath} - 3\hat{k}}{\sqrt{1^2 + 5^2 + 3^2}} \\ &= \frac{1}{\sqrt{35}}(1\hat{\imath} - 5\hat{\jmath} - 3\hat{k}). \end{aligned}$$

1.6 Derivatives of vectors

Suppose we are given a vector ($\vec{A}$) whose components are a function of one variable, say time (t). Then we can express $\vec{A}$ mathematically as

$$\vec{A} = (a_x(t), a_y(t), a_z(t)) = a_x(t)\hat{\imath} + a_y(t)\hat{\jmath} + a_z(t)\hat{k}. \tag{1.13}$$

Then the first derivative of the vector $\vec{A}$ with respect to (w.r.t.) the variable (t) will be given as

[5] Given a particular surface in 3D space, a normal is a vector that is perpendicular to the surface. The normal vector is commonly denoted as $\hat{n}$. When a normal vector has magnitude 1, it is called a unit normal vector.

$$\frac{d\vec{A}}{dt} = \frac{da_x(t)}{dt}\hat{i} + \frac{da_y(t)}{dt}\hat{j} + \frac{da_z(t)}{dt}\hat{k} = \dot{a}_x(t)\hat{i} + \dot{a}_y(t)\hat{j} + \dot{a}_z(t)\hat{k}, \tag{1.14}$$

whereas the second derivative of the vector $\vec{A}$ w.r.t. the variable (t) will be given as

$$\frac{d^2\vec{A}}{dt^2} = \frac{d^2a_x(t)}{dt^2}\hat{i} + \frac{d^2a_y(t)}{dt^2}\hat{j} + \frac{d^2a_z(t)}{dt^2}\hat{k} = \ddot{a}_x(t)\hat{i} + \ddot{a}_y(t)\hat{j} + \ddot{a}_z(t)\hat{k}. \tag{1.15}$$

Concept questions 1.3

1. Can a zero magnitude vector have a direction?
2. Four vectors have the same magnitude. Under what conditions will their sum equal zero?
3. Is it possible to obtain a vector with a negative magnitude?
4. Given that the dot product between vectors $\vec{A}$ and $\vec{B}$ is positive, what can we conclude about the angle between the two vectors?
5. We have talked about vector addition, vector subtraction and vector products. Explain why we have not talked about vector division?

Exercises 1.3

1. Suppose we have a cube whose edges are of unit length. Let one corner coincide with an origin of x, y, z. Find the vector
   (a) that begins at 0 and ends along a major diagonal of the cube; call this vector $\vec{A}$.
   (b) that begins at 0 and ends along the diagonal of the lower face of the cube; call this vector $\vec{B}$.
   (c) Find $\vec{C} = \vec{A} \times \vec{B}$.
   (d) Find the angle between $\vec{A}$ and $\vec{B}$.
2. Given the following vectors, $\vec{a} = 3\hat{i} - 3\hat{j} + 1\hat{k}$ and $\vec{b} = 4\hat{i} + 9\hat{j} + 2\hat{k}$, find:
   (a) $\vec{a} + \vec{b}$.
   (b) $\vec{a} - \vec{b}$.
   (c) The dot product between $\vec{a}$ and $\vec{b}$.
   (d) The cross product between $\vec{a}$ and $\vec{b}$.
3. Given the time-varying vector $\vec{A} = at\hat{i} + bt^2\hat{j} + ct^3\hat{k}$, where a, b and c are constants and $\hat{i}$, $\hat{j}$ and $\hat{k}$ are the unit vectors along the x, y and z coordinate axes, respectively. Find the first and second order derivatives.

1.7 Position vector

A coordinate system uses one or more coordinates (numbers) to uniquely specify the position of a point-particle in space. We employ various coordinate systems to help visualize physics problems in the three-dimensional (3D) space we live in. The three most commonly used coordinate systems are the Cartesian, spherical and cylindrical coordinate systems.

The position vector ($\vec{r}$) of a particle is a vector drawn from a user-defined origin to the particle. It always points radially outwards from the origin and expresses the displacement from the origin. The position vector is also called a *location vector* or *radius vector*;

$$\vec{r} = x\hat{i} + y\hat{j} + z\hat{k}. \tag{1.16}$$

The magnitude of the position vector ($\vec{r}$) of a particle is the distance from the origin and is mathematically expressed as

$$|\vec{r}| = r = \sqrt{x^2 + y^2 + z^2} \tag{1.17}$$

and the unit vector is given as

$$\hat{r} = \hat{\vec{r}} = \frac{\vec{r}}{|\vec{r}|} = \frac{x\hat{i} + y\hat{j} + z\hat{k}}{\sqrt{x^2 + y^2 + z^2}}, \tag{1.18}$$

which is a unit vector pointing radially outwards.

Most of classical physics deals with finding the position vector $\vec{r}$ of things, may it be quarks or planets. The term $\vec{r}$ defines the motion of a particle in a particular coordinate system (figure 1.2).

If we take the origin at (0, 0, 0) in each of the coordinate systems, then the position vector $\vec{r}$ can be defined as follows:

- Cartesian: $(x, y, z) = \vec{r} = x\hat{i} + y\hat{j} + z\hat{k}$.
- Polar: (r, ϕ).
- Spherical: (r, θ, ϕ).
- Cylindrical: (r, ϕ, z).

1.7.1 Separation vector

In classical mechanics, we may sometimes encounter problems where the tail of the displacement vector does not coincide with the user-defined origin. In this case there

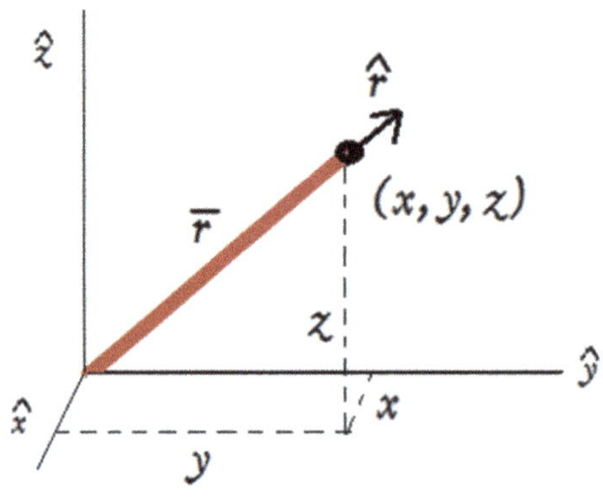

Figure 1.2. Position vector example.

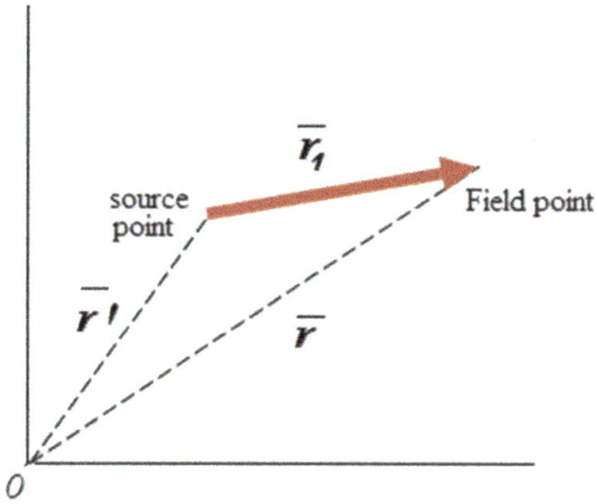

Figure 1.3. A separation vector example, as defined in electrodynamics.

will be two points needed to define the vector, the head-point and tail-point in the coordinate system. This is especially useful in electrodynamics, with the two points being the source point ($\vec{r}'$) and field point ($\vec{r}$). In this case the source point is where the electric charge is located and the field point is where you are calculating the electric or magnetic field strength, as shown in figure 1.3.

In this case the separation vector from source point to field point will be given by

$$\vec{r}_1 = \vec{r} - \vec{r}'. \tag{1.19}$$

Its magnitude is $|\vec{r}_1| = |\vec{r} - \vec{r}'|$, and a unit vector in the direction from $\vec{r}'$ to $\vec{r}'$ is

$$\hat{r}_1 = \frac{\vec{r}_1}{|\vec{r}_1|} = \frac{\vec{r} - \vec{r}'}{|\vec{r} - \vec{r}'|}. \tag{1.20}$$

1.8 Transformation between various coordinate systems

Geometrical figures can be described in many different possible coordinate systems. Hence it is important to understand how different possible coordinate systems are related. Relationships between coordinate systems are defined by using transformation formulas for the coordinates in one system in terms of the coordinates in another system.

1.8.1 Polar coordinates (r, ϕ)

The term polar was used by eighteenth-century Italian writers and is historically attributed to Gregorio Fontana. A polar coordinate system is a two-dimensional (2D) coordinate system where each point is determined by measuring the distance from a user-defined origin (O), also called the *pole*, and an angle from a reference direction, usually taken to lie along the x-axis in a 2D Cartesian coordinate plane. In general we consider polar coordinates to lie on a flat plane, as shown in figure 1.4, and the origin (O) of the system is where all coordinates of the system have zero value.

The magnitude of the distance from the origin (pole) is called the radial coordinate or radius (r), the angle (ϕ) is taken from the reference direction and is called the angular coordinate, polar angle, or azimuth.

In other texts the radial coordinate is also sometimes denoted by ρ, and the angular coordinate by θ. Angles in polar notation are expressed either in degrees or in radians (2π rad being equal to 360°). A positive angular coordinate means that the angle is measured counterclockwise from the reference direction (polar axis). Unit vectors in polar coordinates are $\hat{r}$ and $\hat{\phi}$ (figures 1.4 and 1.5).

Conversion between Cartesian and polar coordinates

In a plane the Cartesian coordinates are given by (x, y) and the polar coordinates are given by (r, ϕ). If we suppose that both coordinates have the same origin, and the polar axis aligns with the positive x-axis, then the coordinate transformation from polar to Cartesian coordinates is given by

$$\begin{cases} x = r\cos\phi \\ y = r\sin\phi \end{cases}, \tag{1.21}$$

whereas the transformation from Cartesian to polar coordinates is given by

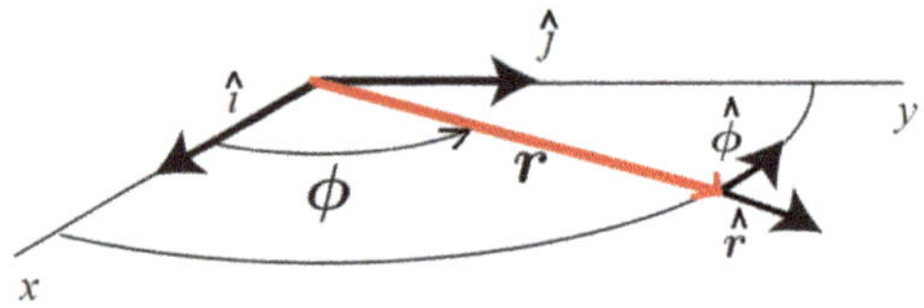

Figure 1.4. Polar coordinates (r, ϕ).

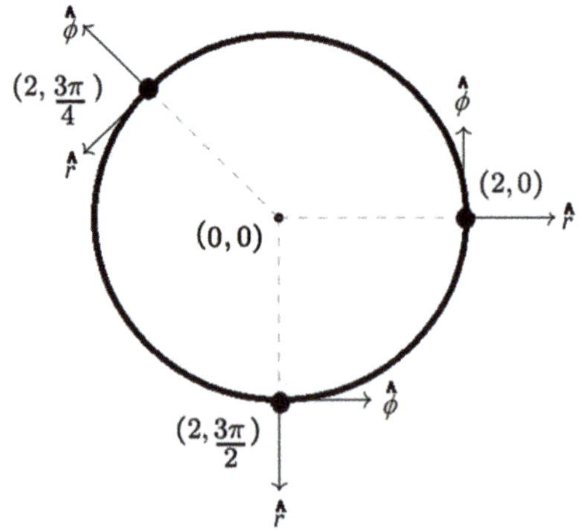

Figure 1.5. Overhead view of various points in polar coordinates (r, ϕ).

$$\begin{cases} r = \sqrt{x^2 + y^2} \\ \phi = \tan^{-1}\left(\dfrac{y}{x}\right) \end{cases}. \tag{1.22}$$

1.8.2 Conversion between Cartesian and spherical coordinates (r, θ, ϕ)

A spherical coordinate system is a way to extend two-dimensional (2D) polar coordinates to three dimensions (3D). It is a coordinate system for 3D space, where the position of a point is specified by three numbers, given as:

- *Radial distance* (r) from a fixed origin, also called the radius, where $r \in [0, \infty)$.
- *Polar angle* (θ) measured from a fixed zenith direction, also called co-latitude, zenith angle, normal angle, or inclination angle, where $\theta \in [0, \pi]$.
- *Azimuth angle* (ϕ) is defined in the in the xy-plane. It is the projection in the xy-plane that passes through the origin and is perpendicular to the zenith and is measured from a fixed reference direction (usually the x-axis) on that plane; $\phi \in [0, 2\pi]$.

It can be seen as the 3D version of the polar coordinate system (figure 1.6). The spherical coordinates on Earth are essentially the *latitude* angle (θ), the *longitudinal* angle (ϕ) and the radial distance (r) from the origin. Unit vectors in spherical coordinates are $\hat{r}$, $\hat{\theta}$ and $\hat{\phi}$.

To convert between Cartesian and spherical coordinates we use the following equations:

$$\text{spherical coordinates } (r, \theta, \phi): \begin{cases} x = r \sin\theta \cos\phi \\ y = r \sin\theta \sin\phi, \\ z = r\cos\theta \end{cases}. \tag{1.23}$$

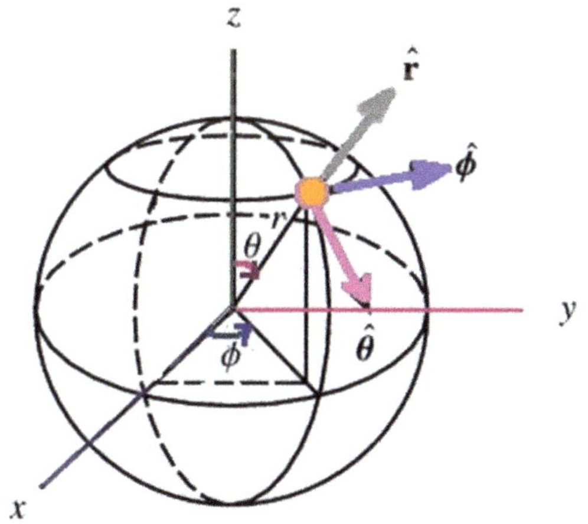

Figure 1.6. Spherical coordinates.

Example 1.9

Convert the point (1, −1, −1) from Cartesian to spherical coordinates.

Solution

Spherical coordinates are given as (r, θ, ϕ). To find the point in spherical coordinates we can take the following steps:

1. Find r:

$$r = \sqrt{x^2 + y^2 + z^2} = \sqrt{1 + 1 + 1} = \sqrt{3}.$$

2. Find θ:

$$z = r\cos\theta \Rightarrow \theta = \cos^{-1}\left(\frac{z}{r}\right) = \cos^{-1}\left(\frac{-1}{\sqrt{3}}\right) = 125^\circ \approx \frac{2\pi}{3}.$$

3. Find ϕ:

$$y = r\sin\theta\sin\phi \Rightarrow \phi = \sin^{-1}\left(\frac{y}{r\sin\theta}\right) = 45^\circ = \frac{\pi}{4} \text{ or}$$

$$x = r\sin\theta\cos\phi \Rightarrow \phi = \cos^{-1}\left(\frac{x}{r\sin\theta}\right) = 45^\circ = \frac{\pi}{4}.$$

Hence, (1, −1, −1) in spherical coordinates are given as $(r, \theta, \phi) = (\sqrt{3}, \frac{2\pi}{3}, \frac{\pi}{4})$.

1.8.3 Conversion between Cartesian and cylindrical coordinates (r, ϕ, z)

Cylindrical coordinates are also an extension of the two-dimensional polar coordinates to three dimensions, also called *cylindrical polar coordinates* or *polar cylindrical coordinates.*

The origin of the cylindrical coordinate system is the point where all three coordinates are zero. This is the intersection between the reference plane and the vertical axis. In the cylindrical coordinate system a point position is specified by the following three points:

- The distance (r) from a chosen reference axis, called the polar axis.
- The direction (ϕ) from the axis relative to a chosen reference direction.
- The distance (z) from a chosen reference plane perpendicular to the axis, called the cylindrical or longitudinal axis. It is also called height or altitude (figure 1.7).

The cylindrical axis is the ray that lies in the reference plane, starting at the origin (0, 0, 0) and pointing in the reference direction. The cylindrical axis can be positive or negative depending on the choice of the origin. The radius and the azimuth correspond to a 2D polar coordinate system and together we can call them polar coordinates.

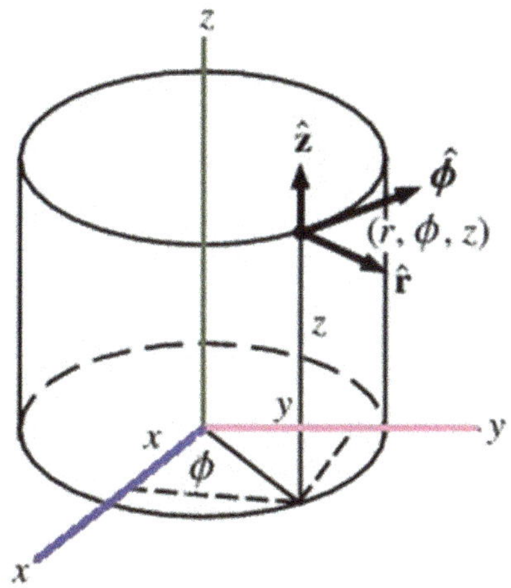

Figure 1.7. Cylindrical coordinates.

The cylindrical coordinates are essentially the *longitudinal* angle (ϕ), the radial distance (r) and height (h) from the origin. The unit vectors in cylindrical coordinates are $\hat{r}$, $\hat{\phi}$ and $\hat{z}$. To convert between Cartesian and cylindrical coordinates we use the following equations:

$$\text{cylindrical coordinates } (r, \phi, z): \begin{cases} x = r\cos\phi \\ y = r\sin\phi\,, \\ z = z \end{cases} \quad (1.24)$$

Example 1.10

Convert the point (2, −1, −1) from Cartesian to cylindrical coordinates.

Solution

Spherical coordinates are given as (r, ϕ, z). To find the point in spherical coordinates we can take the following steps:

1. Find r:

$$r = \sqrt{x^2 + y^2} = \sqrt{4 + 1} = \sqrt{5}.$$

2. Find ϕ:

$$\phi = \tan^{-1}(\frac{y}{x}) = \tan^{-1}(-\frac{1}{2}) = -26.5° \approx -\frac{\pi}{7}.$$

3. Find h:

$$h = -1.$$

Hence, (1, −1, −1) in cylindrical coordinates are given as $(r, \phi, h) = (\sqrt{5}, -\frac{\pi}{7}, -1)$.

1.9 Velocity and acceleration

1.9.1 Velocity and acceleration in Cartesian coordinates

As we saw in section 1.7, the position of a particle can be specified by a single vector, called the position vector or displacement relative to the origin. In Cartesian coordinates, the position vector $(x(t), y(t), z(t))$ is

$$\vec{r} = x\hat{i} + y\hat{j} + z\hat{k}. \tag{1.25}$$

If the vector is the position vector $\vec{r}$ of a moving particle and the parameter is the time (t), then the derivative of $\vec{r}$ with respect to t is called the *velocity*, and is given by

$$\vec{v} = \frac{d\vec{r}}{dt} = \dot{\vec{r}} = \frac{dx}{dt}\hat{i} + \frac{dx}{dt}\hat{j} + \frac{dx}{dt}\hat{k} = \dot{x}\hat{i} + \dot{y}\hat{j} + \dot{z}\hat{k}, \tag{1.26}$$

where the dots indicate differentiation with respect to t. The magnitude of the velocity is called the *speed*. Speed is a scalar quantity and hence only has magnitude but no direction and is mathematically given as

$$\text{speed} = |\vec{v}| = \frac{d|\vec{r}|}{dt}. \tag{1.27}$$

The time derivative of the velocity is called the *acceleration* and is given by

$$\vec{a} = \frac{d\vec{v}}{dt} = \frac{d^2\vec{r}}{dt^2} = \ddot{\vec{r}} = \ddot{x}\hat{i} + \ddot{y}\hat{j} + \ddot{z}\hat{k}, \tag{1.28}$$

where the double dots indicate differentiation with respect to t twice. The magnitude of the acceleration is given by

$$|\vec{a}| = \frac{d|\vec{v}|}{dt}. \tag{1.29}$$

1.10 Velocity and acceleration in various coordinates

As we discussed in section 1.8, the components in each coordinate system are given as:

- Cartesian (x, y, z).
- Polar (r, ϕ).
- Spherical (r, θ, ϕ).
- Cylindrical (r, ϕ, z).

Then, in each coordinate system the velocity and acceleration vectors can be written as:

- Position vector: $\vec{r}$.
- Velocity: $\vec{v} = \frac{d\vec{r}}{dt} = \dot{\vec{r}}$.
- Acceleration: $\vec{a} = \frac{d^2\vec{r}}{dt^2} = \ddot{\vec{r}}$.

Next, let us briefly discuss each coordinate system individually. See appendix B for details of the derivations for the formulas.

1. Cartesian coordinates (x, y, z):

$$\begin{aligned}\vec{r} &= x\hat{i} + y\hat{j} + z\hat{k}\\ \vec{v} &= \dot{x}\hat{i} + \dot{y}\hat{j} + \dot{z}\hat{k}\\ \vec{a} &= \ddot{x}\hat{i} + \ddot{y}\hat{j} + \ddot{z}\hat{k}.\end{aligned} \tag{1.30}$$

2. Polar coordinates (r, ϕ):

$$\begin{aligned}\vec{r} &= r\hat{r}\\ \vec{v} &= \dot{r}\hat{r} + r\dot{\phi}\hat{\phi}\\ \vec{a} &= (\ddot{r} - r\dot{\phi}^2)\hat{r} + (r\ddot{\phi} + 2\dot{r}\dot{\phi})\hat{\phi}.\end{aligned} \tag{1.31}$$

3. Spherical coordinates (r, θ, ϕ):

$$\begin{aligned}\vec{r} &= r\hat{r}\\ \vec{v} &= \dot{r}\hat{r} + r\dot{\theta}\hat{\theta} + r\sin\theta\dot{\phi}\hat{\phi}\\ \vec{a} &= (\ddot{r} - r\dot{\phi}^2\sin^2\theta - r\dot{\theta}^2)\hat{r} + (r\ddot{\theta} + 2\dot{r}\dot{\theta} - r\dot{\phi}^2\sin\theta\cos\theta)\hat{\theta}\\ &\quad + (r\ddot{\phi}\sin\theta + 2\dot{r}\dot{\phi}\sin\theta + 2r\dot{\theta}\dot{\phi}\cos\theta)\hat{\phi}.\end{aligned} \tag{1.32}$$

4. Cylindrical coordinates (r, ϕ, z):

$$\begin{aligned}\vec{r} &= r\hat{r} + z\hat{z}\\ \vec{v} &= \dot{r}\hat{r} + R\dot{\phi}\hat{\phi} + \dot{z}\hat{z}\\ \vec{a} &= (\ddot{r} - R\dot{\phi}^2)\hat{r} + (2\dot{R}\dot{\phi} + R\ddot{\phi})\hat{\phi} + \ddot{z}\hat{z}.\end{aligned} \tag{1.33}$$

Example 1.11

Suppose the position vector of a moving car is given by the equation, $\vec{r}(t) = t^3 + 5t - 3$. Calculate the corresponding velocity and the acceleration functions.

Solution

Velocity function:

$$v(t) = \dot{\vec{r}} = 3t^2 + 5.$$

Acceleration function:

$$a(t) = \ddot{\vec{r}} = 6t.$$

Example 1.12

A rocket is fired vertically from rest from the ground and starts accelerating at the rate of $4t$ m/s^2. If it maintains this rate of acceleration where will the rocket be after 5 s?

Solution

Given the initial conditions, $v_0 = 0\,\text{m s}^{-1}$ and $a(t) = 4t\,\text{m s}^{-2}$, we need to find the displacement $y(t = 5\,\text{s})$. Velocity function:

$$v(t) = \int \vec{r}\,dt = \int 4t\,dt = 2t^2 + v_0.$$

Displacement function:

$$y(t) = \int 2t^2 dt = 2\frac{t^3}{3} + y_0.$$

Then at $t = 5$ s, the rocket starts on the ground, hence $y_0 = 0\,\text{m}$. Put the values into the equation:

$$x(5) = 2\left(\frac{5^3}{3}\right) = 83\ \text{m}.$$

Hence the rocket will be 83 m above the ground.

Concept questions 1.4

1. Does the speedometer in your car give speed or velocity? Does the speedometer care whether you are going north or south at 55 mph?
2. Can you think of a situation where a particle moving at a constant speed experiences an acceleration?
3. Suppose at a particular instance of time the velocity of an object is zero. Can this object have a non-zero acceleration at that instance? Give an example of such a motion.

Exercises 1.4

1. Blue whales are the longest known mammals and they can grow to be the length of a basketball court, about 30 m, and can weigh up to 180 metric tons. If the speed of nerve impulses for a whale is taken to be about $100\,\text{m s}^{-1}$, how long will it take for the whale to know if its tail gets hurt?
2. Suppose a mass is hung by a spring which stretches the spring by 3 cm to some initial state, y_0. We then displace the spring by pulling on the mass at time $t = 0$.

Releasing the mass sets it in motion. Then at any given time the displacement is given by

$$y(t) = 4\sin(\omega t),$$

where ω is the value of angular frequency and can be considered as a constant. Find:

(a) The velocity and acceleration of the mass as a function of time.
(b) The velocity and acceleration of the mass at $t = 3$ s.

3. Given that the acceleration of a certain particle is given by

$$a(t) = \frac{1}{4}t + \cos(2t): \tag{1.34}$$

(a) Find the velocity function and the displacement as a function of time.
(b) Given $v(t = 0) = 2\,\mathrm{m\,s^{-1}}$, find the velocity at $t = 2$ s.

Further reading

Arya A 1990 *Introduction to Classical Mechanics* 2nd edn (Upper Saddle River, NJ: Prentice-Hall)
Chababy R and Sherwood B 2011 *Matter and Interactions* 3rd edn (New York: Wiley)
Fowles G and Cassiday G 2005 *Analytical Mechanics* 7th edn (Belmont, CA: Thomson Learning)
Ohanian H 1985 *Physics* (New York: WW Norton)
Resnick R, Halliday D and Krane K 2002 *Physics* vol 1, 5th edn (New York: Wiley)
Stephenson R 1961 *Mechanics and Properties of Matter* 2nd edn (New York: Wiley)
Symon K 1971 *Mechanics* 3rd edn (Reading, MA: Addison-Wesley)

Chapter 2

Conservation laws

2.1 Introduction

Mechanics deals with objects at rest and objects in motion. It does not necessarily explain why the objects move. It simply describes the motion of an object and helps extrapolate the path of the object. The laws of mechanics apply to both macroscopic and microscopic objects, from planetary motion to the motion of electrons in atoms.

The study of mechanics is generally divided into the following parts:

- *Statics* deals with objects at rest under the influence of balanced external forces.
- *Kinematics* is only concerned with the trajectories of objects. It deals with position, velocity, acceleration and time in any given situation and is not concerned with the forces producing these changes.
- *Dynamics* is concerned with the forces producing the changes in motion of any object or changes in other properties such as the shape or sizes of objects.

In the course of this book we will study all of the above parts of mechanics.

Let us first begin with the fundamental concepts of conservation laws.

2.2 Conservation laws

Conservation laws are very important in physics. They determine which processes are allowed and which are not physically possible. Conservation laws assert that conserved quantities cannot be created or destroyed but are rather just converted from one form to another. There are basically three main conservation laws:

- Conservation of mass.
- Conservation of mechanical quantities, such as energy, angular momentum and linear momentum.
- Conservation of electric charge.

Physicists depend on conservation laws explicitly and take advantage of them in several ways. In mechanics they use conservation laws of energy, momentum and angular momentum to make predictions about the motion of a particle or a system of particles. In sub-atomic and particle physics they use conservation laws to predict possible decays and reactions. In relativistic mechanics, they use the conservation of energy and mass to convert matter into energy.

2.2.1 Conservation of mass

In everyday life and in physics we rely on the conservation of mass, which says that the mass of an object will always remain constant. It would be very strange if the mass of substance could change at will, for example, it would be very disturbing for a sugar merchant if one kilogram of sugar suddenly lost half its mass. Imagine the shock he would receive if he left 100 kg of sugar in his shop one evening before closing and the next day he found only 50 kg there. Because we know that mass cannot spontaneously disappear—the conservation of mass—the merchant will likely conclude that he has been robbed.

Mathematically we write conservation of mass as

$$\Delta m = 0. \tag{2.1}$$

2.2.2 Conservation of energy

Energy is one of the fundamental concepts in physics. This means that energy cannot be further defined in terms of other more fundamental quantities. The best explanation that physicists have come up with is that *energy is the property of matter that must be transferred to an object in order for the object to perform work.*

The law of conservation of energy states that energy cannot be created or destroyed but it can be converted from one form to another. The SI unit of energy is the *joule*, named after James Prescott Joule, an English physicist and mathematician. Mathematically we write conservation of energy as

$$\boxed{\Delta E = 0}, \tag{2.2}$$

where E is the *total energy* of the system, also called the *mechanical energy* of the system.

2.2.3 Conservation of mass and energy

The laws of conservation of energy and conservation of (total) mass both hold true and are equivalent to each other. In other words mass and energy are interchangeable and are related by the formula

$$\boxed{E = mc^2}. \tag{2.3}$$

This formula states that objects that have energy (E) will inevitably have a corresponding mass (m) given by energy divided by the speed of light in a vacuum squared. This is sometimes also called the *mass–energy relationship*. A direct consequence of this relationship is that all matter objects will have intrinsic energies. When the object is at rest we call it *rest energy* and when the object is moving we call it *total energy*.

In general the total energy is always greater than the rest energy of any object since total energy is the sum of the rest energy and the kinetic energy (T), the energy due to the motion of the object.

It is worth mentioning here that conservation of energy follows from Noether's theorem, whose details are mentioned in appendix C, and the fact that the laws of physics are invariant in time.

2.3 Forces that depend on position: energy considerations

2.3.1 What is a force?

We will talk more about forces in chapter 3, but in this chapter we will simply define force as follows: 'If we observe an object to be accelerating, then there must be a force acting on it'.

For rectilinear motion, that is motion in a single straight line when the force is independent of velocity or time, the equation of motion of an object is simply given as a second order differential equation of the form

$$\vec{F}(x) = m\vec{a} = m\frac{d^2x}{dt^2} = m\ddot{x}, \tag{2.4}$$

where the SI unit of force ($\vec{F}$) is called the newton and the acceleration ($\vec{a}$) as a function of position is defined to be

$$\vec{a} = \ddot{x} = \frac{d\dot{x}}{dt} = \frac{d\dot{x}}{dt}\frac{dx}{dx} = v\frac{dv}{dx} = \frac{1}{2}\frac{dv^2}{dx}. \tag{2.5}$$

2.3.2 Kinetic energy

The kinetic energy of an object is the energy that it possesses due to its motion. In classical mechanics, the kinetic energy of a non-rotating object of mass (m) traveling at a speed (v) is called the translational kinetic energy and is mathematically given by

$$T = KE = \frac{1}{2}mv^2. \tag{2.6}$$

As mentioned above, the SI unit for energy is the joule.

The kinetic energy of an object is related to its momentum (defined in section 2.3.4 as $\vec{p} \equiv m\vec{v}$) by the equation

$$T = \frac{1}{2}(mv^2)\frac{m}{m} = \frac{m^2v^2}{2m} = \frac{p^2}{2m}. \tag{2.7}$$

The kinetic energy of any entity depends on the reference frame in which it is measured. However, the total energy of an isolated system, i.e. one in which energy cannot enter or leave, does not change over time.

2.3.3 Force as a function of position

When the force a particle experiences is a function of position ($F = F(x)$), we can use equation (2.5) to rewrite equation (2.4) as

$$F(x) = \frac{dT}{dx}. \tag{2.8}$$

2.3.4 Momentum

Momentum is defined as 'the quantity of motion of a moving body, measured as a product of its mass and velocity'.

If we take the mass of the object as (m) and its velocity as ($\vec{v}$),then its momentum is defined as

$$\vec{p} \equiv m\vec{v}. \tag{2.9}$$

Momentum is a vector quantity, since it has both magnitude and direction, and is measured in kilogram-meters per second in SI units. In terms of force:

$$\boxed{\vec{F}_{\text{net}} = \frac{d\vec{p}}{dt}}. \tag{2.10}$$

This is called the *momentum principle* and it relates the net forces on an object with the observed rate of change of momentum. The momentum principle also tells us that the two vectors in equation (2.10), namely $\vec{F}_{\text{net}}$ and $\vec{p}$, have the same vector components. Momentum is also used to predict the resulting direction and speed of motion of objects after they collide, as discussed in detail in chapter 16.

2.3.5 Conservation of linear momentum

Conservation of momentum is a consequence of the fact that the laws of physics do not depend on position; this is a special case of Noether's theorem. See appendix C for details about Noether's theorem. Conservation of momentum is mathematically given as

$$\boxed{\Delta\vec{p} = 0}. \tag{2.11}$$

We will revisit linear momentum in section 3.3.

2.3.6 Angular momentum

For a point particle

The angular momentum ($\vec{L}$) is a vector quantity that describes an object in circular motion. The angular momentum of a single particle depends on both the momentum of that particle ($\vec{p}$) and its vector location ($\vec{r}$) from some reference point. The angular momentum is defined as a cross-product between $\vec{r}$ and $\vec{p}$ and can be calculated as

$$\vec{L} \equiv \vec{r} \times \vec{p}. \tag{2.12}$$

Angular momentum is a vector quantity, hence it will have both magnitude and direction. The magnitude of angular momentum is equal to (rmv), and its direction is perpendicular to the plane of its circular motion; see section 8.6.1 and figure 8.5 for more details.

For a rigid body

A rigid body in physics is considered as a solid object with a continuous distribution of mass. All particles in a rigid body maintain the same distance relative to each other. For a rigid body the angular momentum is defined as

$$\boxed{\vec{L} \equiv I\omega}, \tag{2.13}$$

where I is moment of inertia and ω is the angular velocity[1] of the object under consideration. For details of the moment of inertia, see chapter 15, section 15.2.

Conservation of angular momentum

Mathematically, conservation of angular momentum is given as

$$\Delta\vec{L} = 0. \tag{2.14}$$

We know that the linear momentum is conserved when there is no net external force acting on the body. The angular momentum will be conserved when the net torque acting on the object is zero.

Conservation of angular momentum results from Noether's theorem and the fact that physical laws are invariant under rotation.

2.3.7 Work

The integral $(\int F(x)dx)$ is defined as the work done on a particle (W) by a variable force, where $\vec{F} = \vec{F}(x)$. If we define $F_{\text{net}, x}$ as the net force applied to the object along the x-direction, then W_{net} which is the net work done by all forces that act on the body will be given as,

$$\boxed{W_{\text{net}} = \int F_{\text{net}, x} \cdot dx}, \tag{2.15}$$

where dx is the displacement[2] of the object. Note that this definition matches our intuitive definition of 'what is work?'. We know that if we push a heavy box over a table, the work that we perform is proportional to the magnitude of the force and the distance we move the box. If the direction of the applied force and the displacement

[1] Angular velocity $(\omega \equiv v/r)$ is a vector quantity that describes the angular speed at which the orientation of the rigid body changes. For details on angular velocity, see chapter 15, section 3.5.

[2] According to the Merriam-Webster dictionary, displacement is defined in physics as the difference between the initial position of something (such as a body or geometric figure) and any later position.

is the same then we say that the work is positive. If the applied force and the displacement are in the opposite directions we say that the work done is negative.

Another way to look at work and energy is to say that, if the work is positive work, then the object will gain energy. If the work is negative work, then the object will lose energy.

The SI unit of work, the joule, is the energy transferred to an object by the work of moving it a distance of one meter by a force of one newton. We can also say that the work is equal to the change in the kinetic energy of the particle.

Example 2.1

Suppose we have a 3 kg box resting on the floor. Also suppose that under the effect of a 2 N force, the box is displaced 2 m to the right (+x-direction) or the left (−x-direction). Then find the work done on the box in the following cases.

Case 1. Moving to the right and speeding up.

The vox moving to right implies $\Delta\ x$ is positive; speeding up implies energy is increasing. Then

$$W_{\text{net}} = \int F_x \cdot dx = (+2\ \text{N})(+2\ \text{m}) = +4\ \text{J}.$$

The work done on the box is positive.

Case 2. Moving to the right and slowing down.

The box moving to the right implies Δx is positive; slowing down implies energy is decreasing, F_x acts opposite to the direction of motion. Then

$$W_{\text{net}} = \int F_x \cdot dx = (-2\ \text{N})(+2\ \text{m}) = -4\ \text{J}.$$

The work done on the box is negative.

Case 3. Moving to the left and speeding up.

The box moving to left implies Δx is negative; speeding up implies energy is increasing, in this case F_x will be negative. Then

$$W_{\text{net}} = \int F_x \cdot dx = (-2\ \text{N})(-2\ \text{m}) = +4\ \text{J}.$$

The work done on the box is positive.

Case 4. Moving to left and slowing down.

The box moving to left implies Δx is negative; slowing down implies energy is decreasing, F_x acts opposite to the direction of motion, in this case F_x will be positive. Then

$$W_{\text{net}} = \int F_x \cdot dx = (+2N)(-2m) = -4\ \text{J}.$$

The work done on the box is negative.

Signs of work done:

- If the object is speeding up, this means force acts in the direction of motion and the work done on the object is positive (cases 1 and 3).
- When the object is slowing down, this means that the force will act opposite to the direction of motion and the work done on the object is negative (cases 2 and 4).

2.3.8 Work–energy theorem

We will now derive a very important identity between the work done on a particle and the change of speed of the particle, called the *work–energy theorem*.

Suppose a force $\vec{F}$ is applied to a particle and the particle moves from point $P_1(x_1)$ to $P_2(x_2)$. If we assume that this force is a function of position and time ($\vec{F} = \vec{F}(x, t)$) then the work done on the particle is given by equation (2.15) and the force is given by equation (2.4). We can write the integral form as

$$W_{\text{net, x}} = \int_{x_1}^{x_2} F_{net} \cdot dx = \int m\frac{dv}{dt}dx = \int mdv\frac{dx}{dt} = \int mvdv = \frac{1}{2}m(v_f^2 - v_o^2). \quad (2.16)$$

Hence we can write equation (2.15) as

$$W_{\text{net}} = T - T_o, \quad (2.17)$$

where we say that $\frac{1}{2}mv_f^2 = T$ is the final kinetic energy value at point P_2 and $\frac{1}{2}mv_o^2 = T_o$ is the initial kinetic energy value at point P_1.

The work–energy theorem is written as

$$\boxed{W_{\text{net}} = \Delta T}. \quad (2.18)$$

2.3.9 Potential energy

Let us define a function $V(x)$ such that

$$\frac{-dV(x)}{dx} = F(x). \quad (2.19)$$

This function is more commonly known as the potential energy (V) of the particle under consideration. The term potential energy was introduced by the nineteenth-century Scottish engineer and physicist William Rankine. Potential energy is associated with forces that act on an object in a way that depends only on the body's position in space.

In a gravitational field the potential energy is the stored energy of an object by the virtue of its position. Suppose a force acting on a body moves the object from an initial position (x_o) to final position (x). If the work done is determined only by the initial and final positions and does not depend on the trajectory of the body, then the potential energy can be evaluated at the two positions to determine this work. We begin by writing equation (2.19) as

$$-dV(x) = F(x) \cdot dx. \tag{2.20}$$

Then in terms of $V(x)$, the work integral is

$$\begin{aligned}\int_{x_o}^{x} F(x)dx &= -\int_{x_o}^{x} dV \\ &= -V(x) + V(x_o) \\ &= W_{\text{net}}.\end{aligned}$$

Note that the potential energy function is defined only to within an arbitrary additive constant. Equation (2.19) remains unaltered if $V(x)$ is changed by adding any constant (c) because

$$-[V(x) + c] + [V(x_o) + c] = -V(x) + V(x_o).$$

There are various types of potential energy, each associated with a particular type of force. Some examples include:

- The work of an elastic force due to a mass–spring system is called elastic potential energy, mathematically given by the formula

$$V(r) = \frac{1}{2}kx^2, \tag{2.21}$$

  where x is the displacement of the mass from its equilibrium position and k is the spring constant.
- The work of the gravitational force on an object of mass m is called gravitational potential energy,

$$V(r) = mgh, \tag{2.22}$$

  where, h is the height of the object and g is acceleration due to gravity.
- The work of the Coulomb force, the force experienced by charged particles (Q and q), is called electric potential energy and is given mathematically as

$$V(r) = \frac{1}{4\pi\epsilon_o}\frac{Qq}{r}. \tag{2.23}$$

Example 2.2

What force is associated with potential energy function, $V = \frac{1}{2}k|\vec{x}|$ in Cartesian coordinates?

Solution

Using equations (2.19) and (2.21) we see that

$$\vec{F} = -\frac{\partial V}{\partial x}$$
$$= \frac{1}{2}k\left(-\frac{dx^2}{dx}\hat{i}\right)$$
$$= -\frac{1}{2}k(2x\hat{i})$$
$$\Longrightarrow \vec{F} = -k\vec{x},$$

which is the formula for elastic force that we will discuss in detail in chapter 7.

2.3.10 Energy equation

We have divided the energy into two classes. Kinetic energy (T), the energy possessed by an object in motion and potential energy (V), the energy an object has because of its position relative to some other object. We can express the sum of the kinetic and potential energies of an object as

$$T_o + V(x_o) = \text{constant} = T + V(x) = E, \tag{2.24}$$

where E is called the *total energy,* or technically the *mechanical energy*, of the system, which remains constant throughout the motion. This is also the statement of the *law of conservation of energy* that we saw in equation (2.2), which I will repeat here because of its importance in physics:

$$\Delta E = 0. \tag{2.25}$$

2.3.11 Conservative versus non-conservative forces

There are two main types of forces, conservative and non-conservative forces.

- *Conservative force.* A force for which the above equation (2.25) holds when the applied force is only the function of position (x) of the particle.

  Another way to define a conservative force is that the work done in moving a particle between two points is independent of the path taken. Or equivalently a particle traveling in a closed loop will have zero net work done.

  A common example of a conservative force is the force of gravity. In a gravitational field, suppose the potential energy (U) for an object at point 1 is (mgh_1) and the potential energy at another point 2 is (mgh_2). When the object moves from point 1 to point 2 the potential energy of the system will change. For a conservative force the mechanical energy of the object will be conserved and the change in potential energy will not depend on the path taken. It will only depend on the initial and final location of the object. A good example is of a child sliding down a frictionless slide. The work done by the gravitational force on the child from the top of the slide to the bottom will be the same irrespective of the shape (straight, spiral or bumpy) of the slide. The amount of work done (and hence the final speed at the bottom of the slide) only depends on the vertical displacement $(h_2 - h_1)$ of the child. For conservative forces the law of conservation of energy is stated as given in equation (2.25).

- *Non-conservative force.* Non-conservative forces are those that can change the total mechanical energy of a system. Friction is a good example for a non-conservative force. In this case the work done around a closed loop will not be zero. For non-conservative forces the law of conservation of energy is stated as

$$\Delta E = \text{work done by non-conservative force}. \tag{2.26}$$

Keep in mind that the total energy does not really change, just that some of it goes where we lose track of it, say for example into heat or sound.

2.3.12 Power

Power is the rate at which the applied force does work on an object. If the force does an amount of work ΔW in an interval of time Δt, then the average power will be given as

$$P_{av} = \frac{\Delta W}{\Delta t}. \tag{2.27}$$

Think of it in terms of elevators; when you want to get to the fortieth floor of a building, how much power you will need will be proportional to how fast you wish to get there. The faster you wish to go the more power you will need to expend.

Instantaneous power is the time derivative of the work, given as

$$P = \frac{dW}{dt}. \tag{2.28}$$

The SI unit of power is the *watt*, named after the Scottish inventor James Watt. One watt is equivalent to one joule per second. In the British system, the unit of power is 'foot-pounds per second'. In practice you may have heard the power of a car referred to in terms of 'horsepower', where

$$1 \text{ horsepower} = 1 \text{ hp} = 550 \text{ ft lbs}^{-1} = 745.7 \text{ W}. \tag{2.29}$$

This is roughly the rate at which a very strong horse will be able to perform work.

2.3.13 Power delivered by a force

During an infinitesimal displacement $d\vec{r}$, a force $\vec{F}$ will perform an amount of work given by

$$dW = \vec{F} \cdot d\vec{r}. \tag{2.30}$$

The instantaneous power delivered by this force will be

$$
\begin{aligned}
P &= \frac{dW}{dt} \\
&= \frac{\vec{F} \cdot d\vec{r}}{dt} \\
&= \vec{F} \cdot \frac{d\vec{r}}{dt} \\
\Longrightarrow P &= \vec{F} \cdot \vec{v}.
\end{aligned}
$$

2.3.14 Power versus watt-hours

Power is the rate at which energy is either generated by power plants or used by consumers. Power is defined as energy per unit time and its SI unit is a watt or joules per second.

The watt-hour is a unit of energy and hence its SI unit is the joule. Electrical energy is often sold in kilowatt hours (symbol kWh) and is used as a billing unit for energy delivered to homes and businesses by electric utility companies. The total energy in kilowatt hours is equal to the power in kilowatts multiplied by the time in hours. For example, a television rated at 100 W operating for 10 h continuously uses one kilowatt hour of energy, whereas a heater rated at 1 kilowatt will use one kilowatt hour of energy in an hour. In general, $1\ \mathrm{J} = 2.77 \times 10^{-7}$ kWh, or vice versa $1\ \mathrm{kWh} = 3.6 \times 10^{6}$ J.

2.3.15 Work done by a force at an angle to the displacement

A box is pulled through a displacement Δx by a rope held at an angle θ, as shown in figure 2.1. In order to find the work done on the box, we have to resolve the force into its horizontal ($F_x = F \cos\theta$) and vertical ($F_y = F \sin\theta$) components.

The work done on the box will be

$$W = \vec{F} \cdot d\vec{r} = |\vec{F}||\Delta\vec{x}|\cos\theta. \tag{2.31}$$

Recall that work is a dot product of two vectors, but itself is a scalar quantity, just a number.

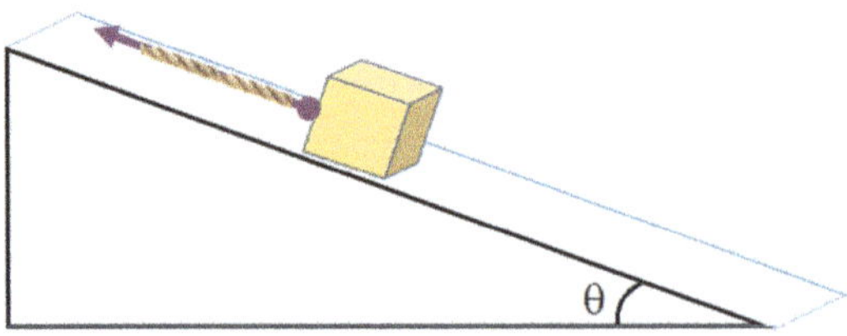

Figure 2.1. Mass on an incline.

Example 2.3

A 2000 kg elevator descends 1300 m in a sky-scraper. Calculate the work done by gravity on the elevator during this displacement.

Solution
Since the elevator is descending, the displacement will be negative and the weight of the elevator will also act downwards. Hence

$$\begin{aligned} \text{work} &= F_y \cdot dy \\ &= (-2000\ \text{kg})(-1300\ \text{m}) \\ &= 2.6 \times 10^6\ \text{J}. \end{aligned} \tag{2.32}$$

Example 2.4

Assuming the same 2000 kg elevator from example 2.3 descending 1300 m in a sky-scraper. Calculate the work done by the tension of the suspension cable during this displacement.

Solution
For a motion with constant velocity the upwards tensional force ($T = +2000$ kg) must balance the downwards weight of the elevator. Hence

$$\begin{aligned} \text{work} &= F_y \cdot dy \\ &= (+2000\ \text{kg})(-1300\ \text{m}) \\ &= -2.6 \times 10^6\ \text{J}. \end{aligned} \tag{2.33}$$

Note in examples 2.3 and 2.4 that the work done by the tensional force is equal in magnitude but opposite in direction to the work done by gravity. Hence, the net work done by both forces together is zero.

But this is a very specific case, as we have assumed motion with a constant velocity. This will not be the case when the motion is accelerated. For example, if the elevator started to move downwards with acceleration (g) then the tensional force and hence the work done from example 2.4 will go to zero, whereas the work done by gravity calculated for example 2.3 will remain the same.

Concept questions 2.1

1. Does the work done by a force on an object depend on the frame of reference? Explain with examples.
2. A ball is tied to a string. Can you rotate it vertically in front of you in a circle with constant speed? What about rotating it horizontally above your head?
3. For the ball in question 2, is the energy constant during both motions?
4. If you hold a rock in your hand while standing, are you doing work? What about if you start walking, are you doing work on the rock?

Exercises 2.1

1. What is the kinetic energy of the Earth due to its motion around the Sun?
2. A spring initially at rest has a force constant of 35 N m^{-1}. How much work do you have to do to compress it by 10 cm? How much more work do you have to do to compress the spring by 10 cm more?
3. A particle of mass m moves along the x-axis from $x = 3$ m to $x = 7$ m. During this motion a force of $\vec{F} = (2x^3 + 8x + 3)N$ acts on the particle. Calculate the work done by the force during the motion.
4. One way to store surplus energy from an electric power plant is to temporarily store it as gravitational potential energy by pumping water from a river into a reservoir at a height (say a water tower) above the level of the river. If the reservoir is 300 m above the river, how much water must we pump in order to store 10^{13} J of energy?

2.4 One-dimensional conservative system: complete solution

The main problem in physics is describing the motion of a particle as a function of time, i.e. find position vector $\vec{x}(t)$, velocity vector $\vec{v}(t)$, and hence $\vec{a}(t)$ and F. The two methods most commonly used are called:

- The *dynamical* or *force method*, which we discuss in chapter 3.
- The *energy method*, which we will talk about here. We will use the energy method to find $\vec{x}(t)$ and $\vec{v}(t)$.

Both methods give identical results.

Assume a 1D system with a force F that depends only on position. Note that in 1D situations, the forces that depend only on position are always conservative. This is not necessarily true in higher dimensional situations.

Associated with this given force F is the potential energy function defined as $V = V(x)$. Then from the definition of mechanical energy, $E = T + V(x)$, and kinetic energy $T = \frac{1}{2}mv^2$, we see that the motion of the particle can be obtained by solving the energy equation for v:

$$v = \pm\sqrt{\frac{2}{m}[E - V(x)]}. \tag{2.34}$$

For any particular value of the total energy (E) of the system, equation (2.34) tells us that the motion is restricted to the regions of the x-axis where $E > V(x)$; this is because we can *never* have a negative kinetic energy or an imaginary velocity.

2.4.1 Qualitative description

One can plot potential energy ($V(x)$) as a function of x to obtain a good qualitative description of the motion based on equation (2.34). As an example, let us consider the potential energy function as shown in the top part of figure 2.2, which represents

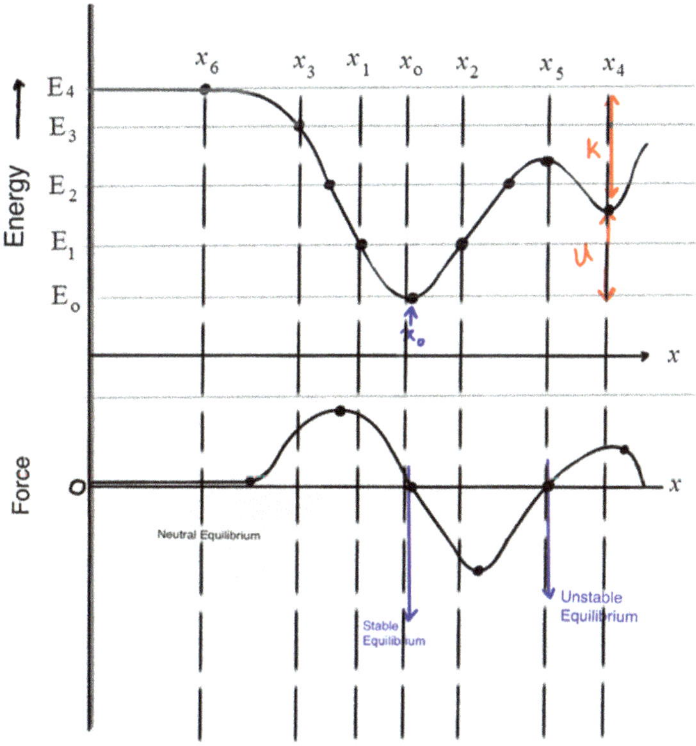

Figure 2.2. Qualitative description.

the potential energy of a particle moving in 1D along the x-axis. The relationship between potential energy and force is determined by

$$F = -\frac{dV}{dx}. \tag{2.35}$$

The force corresponding to this potential is shown in the bottom part of figure 2.2. Let us now consider several different possible choices for the total mechanical energy of the system.

For any particular value of energy, the kinetic energy (T) at any point is found by taking the difference between total energy (E) and potential energy (U). For example kinetic energy at point x_4 is found by $T_4 = E_4 - V_4$ at that point.

- *At energy* E_0. The lowest possible total energy of a system. Here $E = V$, which means that $T = 0$. This implies potential energy (V) is at its maximum value. We see that the particle at this point has zero velocity and hence it is at rest (located at x_o).
- *At energy* E_1. When our particle has this energy it moves in the region between x_1 and x_2. At this energy $T = (E - V)$ and, from figure 2.2, we see the particle has maximum kinetic energy, which means that it has maximum speed at x_o. As the particle approaches points x_1 or x_2, the speed of the particle decreases. At points x_1 and x_2 the particle will stop and reverse its

direction and will start going back towards x_o. Hence points x_1 and x_2 are called *turning points*.

- *At energy* E_2. When the particle has energy E_2, there will be four turning points, namely at points 1, 2, 3, 4 as seen in figure 2.2. The particle moves back and forth in either of the two valleys of the potential energy function.
- *At energy* E_3. At this energy value there is only one turning point located at x_3. If the particle is initially moving in the negative x-direction it will stop at x_3 and then move in the positive direction.
- *At energy* E_4. At energies above E_4 there are no turning points, no place for the particle to reverse direction. The speed changes according to equation (2.34).

2.4.2 Equilibrium

- *Neutral equilibrium*, an interval where $V(x)$ is a constant, such as near point x_6. At this point the slope equals zero and the particle can be displaced slightly without experiencing either a repelling or a restoring force.
- *Stable equilibrium* occurs near point x_0, where $V(x)$ has a minimum value. A particle is at rest here and the slope of the curve is zero. If the particle is displaced to either side, $F(x)$ will tend to return it to the stable equilibrium. For a particle displaced to the left, the force will be positive and for a particle displaced to the right the force will be negative, and both will move the particle towards x_0.
- *Unstable equilibrium* occurs near point x_5, where $V(x)$ has a maximum value. A particle is at rest here as well and the slope of the curve is still zero. However, if the particle is displaced to either side, $F(x)$ will tend to move it away from equilibrium. If the particle is moved to the right, positive force pushes the particles towards even larger x.

2.4.3 General solution for spring force

Suppose you have a mass–spring system made up of a mass (m) attached to the end of a spring (figure 2.3).

The potential energy stored in the spring is given by

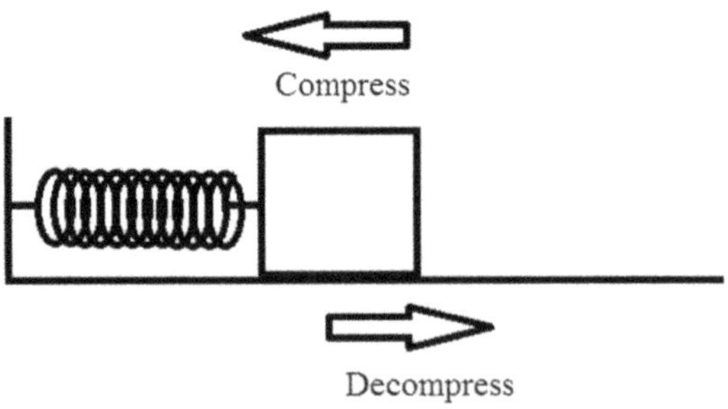

Figure 2.3. Mass–spring system.

$$V_{\text{spring}} = \frac{1}{2}kx^2, \tag{2.36}$$

where x is the displacement of the mass from equilibrium and k is the spring constant measured in newtons per meter (N m^{-1}), or kilograms per second squared (kg s^{-2}). By pulling slightly on the mass and then releasing it, the system will be set in sinusoidal oscillating motion about the equilibrium position, which we will talk more about in chapter 7.

General solution for spring force

Starting from equation (2.36) we can say at $t = 0$ the particle is located at x_o and it is at rest. This implies that the initial velocity (v_o) is zero and the mechanical energy is given by

$$E_0 = T_0 + V_0 = 0 + \frac{1}{2}kx_o^2. \tag{2.37}$$

Then knowing $v = \dot{x}$, we can rewrite equation (2.34) as

$$\int_0^t dt = \int_{x_o}^{x} \frac{1}{\sqrt{\frac{2}{m}(E - V)}} dx. \tag{2.38}$$

Then

$$\begin{aligned}
t &= \int_{x_o}^{x} \frac{1}{\sqrt{\frac{2}{m}(E - v)}} dx \\
&= \int_{x_o}^{x} \frac{1}{\sqrt{\frac{2}{m}\left(\frac{1}{2}kx_o^2 - \frac{1}{2}kx^2\right)}} dx \\
&= \int_{x_o}^{x} \frac{1}{\sqrt{\frac{k}{m}(x_o^2 - x^2)}} dx \\
&= \sqrt{\frac{m}{k}} \int_{x_o}^{x} \frac{1}{\sqrt{x_o^2 - x^2}} dx \\
&= \sqrt{\frac{m}{k}} \left(-\cos^{-1}\frac{x}{x_o}\right)\Bigg|_{x_o}^{x} \\
&= \sqrt{\frac{m}{k}} \left(-\cos^{-1}\frac{x}{x_o} + \cos^{-1}(1)\right) \\
t &= \sqrt{\frac{m}{k}} \cos\left(\frac{x}{x_o}\right).
\end{aligned}$$

This means that we can rewrite

$$\boxed{x(t) = x_o \cos\left(\sqrt{\frac{k}{m}}t\right)}. \tag{2.39}$$

This is a *sinusoidal oscillation.* Remember, $\cos(\pm\theta) \approx \cos\theta$ and for small angles $\cos\theta \approx 1$.

2.4.4 General solution for potential height

Gravitational potential energy is the energy an object has because of its position relative to a gravitating object, such as the Earth. At the top of the stairs you have more potential energy than when you are at the bottom, with respect to the Earth, because the Earth pulls you down. This is called the *gravitational force* and we will talk more about it in chapter 8. For now we will mathematically write gravitational potential energy as

$$V_{\text{height}} = mgh, \tag{2.40}$$

where h is the height of the object with respect to a particular reference. At $t = 0$, if the particle at rest, located at height (h), then the energy is given by $E = T + V = 0 + mgh = mgh$,

$$\begin{aligned} t &= \int_h^x \frac{1}{\sqrt{\frac{2}{m}(mgh - mgx)}}dx \\ &= \frac{1}{\sqrt{2g}}\int_h^x \frac{1}{\sqrt{h - x}}dx \\ &= -\sqrt{\frac{2}{g}}\left[\sqrt{h - x} - \sqrt{h - h}\right]. \end{aligned}$$

Simplifying, we obtain

$$\boxed{t = -\sqrt{\frac{2}{g}(h - x)}}. \tag{2.41}$$

Note that from chapter 3, equation (3.3), the the second equation of motion, we will obtain the same result, when the motion starts from rest ($v_o = 0$):

$$\begin{aligned} \Delta x &= v_0 t + \frac{1}{2}at^2 \\ \Delta x &= -\frac{1}{2}gt^2 \\ -\sqrt{\frac{2\Delta x}{g}} &= t. \end{aligned}$$

Concept questions 2.2

1. Think about a cone in the following situations:
   (a) Lying on its side.
   (b) Standing on its base.
   (c) Standing on its apex.

For which of these positions is the equilibrium stable, unstable or neutral? Explain your reasoning.

2. The potential energy of a system of two identical molecules is proportional to the distance between their centers. Can you explain why a spread out liquid has more potential energy than the same mass of liquid when it is in the shape of a sphere?
3. Air bags are vehicle safety devices and are designed to inflate extremely rapidly then quickly deflate during a collision or impact with a surface or a rapid sudden deceleration. In terms of energy transfers, explain how they do so.

Exercises 2.2

1. (a) Find velocities ($\dot{x}$) as a function of the displacement (x) for a particle of mass m, which starts from rest at $x = 0$ under the influence of forces given by
   $F_x = F_0 + cx$.
   $F_x = F_0\, e^{+cx}$.
   $F_x = F_0 \cos(cx)$.
   (b) Find the potential energy function $V(x)$ for each of the above cases, knowing

$$V(x) = -\int_{x_o}^{x} F_x dx. \tag{2.42}$$

2. Find the force $F(x)$ on a particle of mass (m) moving along a frictionless, horizontal plane with a speed

$$v(x) = \frac{\alpha}{x},$$

   where x is its distance from origin (O) and α is a positive constant.
3. A particle of mass (m) is constrained to a line along a frictionless horizontal plane subject to a force: $F = -kx$. It is projected from $x = 0$ to the right along the positive x-direction with initial kinetic energy

$$T_0 = \frac{1}{2}KA^2,$$

   where K and A are constants. Find the following:
   (a) $V(x)$, the potential energy function for this force.
   (b) $T = T(x)$, the kinetic energy.
   (c) Total energy as a function of its position.

Further reading

Arya A 1990 *Introduction to Classical Mechanics* 2nd edn (Upper Saddle River, NJ: Prentice-Hall)
Baierlein R 1983 *Newtonian Dynamics* (New York: McGraw-Hill)
Chababy R and Sherwood B 2011 *Matter and Interactions* 3rd edn (New York: Wiley)
Fowles G 1986 *Analytical Mechanics* 4th edn (New York: CBS College Publishing)
Fowles G and Cassiday G 2005 *Analytical Mechanics* 7th edn (Belmont, CA: Thomson Learning)
Ohanian H 1985 *Physics* (New York: WW Norton)
Resnick R, Halliday D and Krane K 2002 *Physics* vol 1, 5th edn (New York: Wiley)
Rutherford D E 1951 *Classical Mechanics* (New York: Interscience)
Stephenson R 1961 *Mechanics and Properties of Matter* 2nd edn (New York: Wiley)
Symon K 1971 *Mechanics* 3rd edn (Reading, MA: Addison-Wesley)

Chapter 3

Newtonian mechanics

3.1 Introduction

The ancient civilizations of history, such as the Egyptians, Mesopotamians and the people of the Indus Valley, had a robust understanding of the motion of the stars and planets. However, most of these civilizations believed these celestial objects to represent gods and be worthy of worship. The ancient Greeks were the first to seek natural explanations for these observations. Philosophers such as Thales (624–545 BCE) rejected supernatural explanations for natural phenomena.

Observations by ancient scientists laid the foundations of what we call *celestial mechanics*. Celestial mechanics hence became the study of how things move about the heavens.

By the 16th century, scientists were able to relate celestial mechanics to mathematics. Tycho Brahe (1546–1601 CE) was one of the first astronomers to use clocks capable of counting minutes and seconds, along with quadrants and sextants, to track the movements of celestial objects. Johannes Kepler (1571–1630 CE) based his three laws of planetary motion, known as *Kepler's laws of planetary motion*, on Brahe's data for the motion of Mars.

Around the same time, in his 1638 CE publication, *Dialogues Concerning Two New Sciences*, Galileo Galilei (1564–1642 CE) published the first mathematical proof that uniform acceleration would cause projectiles to move in parabolic trajectories that matched his experimental observations.

In 1687 CE, Newton published *Mathematical Principles of Natural Philosophy*, which describes the motion of objects under the influence of external forces. This work finally unified mathematical reasoning with the motion of objects on the surface of Earth to celestial mechanics and hence developed the field of *Newtonian mechanics*.

3.2 Rectilinear motion under uniform acceleration

In order to talk about motion let us begin with the simplest case of *rectilinear motion*.

doi:10.1088/2053-2563/aae1b9ch3

When a moving particle remains on a single straight line we call it rectilinear motion. If we use the x-axis as the line of motion, we can write the general equation for acceleration of the motion as

$$\vec{a}_x = \frac{d^2\vec{x}}{dt^2} = \ddot{x}. \tag{3.1}$$

Under the condition where acceleration is constant, we can derive equations for the position, displacement and velocity of any particle.

3.2.1 Equations of motion

The easiest way to derive these equations is by using calculus. The first step is to specify initial the conditions of the motion. At time $t = 0$, suppose the initial position vector is given as x_o and the initial velocity is given as v_o. We will also suppose that the acceleration (a) remains constant throughout the rectilinear motion.

Under these initial conditions we can write the equations of motion as follows:

- The first equation of motion:

$$v - v_o = at. \tag{3.2}$$

- The second equation of motion:

$$x - x_o = v_o t + \frac{1}{2}at^2. \tag{3.3}$$

  Here, the displacement vector is given as $\Delta x = x - x_o$.
- The third equation of motion:

$$2a(x - x_o) = v^2 - v_o^2. \tag{3.4}$$

3.2.2 Derivation of the equations of motion

As discussed in the previous section, in order to derive equations of motion we first specify initial conditions of the motion.

At $t = 0$, suppose the initial position is ($x = x_o$), the initial velocity is ($v = v_o$) and, throughout the motion, the acceleration (a) remains constant.

- The first equation of motion:

  To derive the first equation of motion, start from the definition of acceleration and solve as follows,

$$a = \frac{dv}{dt}$$
$$dv = adt$$
$$\int_{v_o}^{v} dv = \int_{0}^{t} adt$$
$$v - v_o = at.$$

- The second equation of motion:

  To derive the second equation of motion, start from the first equation of motion and solve,

$$
\begin{aligned}
v - v_o &= at \\
\frac{dx}{dt} &= v_o + at \\
\int_{x_o}^{x} dx &= \int_0^t (v_o + at)dt \\
x - x_o &= v_o t + \frac{1}{2}at^2 \\
\Delta x &= v_o t + \frac{1}{2}at^2.
\end{aligned}
$$

- The third equation of motion:

$$
\begin{aligned}
a &= \frac{dv}{dt} \\
a &= \frac{dv}{dt}\frac{dx}{dx} \\
a &= v\,\frac{dv}{dx}
\end{aligned}
$$

$$
\begin{aligned}
a\ dx &= v\ dv \\
\int_{x_o}^{x} adx &= \int_{v_o}^{v} v\ dv \\
a\ (x - x_o) &= \frac{v^2 - v_o^2}{2} \\
2a\Delta x &= v^2 - v_o^2.
\end{aligned}
$$

3.2.3 Free-falling objects

A free-falling object is accelerated toward the Earth with a constant acceleration numerically equal to $g = 9.8\ \mathrm{m\ s^{-2}}$. The values of various useful parameters can be related through the kinematic equations with the following few equations,

$$
\begin{aligned}
v - v_o &= -gt \\
y - y_o &= v_o t - \frac{1}{2}gt^2 \\
2a(y - y_o) &= v^2 - v_o^2,
\end{aligned}
\tag{3.5}
$$

where initial velocity is v_o, initial position is x_o and g is the acceleration due to gravity. The negative sign is added in these equations to show that g is always towards the center of the Earth. The displacement vector is given as ($\Delta y = y - y_o$).

3.3 Linear momentum

We introduced linear momentum in section 2.3.4—now let us talk about it in detail. Linear momentum, also called *momentum*, is a vector quantity used in classical mechanics. It is defined as the product of the mass (m) and velocity ($\vec{v}$) of a particle. Linear momentum is denoted as $\vec{p}$. The SI units of momentum are kg m s^{-1}. Since momentum is a vector quantity it will have both magnitude and direction. The momentum of a particle is always in the same direction as its velocity vector. For example, a 1 kg toy car, traveling due north at 1 m s^{-1} in a straight line, has a momentum of 1 kg m s^{-1} due north,

$$\vec{p} = m\vec{v} = m\frac{d\vec{x}}{dt}. \tag{3.6}$$

The dimensionality of linear momentum depends on the dimensionality of the velocity vector. For a particle with a 3D position vector $\vec{r} = x\hat{\imath} + y\hat{\jmath} + z\hat{k}$, the linear momentum is also a three-dimensional vector.

A system is a collection of two or more objects. Linear momentum is a conserved quantity, which means that for a closed system[1] its total linear momentum cannot change. A Newton's cradle demonstrates conservation of momentum.

For an isolated system[2], linear momentum depends on the frame of reference. Observers in different frames would find different values of linear momentum of a system, but each would observe that the value of linear momentum does not change with time.

For example, a person, say Jane, is standing (at rest) in a non-moving glass elevator. She will have zero velocity regardless of the location of the observer. However, if the elevator starts ascending with velocity (v), an outside observer, say Mary standing outside, will see that Jane is now moving with velocity (v) and will now have a non-zero momentum according to Mary. However, Jane in the elevator is just standing and not moving inside the elevator. Hence in Jane's frame of reference, her own momentum is still zero.

The two observers, Jane and Mary, each have a frame of reference, in which they observe the same motion with different observations. So who is incorrect? Jane with zero velocity in her frame or Mary in her frame who sees Jane with velocity v?

Suppose the position of Mary (in a stationary frame of reference) is x. Then from the point of view of Jane in the moving frame of reference Mary's position will be ($\hat{x}$). The relationship between the two frames changes with time and is given by a Galilean transformation as

[1] In classical mechanics, a closed system is a particular physical system that is not affected by net force external to the system and is a system that does not exchange any matter with its surroundings.

[2] An isolated system is a physical system so far removed from other systems that it does not interact with them.

$$\hat{x} = x - vt, \tag{3.7}$$

where v is the speed of the elevator. The speed of the elevator in Mary's unmoving frame of reference will be given as

$$v = \frac{dx}{dt} \tag{3.8}$$

and the speed of elevator in Jane's moving frame of reference will be given as

$$\hat{v} = \frac{d\hat{x}}{dt} = \frac{d(x - vt)}{dt} = \frac{dx}{dt} - \frac{dv}{dt}t - v\frac{dt}{dt} = v - at - v = at. \tag{3.9}$$

Then acceleration will be, given that $\hat{v} = at$,

$$\hat{a} = \frac{d\hat{v}}{dt} = \frac{d(at)}{dt} = a. \tag{3.10}$$

Hence we have shown $\hat{a} = a$, which means momentum is conserved in both reference frames as

$$\frac{d\hat{p}}{dt} = \frac{dp}{dt}. \tag{3.11}$$

3.4 Newton's laws of motion

In his 1687 CE *Mathematical Principles of Natural Philosophy*, Isaac Newton defined what are now called Newton's laws of motion as:

1. An object at rest will stay at rest and an object in motion stays in motion, unless it is compelled to change that state by an external force applied on it.
2. The change of motion is proportional to the force acting on it and is in the direction of the applied force.
3. The mutual actions of two bodies upon each other are always equal and opposite in direction.

3.4.1 Newton's first law: the inertial reference system

Newton's first law is also called the inertial law. This law describes an inherent property of matter called *inertia*. Inertia is the resistance of all matter to motion change. A particle at rest will resist being moved and a particle in motion resists being brought to rest. In both cases, an external force ($\vec{F}_{\text{ext}}$) is required to bring change in state. It is as though matter has been endowed with an innate abhorrence of acceleration. It always takes ($\vec{F}_{\text{ext}}$) to accelerate matter. When the force is in the direction of motion it makes the object speed up or accelerate and when the force is opposite to the direction of motion it makes the object slow down or decelerate.

3.4.1.1 Inertial frame of reference

A mathematical description of the motion of a particle requires the selection of a frame of reference. A frame of reference is a set of coordinates in *configuration space* (see appendix D for more details) that can be used to specify the position, velocity and acceleration of a particle at any instant of time. The frame of reference in which Newton's laws hold is called an *inertial frame of reference.*

Accelerated frames of reference will not be 'inertial frames', because an object at rest or moving at constant velocity in an accelerated frame as seen from the inertial frame of reference would *appear* to be accelerated.

3.4.1.2 Does a perfect or a true inertial reference frame exist in the Universe?

In an inertial frame, Newton's first law states that an object at rest remains at rest, and an object in motion keeps going, until acted upon by an external force. Newton's second law states that the net external force acting on a particle equals its mass times its acceleration. This means that Newton's first and second laws have to hold in all inertial frames.

Consider your textbook sitting on your table. It is at rest with respect to you. However, if you were on the space station looking at it, you will need to account for the rotation of the Earth. So in the frame of reference of the space station the book will appear to be moving with relative speed of space station and Earth's rotation. Hence, the space station frame of reference is not an inertial frame[3].

To account for these changes in frames of references, we sometimes talk about the *Earth-centered, Earth-fixed* (ECEF) frame or the *Earth-centered inertial* (ECI) frame. In both frames the origin is located at the center of the Earth. However, in the ECEF-frame the x-, y- and z-coordinates are aligned with the international reference pole (IRP) and international reference meridian (IRM) that are fixed with respect to the surface of the Earth. In contrast, in the Earth-centered inertial (ECI) coordinate frame the xy-plane coincides with the Earth's equatorial plane, but the x- and y-axes do not rotate with the Earth, and the z-axis extends through the North Pole. Keep in mind that as the Earth rotates the ECI does not rotate and remains fixed with respect to the celestial sphere (which does not rotate as the Earth rotates).

In summary, the Earth is not an inertial frame of reference since:

- The Earth is rotating about its axis, which gives it *centripetal acceleration* (a_c).[4]
- The Earth is revolving around the Sun at a certain speed.

However, in dealing with classical mechanical problems on the surface of the Earth we can neglect the Earth's motion. The Earth can be assumed to be approximately stationary and we can consider Earth as an inertial reference frame without any loss of generalization.

[3] Keep in mind that it is not the speed of rotation but the acceleration inherent in rotation that makes a non-inertial frame.

[4] An object moving in a circular path at constant speed experiences an acceleration due to the change in direction. Centripetal means 'toward the center' or 'center seeking'. Mathematically it is given as $a_c = \frac{v2}{r} = \omega^2 r$, where r is the radius of the circle, v is velocity and ω is the angular velocity.

However, there are effects such as Coriolis forces[5] that cause torques in projectile motion over long distances that are directly due to the spin of the Earth. Coriolis force have no sources and hence they violate Newton's laws and prove that Earth is not an inertial frame.

Example 3.1

Calculate the centripetal acceleration of a point at the equator on the Earth, due to the rotation of the Earth, given a radius of the Earth of $R_E = 6.4 \times 10^6$ m.

Solution

The centripetal acceleration of Earth is

$$a_c = \omega^2 R_E = (2\pi f)^2 R_E = \left(\frac{2\pi}{T}\right)^2 R_E.$$

We know that T = period = 24 h = 86 400 s. Hence,

$$a_c = \left(\frac{2\pi}{86\,400}\right)^2 (6.4 \times 10^6) = 0.0338 \text{ ms}^{-2}.$$

Example 3.2

Calculate the centripetal acceleration of Earth in its orbit around the Sun, given the average distance between the Earth and Sun, 92.96 million miles or $R_{SE} = 1.49 \times 10^{11}$ m.

Solution

The centripetal acceleration of the Earth around the Sun is

$$a_c = \omega^2 R_{SE} = (2\pi f)^2 R_{SE} = \left(\frac{2\pi}{T}\right)^2 R_{SE}.$$

We know that T = period = 365 days = 3.15×10^7 s. Hence,

$$a_c = \left(\frac{2\pi}{3.15 \times 10^7}\right)^2 (1.49 \times 10^{11}) = 0.0059 \text{ ms}^{-2}.$$

Example 3.3

Compare the centripetal acceleration calculated in example 3.1 to the centripetal acceleration calculated in example 3.2.

[5] Coriolis force is a virtual force that is a direct consequence of inertia. Coriolis forces exist only in a rotating reference frame, where they behave like a real force, that can cause acceleration and has real effects. However, the Coriolis force cannot be identified to be caused by a particular body and hence we call it a virtual force. Since the Earth is rotating, the Coriolis effect tends to deflect moving objects to the right in the Northern Hemisphere and to the left in the Southern Hemisphere.

Solution

$$\frac{\text{centripetal acceleration of Earth}}{\text{centripetal acceleration of Earth around Sun}} = \frac{a_{cE}}{a_{cES}} = \frac{0.0338}{0.0059} = 5.7.$$

This implies that the centripetal acceleration of the Earth is 5.7 times the centripetal acceleration of the Earth around the Sun.

3.4.2 Newton's second and third laws: mass and force

The quantitative measure of inertia is called *mass*. It is a common everyday experience that objects with more mass require more force to accelerate them. Think of a MINI Cooper and an eighteen-wheeler truck. If both start from rest, it requires less force to accelerate a MINI to 60 mph than an eighteen-wheeler truck.

Consider two masses (m_1) and (m_2) initially at rest in an inertial frame attached to each other with a spring placed on an air-track, as shown in figure 3.1. Let m_2 equal two times m_1. If we push the masses together by applying an external force, the spring is compressed. When the force holding the masses is removed, both masses spring away from each other.

Suppose they attain speeds v_1 and v_2, respectively. We know that since initial velocities are zero for both masses, the final velocities must be opposite in direction. We can define the ratio of masses as

$$\frac{m_2}{m_1} = \left| \frac{v_1}{v_2} \right|. \tag{3.12}$$

Rewriting the above equation $\Delta(m_1 v_1) = -\Delta(m_2 v_2)$ and dividing by Δt will give us the above equation in terms of linear momentum:

$$\frac{dp_1}{dt} = -\frac{dp_2}{dt}. \tag{3.13}$$

Hence mathematically Newton's second law is

$$F = k\frac{d}{dt}(m\vec{v}), \tag{3.14}$$

where k is a constant of proportionality, which is usually taken to be unity for SI units. Hence the statement of Newton's second law is:

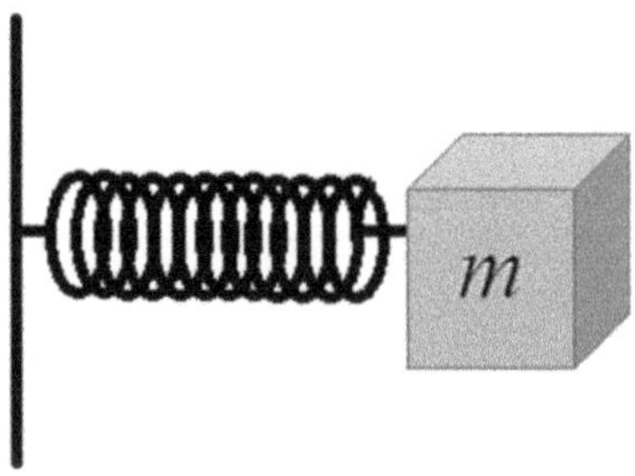

Figure 3.1. Two masses connected with a spring.

'The time rate of change of an object's linear momentum is proportional to the external or applied force, $\vec{F}$'.

When we talk about situations where the mass does not change, the equation (3.14) can be rewritten as

$$F = k\frac{d}{dt}(m\vec{v}) = k\frac{d\vec{p}}{dt} = km\frac{d\vec{v}}{dt} = km\vec{a}. \tag{3.15}$$

Then in SI units, Newton's second law for a constant mass is written as

$$F = \frac{d\vec{p}}{dt} = m\frac{d\vec{v}}{dt} = m\vec{a}, \tag{3.16}$$

then from equations (3.13) and (3.16) we can easily see

$$F_1 = -F_2, \tag{3.17}$$

which is better known as Newton's third law and states:

'Two interacting bodies exert equal and opposite forces upon one another.'

From equation (3.13) we can also see that

$$p_1 + p_2 = \text{const}. \tag{3.18}$$

Equation (3.18) is a special case of the more general situation in which the total linear momentum of an isolated system is a conserved quantity.

$$\Delta p = 0. \tag{3.19}$$

Remember, conservation of linear momentum is one of the fundamental laws of physics. It is always guaranteed to hold, even in cases where Newton's laws may fail.

Example 3.4

A spaceship of mass (M) is traveling in space with velocity $v_i = 40\,\text{km min}^{-1}$ relative to the Sun. It ejects a rear-stage rocket of mass $0.4M$ with a relative speed $u = 10\,\text{km min}^{-1}$ at time t, as shown in figure 3.2. What is the final velocity ($\vec{v}_f$) of the spaceship after ejection of the rear-stage rocket?

Solution

The system is made up of the spaceship and the rear-stage rocket together and we can assume that they make a closed system, which implies there is no external force. Hence we can say conservation of momentum holds and the initial momentum will equal the final momentum of the system.

Assume that we take the direction of the spaceship as positive, then momentum before ejection will be given as

$$p_i = Mv_i.$$

If we say that the velocity of the ejected rear-stage is given as $\vec{U}$, which is the difference in velocities of the spaceship and the rear-stage rocket and can be written as ($\vec{U} = v_f - u$). Then we can write total momentum after ejection as

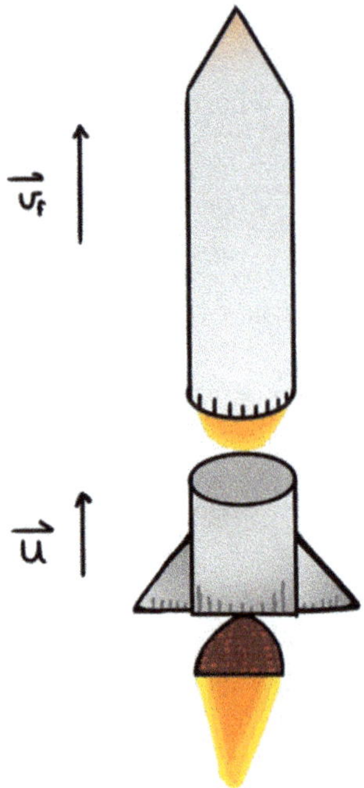

Figure 3.2. Spaceship example.

$$\vec{p}_f = 0.40M\vec{U} + 0.60M\vec{v}_f$$
$$\vec{p}_i = 0.40M(\vec{v}_f - \vec{u}) + 0.60M\vec{v}_f$$
$$M\vec{v}_i = (\vec{v}_f - 0.40\vec{u})M.$$

Putting in values we will obtain the final velocity of the spaceship as

$$\Longrightarrow \vec{v}_f = 44\,\frac{\text{km}}{\text{min}}.$$

3.4.3 Motion of a particle

Equation (3.16) is the most fundamental equation of classical mechanics when considering SI units. It describes the motion of a particle subjected to a net external force, F_{net}:

$$\Sigma F = F_{\text{net}} = m\vec{a} = m\frac{d\vec{v}}{dt} = m\frac{d^2\vec{r}}{dt^2} = m\ddot{r}. \tag{3.20}$$

For objects that are rest or moving with constant velocity

$$\Sigma F = 0, \tag{3.21}$$

whereas for accelerating objects

$$\Sigma F = m\vec{a}. \tag{3.22}$$

3.4.4 Problem solving: the motion of a particle

A common classical mechanics problem is to predict the motion of a particle as a function of time given some initial conditions. The usual solution involves using equation (3.20) along with the equations of motion we derived in section 3.2. A good idea is to follow this prescription:

1. Choose a suitable inertial frame of reference.
2. Set up a suitable coordinate system with a user-defined origin.
3. Draw a free-body diagram for each object in the problem. The free-body diagram should include all the forces acting on the object and their directions.
4. Where needed, find the components of forces along the coordinate axes.
5. For each object, write the Newton's force equation, which is the vector sum of all the forces acting on the object. Usually, this means adding the x-, y- and z-components of the forces separately.

Example 3.5

Ron pushes two boxes of mass m_1 and m_2 with a push force equal to $\vec{F}$, as shown in figure 3.3. Find the accelerations of each box in the following cases:

(a) Suppose both boxes slide over a floor without friction.
(b) Suppose both boxes slide over a carpeted floor with friction f_f.

Solution
(a) For m_1, Newton's force equation will be written as

$$\begin{cases} \Sigma F_x &= F - P_{12} \\ \Sigma F_y &= N_1 - m_1 g \end{cases}, \tag{3.23}$$

where P_{12} is the force exerted by m_2 on m_1.

For m_2, Newton's force equation will be written as

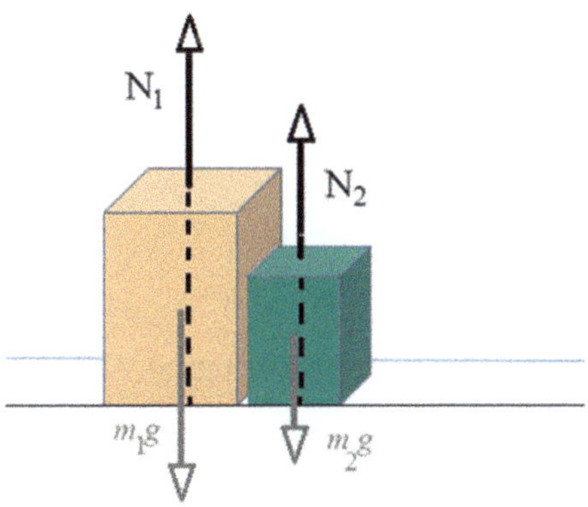

Figure 3.3. Ron pushes two boxes of mass m_1 and m_2.

$$\begin{cases} \Sigma F_x &= P_{21} \\ \Sigma F_y &= N_2 - m_2 g \end{cases}, \tag{3.24}$$

where P_{21} is is the force exerted by m_1 on m_2.

For m_1, $\Sigma F_x = m_1 a_1$ and for m_2, $\Sigma F_x = m_2 a_2$, and since both boxes are connected we say $a_1 = a_2 = a$. There is no motion in the y-axis, so ΣF_y for both masses will equal zero. Then the above equations can be rewritten as follows.

For m_1, Newton's force equation will be written as

$$\begin{cases} m_1 a &= F - P_{12} \\ 0 &= N_1 - m_1 g \end{cases}. \tag{3.25}$$

For m_2, Newton's force equation will be written as

$$\begin{cases} m_2 a &= P_{21} \\ 0 &= N_2 - m_2 g \end{cases}. \tag{3.26}$$

Adding F_x components for m_1 and m_2 and knowing $P_{12} = P_{21}$ we find the acceleration as

$$\begin{aligned} (m_1 + m_2)a &= F \\ a &= \frac{F}{m_1 + m_2}. \end{aligned} \tag{3.27}$$

(b) For m_1, Newton's force equation will be written as

$$\begin{cases} \Sigma F_x &= F - P_{12} - f_f \\ \Sigma F_y &= N_1 - m_1 g \end{cases}. \tag{3.28}$$

For m_2, Newton's force equation will be written as

$$\begin{cases} \Sigma F_x &= P_{21} - f_f \\ \Sigma F_y &= N_2 - m_2 g \end{cases}. \tag{3.29}$$

For m_1, $\Sigma F_x = m_1 a_1$, and for m_2, $\Sigma F_x = m_2 a_2$, and since both boxes are connected we say $a_1 = a_2 = a$. Again, there is no motion in the y-axis, so ΣF_y for both masses will equal zero. Then the above equations will be rewritten as follows. For m_1, Newton's force equation will be written as

$$\begin{cases} m_1 a &= F - P_{12} - f_f \\ N_1 &= m_1 g \end{cases}. \tag{3.30}$$

For m_2, Newton's force equation will be written as

$$\begin{cases} m_2 a &= P_{21} - f_f \\ N_2 &= m_2 g \end{cases}. \tag{3.31}$$

The frictional force is given as ($f_f = \mu N$), where N is the normal force which implies that we can rewrite $f_f = \mu mg$ in the above equations,

$$\begin{aligned} m_1 a &= F - P_{12} - \mu m_1 g \\ m_2 a &= P_{21} - \mu m_2 g. \end{aligned} \tag{3.32}$$

Adding F_x components for m_1 and m_2 and knowing $P_{12} = P_{21}$ we find the acceleration as

$$(m_1 + m_2)a = F - \mu g(m_1 + m_2)$$

$$\boxed{a = \frac{F - \mu(m_1 + m_2)g}{m_1 + m_2}}. \tag{3.33}$$

3.4.5 Force as a function of time: impulse

If the force on a particle is known explicitly as a function of time, then the equation of motion for a constant mass can be written as

$$\int_0^t F(t)dt = mv - mv_o \tag{3.34}$$

$$\boxed{\int_0^t F(t)dt = \Delta p}. \tag{3.35}$$

Hence we can define the impulse as:

'Change of momentum imparted to a body by the force acting over a certain time interval'.

We can find the velocity as a function of time starting from equation (3.34),

$$v = v_o + \frac{1}{m}\int_0^t F(t)dt. \tag{3.36}$$

The position as a function of time ($dx = vdt$) will be given by integration as

$$x - x_o = v_o t + \frac{1}{m}\int_0^t \int_0^{t'} [F(t')dt']dt. \tag{3.37}$$

Next let us talk about various kinds of forces that can act on objects.

- *Constant force* means that the applied force is not a function of time. Then from equation (3.36) we find the velocity as a function of time:

$$v(t) = v_o + \frac{F_o}{m}\int_o^t dt$$

$$v(t) = v_o + \frac{F_o}{m}t, \tag{3.38}$$

which is just the equation of motion we found in equation (3.2). If we take $\frac{F_o}{m} = a$, then from equation (3.37) we find the position as a function of time as

$$x(t) = x_o + v_o t + \int_0^t \frac{Ft}{m} dt$$

$$x(t) = x_o + v_o t + \frac{Ft^2}{2m}, \tag{3.39}$$

which is just the second equation of motion we found in equation (3.3) taking $\frac{F_o}{m} = a$ again.

- *Step force*. Given a body of mass (m) which is first subject to a constant force F_1 that acts for time t_1. Then suddenly the force is changed to another constant force (F_2) that acts until time t_2. Then for time interval $t = [0, t_1]$ we will have velocity and position

$$v(t) = v_o + \frac{F_1 t}{m} \tag{3.40}$$

and

$$\Delta x = v_o t + \frac{F_1 t^2}{2m}, \tag{3.41}$$

and for the time interval $t = [t_1, t_2] = [t, t - t_1]$ the velocity and position will be given as

$$v = v_o + \frac{F_1 t_1}{m} + \frac{F_2}{m}(t - t_1) \tag{3.42}$$

and

$$\Delta x = v_o t_1 + \frac{F_1 t_1^2}{2m} + \left(v_o + \frac{F_1 t_1}{m}\right)(t_1 - t_2) + \frac{F_2}{2m}(t_1 - t_2)^2. \tag{3.43}$$

- *Uniformly increasing force (the jerk)*. Suppose a body of mass (m) initially at rest at $t = 0$, is subjected to a constantly increasing force of the form

$$F(t) = ct, \tag{3.44}$$

where c is a constant of proportionality. Then we can write the equation of motion as

$$m\frac{dv}{dt} = ct.$$

We can solve the above equation to obtain velocity as

$$v = \int_0^t \frac{ct}{m} dt = \frac{ct^2}{2m} \tag{3.45}$$

and the position will be given as

$$x = \int_0^t \left(\frac{ct^2}{2m}\right) dt = \frac{ct^3}{6m}. \tag{3.46}$$

Note that the third derivative of x w.r.t. time is constant. Here, $\frac{d^3x}{dt^3} = \frac{c}{m}$ represents the time rate of change of acceleration and is called the *jerk*.

Concept questions 3.1

1. As the Moon goes around the Earth, it causes the tides in the oceans to rise and fall, through which the continents move as the Earth turns. This causes friction by which the Earth's rotation slows. Explain why this does not violate the law of conservation of energy.
2. Why does a heavy gun kick less than a light one?
3. Discuss the energy changes in the water-evaporation cycle, where the water flows over Niagara falls, evaporates and returns to the head of the falls.

Exercises 3.1

1. The speed of a car traveling due east is reduced from 45 mph to 30 mph in about 80 m.
   (a) Find the magnitude and direction of the constant acceleration.
   (b) How much time has elapsed during this deceleration?
   (c) If we continue to decelerate at the same rate, how much time would elapse to bring it to rest from 45 mph?
   (d) What total distance is required to bring the car to rest from 45 mph?
2. A rocket is projected vertically upwards from the ground with an initial speed of 80 m s^{-1}. Assume it is not subjected to air drag and find the following:
   (a) How long will it take to reach the highest point?
   (b) How high will the rocket rise?
   (c) At what time will the rocket be 100 m above the ground?
3. Suppose a particle of mass (m) starts from rest at $x = 0$ and $t = 0$ and is subject to the following force functions. Find the velocity and position as a function of time for each of the following cases:
   (a) $F(t) = F_0 t$,
   (b) $F(t) = F_0 \sin(\omega t)$,

(c) $F(t) = F_0 e^{\omega t}$,
where F_0 is force at $t = 0$ and ω is called angular frequency. Consider both F_0 and ω to be constants for this question.

3.5 Torque

The moment of a force $\vec{F}$ about a point, called torque ($\vec{\tau}$), is its turning effect about the point. It is quantified as the product of the force and the perpendicular distance from the point to the line of action of the force. Moments of force, or torques, are represented by cross products. Let a force $\vec{F}$ act at a point P(x, y, z), as shown in figure 3.4, and let the position vector be defined as $\vec{r} = x\hat{\imath} + y\hat{\jmath} + z\hat{k}$; then the torque $\vec{\tau}$ about a given point 0 is defined as

$$\boxed{\vec{\tau} = \text{(moment arm) (force)} = \vec{r} \times \vec{F}}. \tag{3.47}$$

Thus, the moment of a force about a point is a vector quantity having a magnitude and a direction. Suppose we have an object initially at rest but free to turn about a fixed point 0 acting as a pivot, then the body tends to rotate under the effect of a single force. The axis of this rotation is perpendicular to the plane containing both the position vector ($\vec{r}$) and the force ($\vec{F}$). In figure 3.4 both $\vec{r}$ and $\vec{F}$ are in the xy-plane, hence the direction of the torque vector ($\vec{\tau}$) will be along the perpendicular (z) axis, called the *axis of rotation*. The magnitude of the torque is given by

$$\boxed{|\vec{\tau}| = |\vec{r} \times \vec{F}| = rF \sin\theta}, \tag{3.48}$$

in which (θ) is the angle between ($\vec{r}$) and ($\vec{F}$). Thus, $|\tau|$ can be regarded as the product of the magnitude of the force and the quantity ($r \sin\theta$), which is just the perpendicular distance from the line of action of the force to the point 0. The SI unit of torque is the newton-meter.

When several forces are applied to a single body at different points, from the distributive law of vector multiplication we know the moments will add vectorially:

$$\vec{\tau}_{\text{net}} = \vec{\tau}_1 + \vec{\tau}_2 + \vec{\tau}_3 + \cdots. \tag{3.49}$$

3.5.1 For rotating objects

3.5.1.1. Work done by a constant torque

If a force acts through a particular distance, the force does mechanical work mathematically given as force × distance.

Similarly, if a rotating force, torque, acts through a rotational distance (angle), it does work mathematically given as

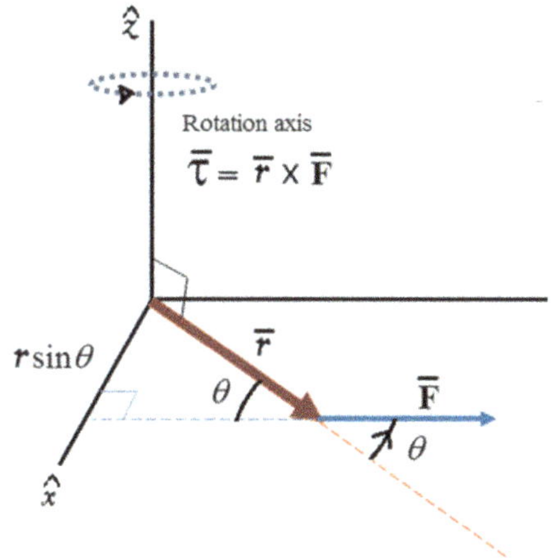

Figure 3.4. Torque or moment of force about O.

$$\boxed{W = \int_{\theta_1}^{\theta_2} \tau d\theta}, \tag{3.50}$$

where W is work done, τ is torque, and θ_1 and θ_2 represent (respectively) the initial and final angular positions of the body. In other words, the work done by a torque is to move the object through a particular angle.

3.5.1.2. Rotational kinetic energy (T_r*)*

From the work–energy theorem W also represents the change in the rotational kinetic energy (T_r) of the body, given mathematically as

$$\boxed{T_r = \frac{1}{2} I\omega^2}. \tag{3.51}$$

*3.5.1.3. Power (*P*)*

Power (with SI units of watts) is the work per unit time, given by

$$\boxed{P = \vec{\tau} \cdot \vec{\omega}}, \tag{3.52}$$

where $\vec{\tau}$ is torque, $\vec{\omega}$ is the angular velocity and $(\cdot)$ represents the scalar product. Note that:

- Equation (3.52) may be rearranged to find torque for a given angular speed and power output.
- Power injected by the torque depends only on the instantaneous angular speed and not on whether the angular speed increases, decreases, or remains constant while the torque is being applied.

3.5.1.4. Principle of moments

The principle of moments, also known as Varignon's theorem, states that:

'The sum of torques due to several forces applied to a single point is equal to the torque due to the sum (resultant) of the forces'.

Mathematically we write Varignon's theorem as

$$(\vec{r} \times \vec{F_1}) + (\vec{r} \times \vec{F_2}) + \cdots = \vec{r} \times (\vec{F_1} + \vec{F_2} + \cdots). \tag{3.53}$$

3.5.1.5. For objects in equilibrium

The vector sum of all the moments under equilibrium is zero, and is mathematically expressed as

$$\Sigma\tau = 0. \tag{3.54}$$

3.5.1.6. Torques due to inertia

Torque in terms of the moment of inertia is discussed in detail chapter 15. Here I will just use it to define torque (τ) as

$$\boxed{\vec{\tau} = I\alpha}, \tag{3.55}$$

where I is the moment of inertia and α is the angular acceleration.

Example 3.6

A spanner that is used to tighten a nut is 30 cm long. The force exerted on the end of a spanner is 100 N.

(a) What is the torque exerted on the nut?
(b) What is the work done when the nut turns through 45°?

Solution

(a) The torque exerted on the nut will be

$$\tau = F \times r = 30 \text{ N m}.$$

(b) The work done when the nut turns through 45° will be

$$W = \tau\theta = 30\frac{\pi}{4} \text{ J}.$$

Concept questions 3.2

1. A commonly held belief is that 'buttered toast always lands buttered side down'. Can you prove if this statement is true or not?
2. The wheel is one of the most important inventions of all time. Explain.
3. Sit in a straight-backed chair without arm rests and try to stand up without leaning forward. Why can you not do this?

Exercises 3.2

1. A car repair manual states that certain bolts are to be tightened to a torque of 50 N m. If you use a wrench 8 inches in length, what perpendicular force must be exerted on the end of the wrench to achieve the correct torque?
2. What is the torque about the origin on an object located at (1, −2, 2) m due to a force given by, $\vec{F} = (3, 3, 4)$ N. Express your answer in the vector notation.

Further reading

Arya A 1990 *Introduction to Classical Mechanics* 2nd edn (Upper Saddle River, NJ: Prentice-Hall)

Baierlein R 1983 *Newtonian Dynamics* (New York: McGraw-Hill)

Chababy R and Sherwood B 2011 *Matter and Interactions* 3rd edn (New York: Wiley)

Fowles G and Cassiday G 2005 *Analytical Mechanics* 7th edn (Belmont, CA: Thomson Learning)

Ohanian H 1985 *Physics* (New York: WW Norton)

Resnick R, Halliday D and Krane K 2002 *Physics* vol 1, 5th edn (New York: Wiley)

Stephenson R 1961 *Mechanics and Properties of Matter* 2nd edn (New York: Wiley)

Symon K 1971 *Mechanics* 3rd edn (Reading, MA: Addison-Wesley)

Chapter 4

Lagrangian mechanics

4.1 Lagrangian mechanics

Lagrangian mechanics was first introduced by the Italian-French mathematician and astronomer Joseph-Louis Lagrange in 1788 CE, almost 120 years after Newton's *Mathematical Principles of Natural Philosophy*. Lagrange did not introduce any new physics in his Lagrangian mechanics, he just made a mathematical change to the formalism of Newtonian mechanics. Basically, Lagrangian mechanics is just another way of looking at Newtonian mechanics.

Hence we will begin the discussion of Lagrangian mechanics by recalling fundamental equations from Newtonian mechanics. In chapter 2, we expressed force as a function of position in equation (2.19) as

$$\vec{F}(x) = -\frac{dV(x)}{dx}. \tag{4.1}$$

Most forces we deal with every day, with the possible exception of the force of friction, are conservative. A conservative force can be represented as the gradient of a potential, as in equation (4.1). So we can rewrite Newton's second law in 3D vector form for objects, whose mass remains constant, under the influence of conservative forces as

$$\vec{F}(x) = -\frac{\partial V}{\partial x}\hat{i} - \frac{\partial V}{\partial y}\hat{j} - \frac{\partial V}{\partial z}\hat{k} = -\nabla V, \tag{4.2}$$

where the potential $V = V(x, y, z)$ is a scalar quantity and $\nabla \equiv \frac{\partial}{\partial x}\hat{i} - \frac{\partial}{\partial y}\hat{j} - \frac{\partial}{\partial z}\hat{k}$.

In chapter 3 we expressed, in the following two equations,

$$\begin{aligned} &\text{Newton's second law:} \quad \vec{F} = \frac{d\vec{p}}{dt} \\ &\text{Newton's third law:} \quad \vec{F}_{12} = -\vec{F}_{21}. \end{aligned} \tag{4.3}$$

doi:10.1088/2053-2563/aae1b9ch4

4.2 From Newtonian to Lagrangian formalism

Newtonian mechanics defines and evaluates the motion of particles and is based on Newton's laws given by equation (4.3) along with the equation of motion. There are two main issues with this formalism:

- One is that Newton's second law has to apply to each particle and each coordinate of each particle in a given system individually. This means we need three equations to completely describe a system of one object moving in 3D space. A two-object system moving in 3D space will need six individual equations to describe it. Hence, for an n-particle system moving in 3D space, we will need $3N$ second order ordinary differential equations to describe it completely. Which is no easy feat for any system made of more than three particles.
- The second issue with Newtonian mechanics is that Newton's laws are relationships among vectors, which becomes complicated and messy when we try to change coordinate systems.

4.2.1 Advantage of Lagrangian mechanics

The *Lagrangian formulation* of classical mechanics is based on equation (4.2), which is just an alternative form of Newton's laws applicable where forces are conservative. Some advantages of Lagrangian mechanics can be summed up as:

- Lagrangian mechanics uses the energies in the system instead of forces. Energy is a scalar quality, hence we do not have to deal with vectors in this formalism, which makes it easier to deal with many complicated systems, as we will see in this book.
- Additionally, since the Lagrangian formulation just uses scalars, coordinate transformations tend to be much easier.

The central quantity of Lagrangian mechanics is the *Lagrangian*, a function which summarizes the dynamics of the entire system and is defined mathematically as

$$\boxed{L(\dot{\vec{r}}, \vec{r}) \equiv T(\dot{\vec{r}}) - V(\vec{r}),} \tag{4.4}$$

where $T = \frac{1}{2}\sum_{k=1}^{N} m_k v_k^2$ is the total kinetic energy and V is the potential energy of the system under consideration.

Overall, the Lagrangian has units of energy, but no single expression for all physical systems. Any function which generates the correct equations of motion, and is in agreement with physical laws, can be taken as a Lagrangian.

While writing the Lagrangian, keep in mind that the Lagrangian varies with the potential energy of the system. The potential energy of the system will give the amount of energy on any particle due to all the other particles and other external influences in a system:

- For conservative forces, potential is a function of the position vectors of the particles, $V = V(r_1, r_2, \ldots)$. For example, gravitational potential is given as $V(z) = mgz$.
- For non-conservative forces, such as the electromagnetic potential, potential is a function of the position and velocities of the particles, $V = V(r_1, r_2, \ldots, v_1, v_2, \ldots)$.
- If an external force acts on the system which changes with time, the potential of the system will also change with time; in this case, $V = V(r_1, r_2, \ldots, v_1, v_2, \ldots, t)$.

4.2.2 Principle of least action

In my definition of the Lagrangian in equation (4.4), note the minus sign between T and V. The negative sign is there to describe the *principle of least action.* The principle of least action states that:

'The actual path taken by the system is an extremum (whether it is a maxima or minima) of action (S)'.

In physics, *action* (say, S) is a quantity that describes the overall quantity of motion of a physical system. In most physics problems we are concerned with the initial and final states of the systems and do not necessarily concern ourselves with instantaneous changes in the system.

Think about it—when we wrote the equations of motion in chapter 3 we only talked about the initial and final velocities of a moving object and we only talked about Δt, which is the difference between the initial (t_i) and final (t_f) times. We hardly ever talked about the path the object actually takes. However, in order to get from the initial to the final state, the system has many options and many paths it may follow. We can call each route a system is allowed to take between initial and final states to be characterized by a specific numerical quantity which we call its *action.* According to Encyclopedia Britannica, we can think of action (S) as either

$$\begin{aligned} \text{action } (S) &= (2 \times \text{average kinetic energy of the system}) \\ &\quad \times (\text{time interval between the initial and final states}), \\ \text{action } (S) &= 2 \times T_{av} \times \Delta t \end{aligned} \tag{4.5}$$

or equivalently as

$$\begin{aligned} \text{action } (S) &= (\text{average momentum of the system}) \\ &\quad \times (\text{length of the path between the initial and final states}). \\ \text{action } (S) &= p_{av} \times \Delta \ell \end{aligned} \tag{4.6}$$

Feynman's lectures

Richard Feynman in his *Lectures*[1], volume 2, chapter 19, gives an example of an object in a gravitational field. Suppose the object goes from point A to point B in, say, time Δt. Now, you try a different motion, however, you make it a condition that

[1] http://www.feynmanlectures.caltech.edu/II_19.html.

it has to still get there in time Δt. If you calculate the kinetic energy at every moment on the path, take away the potential energy and integrate it over the time during the whole path, you will find that the number you get is bigger than that for the actual motion. According to Feynman:

> 'The laws of Newton could be stated not in the form $F = ma$ but in the form: the average kinetic energy less the average potential energy is as little as possible for the path of an object going from one point to another'.

We observe that when the object is projected upwards it first rises with the initial speed of projection, then as times passes it slows down until it stops at the top of the motion. This happens because we have both kinetic (T) and potential (V) energies to contend with during the motion and the system on average must have the least difference between kinetic and potential energies. Then the integral can be written as

$$S(x(t)) = \int_{t_i}^{t_f} \left(\frac{1}{2}mv^2 - mgy\right)dt, \tag{4.7}$$

where $(\frac{1}{2}mv^2)$ is the kinetic and (mgy) is the potential energy of the object.

The motion of the object is a parabola and will give a certain numerical result for the value of action. But if we pick another curved path, not a parabola, and calculate the integral we will find that it will give another value of action which will be more than that for a parabolic *true* path. Hence we define action as

$$S(x(t)) \equiv \int_{t_i}^{t_f} (T - V)dt \tag{4.8}$$

and our mathematical problem is to find the correct curve for which the action number is the least. This will give us the *true* path of any system under consideration.

The value of action (S) for any actual motion of a system between two configurations is always a minimum or a maximum. In most instances, the behavior of the system follows the path of minimum, or the *least action*, hence we call it the *principle of least action*.

From equation (4.8) we see that action has dimensions of [energy][time], and its SI unit will be joule second, which is the same as the units for angular momentum, which is a happy coincidence.

4.2.2.1 Proof: principle of least action

To describe the principle of least action, we consider two fixed end points from some initial point say x_i to some final point x_f as shown in figure 4.1. There are infinitely many smooth paths $x(t)$ from $x_i(t_i)$ to $x_f(t_f)$, but of all these possible paths, only one is the true path taken by the system. The question is, which one?

To each path, let us assign a number called the action (S) defined as

$$S(x(t)) = \int_{t_i}^{t_f} (T - V)dt = \int_{t_i}^{t_f} L(x, \dot{x})dt, \tag{4.9}$$

where we call $L = L(x, \dot{x})$ the *Lagrangian*.

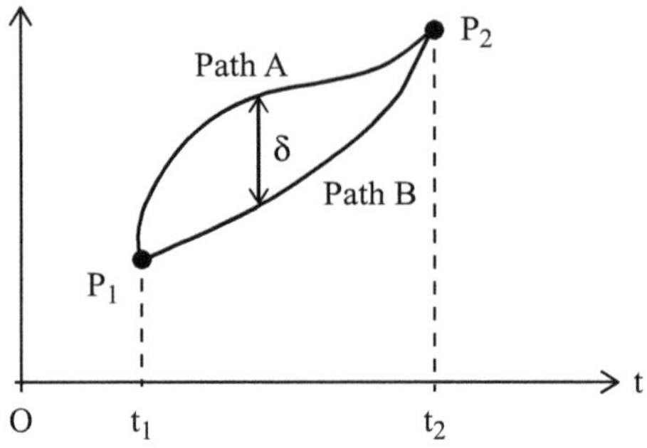

Figure 4.1. Different paths of integration.

Consider varying a given path slightly, such that $x(t) \to x(t) + \delta x(t)$, where we keep the end points of the path fixed by demanding $\delta x(t_i) = \delta x(t_f) = 0$. Then the change in the action will be

$$\begin{aligned}\delta S(x(t)) &= \delta\left[\int_{t_i}^{t_f} L(x, \dot{x})dt\right] \\ &= \int_{t_i}^{t_f} \delta L(x, \dot{x})dt \\ &= \int_{t_i}^{t_f} \left(\frac{\partial L}{\partial x}\delta x + \frac{\partial L}{\partial \dot{x}}\delta \dot{x}\right)dt.\end{aligned} \tag{4.10}$$

Integrating by parts ($\int udw = uw - \int wdu$), we obtain

$$\delta S(x(t)) = \int_{t_i}^{t_f} \left(\frac{\partial L}{\partial x} - \frac{d}{dt}\left(\frac{\partial L}{\partial \dot{x}}\right)\right)\partial x dt + \frac{\partial L}{\partial \dot{x}}\partial x \Big|_{t_i}^{t_f}. \tag{4.11}$$

The final term ($\frac{\partial L}{\partial \dot{x}}\partial x$) vanishes since we have fixed the end points of the path. So changes in initial and final points will be given by $\delta x(t_i) = \delta x(t_f) = 0$.

The requirement that the action is an extremum says that $\delta S = 0$ for all changes in the path $\delta x(t)$. We see that this holds if and only if

$$\boxed{\frac{d}{dt}\left(\frac{\partial L}{\partial \dot{x}}\right) - \frac{\partial L}{\partial x} = 0.} \tag{4.12}$$

This is known as *Lagrange's equation* (or sometimes as the Euler–Lagrange equation).

To finish the proof, we need only show that Lagrange's equations are equivalent to Newton's equations. From the definition of the Lagrangian from equation (4.4) $L = T - V = \frac{1}{2}m\dot{x}^2 - V(x)$, we have

$$\frac{\partial L}{\partial x} = -\frac{\partial V}{\partial x}$$

and

$$\frac{\partial L}{\partial \dot{x}} = \frac{1}{2}m(2\dot{x}) = m\dot{x} = p,$$

where p is the linear momentum. Put these values in equation (4.12) to obtain

$$\begin{aligned} \frac{d}{dt}(p) + \frac{\partial V}{\partial x} &= 0 \\ \frac{dp}{dt} &= -\frac{\partial V}{\partial x} \\ F &= -\frac{\partial V}{\partial x}, \end{aligned} \quad (4.13)$$

which is clearly the same as Newton's law from equation (4.3).

4.2.3 Free-particle Lagrangian

A free particle is defined as a particle that is not bound by any external force or is defined as a particle in a region of space where its potential energy does not vary. The Lagrangian for a free particle in one dimension is given by

$$L_{\text{free-particle}}(\dot{x}) = \frac{1}{2}m\dot{x}^2. \quad (4.14)$$

The Lagrangian for a free particle in three dimensions is given by

$$L_{\text{free-particle}}(\dot{r}) = \frac{1}{2}m\dot{r}^2 = \frac{1}{2}m(\dot{x}^2 + \dot{y}^2 + \dot{z}^2). \quad (4.15)$$

4.2.4 Generalized coordinates

Generalized coordinates are usually selected to provide the minimum number of independent coordinates that completely define a system. Although there are many choices for generalized coordinates for a physical system, the parameters usually selected for the specification of the configuration of the system are those that make the solution of equations of motion easier.

If these parameters are independent of each other, the number of independent generalized coordinates is given by the *number of degrees of freedom* of the system. The degrees of freedom of a system is the number of independent physical parameters or pieces of information that can be used to completely define the system. The simplest example to consider is that in order to locate a particle in three-dimensional space we require three position coordinates, hence this system has three degrees of freedom. Now if the object starts to move with a velocity $\vec{v} = \vec{v}(t)$, velocity is a function of time, then such a system will have six degrees of freedom, three corresponding to position coordinates and three to velocity coordinates. If the motion of the particle is constrained, say the particle is only allowed to move on a wire, then the system has fewer than six degrees of freedom. On the other hand, an extended object that can rotate or vibrate can have more than six degrees of freedom.

Lagrange's equation in generalized coordinates will be written as

$$\boxed{\frac{d}{dt}\left(\frac{\partial L}{\partial \dot{q}}\right) - \frac{\partial L}{\partial q} = 0.} \tag{4.16}$$

These equations are called *generalized equations of motion*. There are a lot of generalized coordinates and there may be multiple ones used in a single question. In general, the term *coordinate* is used to define a system in Cartesian coordinates. For example to locate a point (P) on a sphere we can use the x-, y- and z-coordinates. However, we can also use radius (r), angle θ and angle(ϕ to locate the point P, these will be called generalized coordinates. Basically, the adjective *generalized* distinguishes these parameters from the traditional use of the term coordinate, which refers to Cartesian coordinates.

Example 4.1

Starting from the Cartesian coordinates (x, y, z), find the kinetic energy in 3D for cylindrical coordinates (r, ϕ, z).

Solution

We know that the kinetic energy in 3D in Cartesian coordinates (x, y, z) is given as

$$T = \frac{1}{2}m(v_x^2 + v_y^2 + v_z^2) = \frac{1}{2}m(\dot{x}^2 + \dot{y}^2 + \dot{z}^2). \tag{4.17}$$

The conversion from Cartesian to cylindrical coordinates is

$$\begin{cases} x &= r\cos\phi \\ y &= r\sin\phi\,\cdot \\ z &= z \end{cases} \tag{4.18}$$

Differentiating the above equations we obtain

$$\begin{cases} \dot{x} &= \dot{r}\cos(\phi) - r\sin(\dot{\phi})\phi \\ \dot{y} &= \dot{r}\sin(\phi) + r\cos(\dot{\phi})\phi\cdot \\ \dot{z} &= \dot{z} \end{cases} \tag{4.19}$$

To obtain the kinetic energy we square the above equations:

$$\begin{cases} \dot{x}^2 &= \dot{r}^2\cos^2(\phi) + r^2\sin^2(\phi)\dot{\phi}^2 - 2r\dot{r}\cos(\phi)\sin(\phi)\dot{\phi} \\ \dot{x}^2 &= \dot{r}^2\sin^2(\phi) + r^2\cos^2(\phi)\dot{\phi}^2 + 2r\dot{r}\cos(\phi)\sin(\phi)\dot{\phi}\cdot \\ \dot{z}^2 &= \dot{z}^2 \end{cases} \tag{4.20}$$

Adding the above equations and putting values in equation (4.17) we obtain

$$\dot{x}^2 + \dot{y}^2 + \dot{z}^2 = \dot{r}^2 + r^2\dot{\phi}^2 + \dot{z}^2. \tag{4.21}$$

Hence the kinetic energy in 3D in cylindrical coordinates is given as

$$\boxed{T = \frac{1}{2}m(\dot{r}^2 + r^2\dot{\phi}^2 + \dot{z}^2).} \tag{4.22}$$

Example 4.2

Starting from the Cartesian coordinates (x, y, z), find the kinetic energy in 3D for spherical coordinates (r, θ, ϕ).

Solution

Starting from the definition of kinetic energy in Cartesian coordinates in equation (4.17) and knowing the conversions to spherical coordinates, we can find the kinetic energy in 3D in spherical coordinates coordinates as follows:

$$\begin{cases} x &= r \sin\theta \cos\phi \\ y &= r \sin\theta \sin\phi \\ x &= r \cos\theta \end{cases} \tag{4.23}$$

$$\begin{cases} \dot{x} &= \dot{r} \sin\theta \cos\phi + r \cos\theta\dot{\theta} \cos\phi - r \sin\theta \sin\theta \sin\phi\dot{\phi} \\ \dot{y} &= \dot{r} \sin\theta \cos\phi + r \cos\theta\dot{\theta} \sin\phi - r \sin\theta \sin\theta \cos\phi\dot{\phi} \cdot \\ \dot{z} &= \dot{r} \cos\theta - r \sin\theta\dot{\theta} \end{cases}$$

Square the above equations by using the algebraic identities

$$(a + b)^2 = (a^2 + b^2 + 2ab) \text{ and } (a + b + c)^2 = a^2 + b^2 + c^2 + 2(ab + ac + bc).$$

We obtain

$$\begin{cases} \dot{x}^2 &= \dot{r}^2 \sin^2\theta \cos^2\phi + r^2 \cos^2\theta\dot{\theta}^2 \cos^2\phi + r^2 \sin^2\theta \sin^2\phi\dot{\phi}^2 + 2(\ldots) \\ \dot{y}^2 &= \dot{r}^2 \sin^2\theta \sin^2\phi + r^2 \cos^2\theta\dot{\theta}^2 \sin^2\phi + r^2 \sin^2\theta \cos^2\phi^2\dot{\phi}^2 + 2(\ldots) \cdot \\ \dot{z}^2 &= \dot{r}^2 \cos^2\theta + r^2 \sin^2\theta\dot{\theta}^2 - 2r\dot{r} \cos\theta \sin\theta\dot{\theta} \end{cases}$$

Adding the above equations and using the trigonometric identity $\cos^2(\theta) + \sin^2(\theta) = 1$, we can solve to obtain

$$\dot{x}^2 + \dot{y}^2 + \dot{z}^2 = \dot{r}^2 + r^2\dot{\theta}^2 + r^2 \sin^2\theta\ \dot{\phi}^2. \tag{4.24}$$

Note that all other terms will cancel out. It is left as an algebraic exercise for you to attempt. Hence, we find the kinetic energy in 3D in spherical coordinates by putting values in equation (4.17) as

$$\boxed{T = \frac{1}{2}m(\dot{r}^2 + r^2\dot{\theta}^2 + r^2 \sin^2\theta\ \dot{\phi}^2).} \tag{4.25}$$

4.2.5 Summary kinetic energy in various coordinates

Kinetic energy in 3D in various coordinates is given as:

- KE in Cartesian coordinates: $T = \frac{1}{2}m(\dot{x}^2 + \dot{y}^2 + \dot{z}^2)$.
- KE in cylindrical coordinates: $T = \frac{1}{2}m(\dot{r}^2 + r^2\dot{\phi}^2 + \dot{z}^2)$.
- KE in spherical coordinates: $T = \frac{1}{2}m(\dot{r}^2 + r^2\dot{\theta}^2 + r^2 \sin^2\theta\ \dot{\phi}^2)$.

4.3 Choosing Lagrange's formalism—when and where?

Back to Lagrangian mechanics, there are two very important reasons for working with Lagrange's equations:

1. Lagrange's equations hold in any coordinate system, while Newton's equations are limited to only inertial frames.
2. In Lagrange's formalism, dealing with constraints present in a system is relatively easier.

We will look at these two aspects in the next two subsections.

4.3.1 Lagrange's equations hold in any coordinate system

From the principle of least action, as discussed in section 4.2.2, we know that Lagrange's equations hold in any coordinate system. We can show this with the help of the following example.

Example 4.3

Consider a free particle with the Lagrangian given by

$$L = \frac{1}{2}mr^2, \tag{4.26}$$

with $r = (x, y, z)$. Now measure the motion of the particle with respect to a coordinate system which is rotating with angular velocity $\vec{\omega} = (0, 0, \omega_z)$ about the z-axis. If $r^1 = (x^1, y^1, z^1)$ are the coordinates in the rotating system, we have the relationship

$$\begin{cases} x^1 = & x\cos(\omega t) + y\sin(\omega t) \\ y^1 = & y\cos(\omega t) + x\sin(\omega t)\cdot \\ z^1 = & z \end{cases} \tag{4.27}$$

Then we can substitute these expressions into the Lagrangian to find L in terms of the rotating coordinates as

$$\begin{aligned} L &= \frac{1}{2}m[(\dot{x}^1\omega y^1)^2 + (\dot{y}^1 + \omega x^1)^2 + \dot{z}^{1^2}] \\ L &= \frac{1}{2}m[(\dot{\vec{r}}^1 + \omega \times \vec{r}^1)^2. \end{aligned} \tag{4.28}$$

In this rotating frame, we can use Lagrange's equations to derive the equations of motion. Taking the derivatives, and putting values in Lagrange's equation (4.12), we obtain

$$\begin{aligned} \frac{d}{dt}\frac{\partial L}{\partial \dot{r}^1} - \frac{\partial L}{\partial r^1} = 0 \\ m(\ddot{r}^1) + m(\omega \times (\omega \times r^1)) + m(2\omega \times \dot{r}^1) = 0. \end{aligned} \tag{4.29}$$

The second and third terms in equation (4.29) are the centrifugal[2] and Coriolis forces[3], respectively. These are examples of fictitious forces and are called fictitious because they are a consequence of the reference frame, rather than actual forces acting on each other. Note that we cannot ignore the importance of these fictitious forces. According to Einstein's theory of general relativity, these fictitious forces and the force of gravity are on equal footing and we certainly cannot ignore the force of gravity while we are on Earth.

4.3.2 Constraints

Another advantage of the Lagrangian formulation is that we can ignore the *constraints* that are inherently found in Newton's formalism. What do we mean by constraints? Constraints on a system are conditions that a system must obey. These are things such as the motion of a bead constrained to move along a wire or on the surface of a sphere, or the fact that a box sliding on a slope must remain fixed on the slope, etc.

In Newtonian mechanics, we introduce *constraint forces* into our second order differential equations. In the Lagrangian formulation, we do not have to worry about such things. There are two main kinds of constraints and we will talk about them in this section.

4.3.2.1 Holonomic constraints

Holonomic constraints depend only on the coordinate (x_i) and time (t) and do not depend on velocities or accelerations. The holonomic constraint equation is of the form

$$f = \sum_{1}^{3} f(\vec{r}_i, t) = 0 \tag{4.30}$$

and it connects the three spatial coordinates of the particle under consideration. For holonomic constraints the three spatial coordinates are not independent,

[2] Centrifugal force is a virtual force that acts outward on an object moving around a center when the frame of reference is rotating (e.g. Earth is a rotating frame). It arises from the inertia of the object and is directed away from the axis of rotation. Sometimes engineers use the term 'reactive centrifugal forces' since it is a reaction to a real force called 'centripetal force'.

[3] The Coriolis force acts on objects in motion relative to a rotating reference frame. Since Earth is rotating, on Earth the Coriolis effect tends to deflect moving objects to the right in the Northern Hemisphere and to the left in the Southern Hemisphere. The Coriolis force is also responsible for large scale weather patterns.

additionally the constraint may change with time, so time (t) cannot be ignored in the constraint equation. This means that at all times, any one coordinate will be determined from knowing the other coordinates, e.g. if coordinates x_i and y_i are given, then so is z_i.

Consider the example of a simple pendulum. A simple pendulum is made up of a string attached to a pivot on one end and a mass hanging on the other end, see figure 7.2. If we consider the string to be in-extendable with length (L) then the string's length is a constant. Therefore, this system is holonomic; it obeys the holonomic constraint

$$(x^2 + y^2) - L^2 = 0, \tag{4.31}$$

where (x, y) is the position of the mass at time t.

Keep in mind that one constraint equation accounts for one constraint. If there are n constraints on the system, there will be n constraint equations. If there are zero constraints on the system the then there are zero constraint equations.

4.3.2.2 Non-holonomic constraints

A non-holonomic system is a mechanical system whose state depends on the path taken. A non-holonomic system has constraints on both the generalized coordinates and derivatives of generalized coordinates. In general, non-holonomic constraints are considered to depend on the velocities and accelerations of particles.

First-order non-holonomic constraints have the form:

$$g = \sum_{1}^{3} g(\vec{q}, \dot{\vec{q}}, t) = 0. \tag{4.32}$$

In a non-holonomic system, the path integrals depend on the intermediate values the path follows and we cannot say that it is a conserved system. Hence when the system returns to the beginning of the trajectory, the system may not return to its original state. Hence we say that the non-holonomic system is *non-integrable*. A classic example of a non-holonomic system is the Foucault pendulum. In contrast, in the holonomic system the path integral depends only on the initial and final state of the system and are completely independent of the path.

4.4 Lagrangian formalism for non-conservative forces

Certain non-conservative forces can be integrated to obtain a *pseudo-potential* function, often called a *generalized potential.* The generalized potential can then be added to the potential energy (V) in the Lagrangian formulation and we can solve Lagrange's equation of the form (4.16) to find the motion of any system.

However, in certain cases this is not easy to do, for example, velocity-dependent non-conservative forces (such as kinetic frictional forces) are especially difficult to incorporate into the Lagrangian formalism. For cases where we cannot easily find a generalized potential function to add to the Lagrangian, we can simply add the force to equation (4.16), which then becomes

$$\boxed{\frac{d}{dt}\left(\frac{\partial L}{\partial \dot{q}_i}\right) - \frac{\partial L}{\partial q_i} = F_i,} \tag{4.33}$$

where F_i is the non-conservative force. The simplest example of a non-conservative force is the force of friction (f_f). In most cases the force of friction is considered a constant. However, friction can also be velocity-dependent, say for the case of viscous drag, which we will address in chapter 12, section 12.2.

4.5 The Lagrangian formalism in a nutshell

A system is described by n generalized spatial coordinates q_i, which define a point in an n-dimensional configuration space. The Lagrangian function in terms of generalized coordinates is defined as

$$L = T(\dot{q}) - V(q). \tag{4.34}$$

The time evolution is a curve in the configuration space governed by the Lagrangian (L) such that the coordinates q_i obey the Lagrangian equations given by

$$\frac{d}{dt}\left(\frac{\partial L}{\partial \dot{q}_i}\right) - \frac{\partial L}{\partial q_i} = 0. \tag{4.35}$$

These are n coupled second order (usually) non-linear differential equations. Equation (4.35) is also the Lagrangian equation of motion for a conservative system under holonomic constraint. For example, if we take the generalized coordinates (q_i) to be position or displacement of the particle under consideration in the x-direction, then $\dot{q}_i$ will be the corresponding velocity in the x-direction.

The Lagrangian equation of motion that includes frictional forces is given as

$$\frac{d}{dt}\left(\frac{\partial L}{\partial \dot{q}_i}\right) - \frac{\partial L}{\partial q_i} = F_i. \tag{4.36}$$

Here, F_i represents the resistive force that the particle under consideration is subjected to. Note that the generalized momentum conjugate to q_i is given as

$$p_i = \frac{\partial L}{\partial \dot{q}_i}, \tag{4.37}$$

where $_{pi}$ coincides with the real momentum only in Cartesian coordinates. In this case, equation (4.35) for a conservative system can be written as

$$\frac{d}{dt}(p_i) - \frac{\partial L}{\partial q_i} = 0$$

$$\dot{p}_i = \frac{\partial L}{\partial q_i}. \tag{4.38}$$

The interesting thing to note here is that equation (4.38) is equivalent to equation (4.1),

$$\frac{dp}{dt} = \frac{\partial L}{\partial q_i} = \frac{\partial(T - V)}{\partial x} = \frac{\partial T(\dot{x})}{\partial x} - \frac{\partial V(x)}{\partial x} = -\frac{\partial V}{\partial x}.$$

4.5.1 General problem-solving techniques—Lagrange's equations

The general problem-solving strategy using Lagrange's formulation can be summed up as follows:

1. Select a suitable set of generalized coordinates that uniquely specifies the configuration of the system.
2. Find the equations to relate the dependent Cartesian coordinates to the independent generalized coordinates.
3. Find the kinetic energy as a function of generalized coordinates $[T = T(\dot{q}_i)]$ and their derivatives, using $T = \frac{1}{2}m(\vec{v} \cdot \vec{v})$, where $\vec{v}$ is expressed in terms of appropriate vectors in terms of selected generalized coordinates. In general,

$$T = \sum_{i=1}^{3N} \frac{1}{2} m_i \dot{q}_i^2. \tag{4.39}$$

4. Find potential energy as a function of the generalized coordinates, $[V = V(q_i)]$.
5. The equation of motions are then given by equations (4.35) or (4.36), depending on the system under consideration.

Example 4.4

Two blocks of equal mass (m) are connected with a flexible cord. One block is placed on a smooth horizontal table and the other is hanging over a massless pulley, as shown in figure 4.2. Find the acceleration of the blocks using the (a) Newtonian and (b) Lagrangian formalisms.

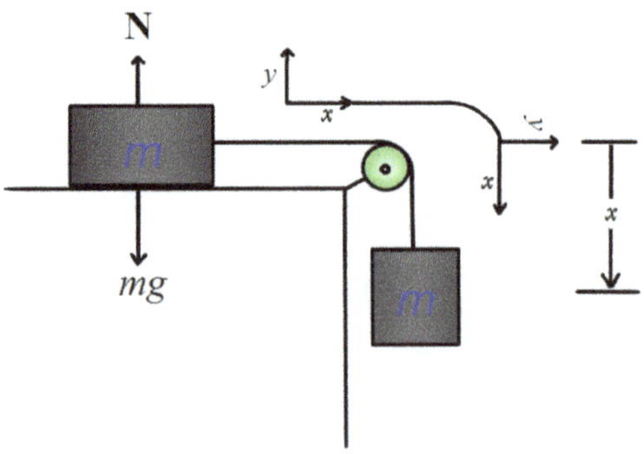

Figure 4.2. Two masses connected over a pulley.

Solution
(a) Lets first solve using Newtonian mechanics, writing Newton's force equations for both mass 1 and mass 2:

mass 1:

$$\Sigma F_x = T \longrightarrow ①$$
$$\Sigma F_y = N - m_1 g$$

mass 2:

$$\Sigma F_x = 0$$
$$\Sigma F_y = -T + m_2 g \longrightarrow ②.$$

Then simplifying ① and ②, knowing $m_1 = m_2$, we obtain

$$ma = T$$
$$ma = -T + mg.$$

Adding the above equations,

$$2ma = mg$$
$$a = \frac{g}{2}$$

Hence the acceleration is given by

$$\boxed{\ddot{x} = \frac{g}{2}.} \tag{4.40}$$

(b) Next we will solve the same problem using the Lagrangian method. In order to do that we will first write the kinetic and potential energies in terms of generalized coordinates ($q = x$ and $\dot{q} = \dot{x}$) as follows.

Step 1. Find the kinetic and potential energies,

$$T = \frac{1}{2}m_1\dot{x}^2 + \frac{1}{2}m_2\dot{x}^2,$$

since $m_1 = m_2$,

$$T = m\dot{x}^2$$
$$V = -m_2 gx = -mgx$$

Step 2. Write the Lagrangian,

$$L = T - V$$
$$L = m\dot{x}^2 + mgx.$$

Step 3. Write the equations of motion,

$$\frac{d}{dt}\left(\frac{2L}{2\dot{x}}\right) - \frac{2L}{2x} = 0$$
$$\frac{d}{dt}(2m\dot{x}) - mg = 0$$
$$2m\ddot{x} - mg = 0$$

$$\boxed{\ddot{x} = \frac{g}{2}.} \tag{4.41}$$

The answer in equation (4.41) is identical to the answer of equation (4.40) above.

Example 4.5

Find the velocity as a function of time for (a) a freely falling object dropped from an initial height (h) subject to (b) a linear drag force caused by air resistance using the Lagrangian formalism.

Solution

(a) Freely falling object (non-conservative system).

Starting from rest, we can write the kinetic (T) and potential (V) energies as

$$T = \frac{1}{2}mv^2 = \frac{1}{2}(\dot{x}^2 + \dot{y}^2 + \dot{z}^2)$$
$$V = -\,mgh.$$

Hence the Lagrangian will be written as

$$\begin{aligned} L &= T - V \\ &= \frac{1}{2}m(\dot{x}^2 + \dot{y}^2 + \dot{z}^2) + mgh \\ L &= \frac{1}{2}m\dot{r}^2 + mgr. \end{aligned}$$

(b) Drag force.

Under the influence of a linear drag force of the form ($-bv$) the Lagrangian equation will be written as

$$\begin{aligned} \frac{d}{dt}\left(\frac{2L}{2\dot{r}}\right) - \frac{2L}{2r} &= -\,bv \\ \frac{d}{dt}(m\dot{r}) - mg &= -\,b\dot{r} \\ m\ddot{r} + b\dot{r} - mg &= 0 \\ \ddot{r} + \frac{b}{m}\dot{r} - g &= 0. \end{aligned}$$

Hence the acceleration will be

$$\boxed{a = \ddot{r} = \left(g - \frac{b\dot{r}}{m}\right).} \tag{4.42}$$

To find the velocity we will start from equation (4.42) and integrate both sides,

$$\frac{dv}{dt} = g - \frac{b}{m}v.$$

Rearranging, $$\frac{1}{g - \frac{b}{m}v}dv = dt.$$

Integrating, $$\int_0^v \frac{1}{g - \frac{b}{m}v}dv = \int_0^t dt$$

$$\frac{\ln(g - \frac{b}{m}v)}{-\frac{b}{m}}\Big|_o^v = t$$

$$\ln\left(g - \frac{bv}{m}\right) - \ln(g) = -\frac{b}{m}t$$

$$\ln\left[\frac{g - \frac{bv}{m}}{g}\right] = \frac{b}{m}t$$

$$1 - \frac{bv}{m} = e^{-\frac{b}{m}t}$$

$$-\frac{b}{mg}v = e^{-\frac{b}{m}t} - 1.$$

Hence the velocity will be given by

$$\boxed{v = \frac{mg}{b}(1 - e^{-\frac{b}{m}t}).} \tag{4.43}$$

4.5.2 Advantages of Lagrangian formulation

So why use Lagrangian mechanics or the Lagrangian formulation?

Good question, because…

- The Lagrangian formulation can easily handle situations, such as electromagnetism and relativity, that are very tedious to do using Newtonian mechanics.
- Newtonian mechanics works well in Cartesian coordinates, but it is sometimes complicated to switch coordinate systems. In contrast, the Lagrangian formulation is coordinate-independent, and equations of motion for any non-Cartesian coordinate system can be easily handled.

Concept questions 4.1

1. How many generalized coordinates do we need to describe the motion given in example 4.4 versus the motion described in example 4.5?
2. The Lagrangian says to treat velocity and position as independent variables. However, we know velocity is the derivative of position, so how can you treat them as independent variables?
3. If we shift the potential energy (V) by a constant (c) for a particle moving in 1D what happens to the action (S) and the physical path $x(t)$ followed by the particle?

Exercises 4.2

1. Find the Lagrangian of a simple pendulum made up of a hanging mass (m_1) attached to a string of length (ℓ) in terms of its length and the angle (θ).
2. Find the equations of motion of a particle in rotating polar coordinates (r, θ), with a conservative central force acting on it. *Hint*: the Cartesian coordinates are related to the rotating coordinates using rotating angular velocity (ω_0),

$$\begin{aligned} x &= r\cos(\theta + \omega_0 t) \\ y &= r\sin(\theta + \omega_0 t). \end{aligned} \tag{4.44}$$

3. Suppose we have two weights, m_1 and m_2, attached together with a string of length (ℓ). Let m_1 be sitting on a ramp with angle (θ) while the other mass m_2 hangs over a pulley. Both masses are initially at rest. Find the acceleration and velocities of each mass for the following conditions:
   (a) When the ramp is frictionless.
   (b) When the friction between m_1 and the ramp is given by f_f.

Further reading

Baierlein R 1983 *Newtonian Dynamics* (New York: McGraw-Hill)
Fowles G 1986 *Analytical Mechanics* 4th edn (New York: CBS College Publishing)
Fowles G and Cassiday G 2005 *Analytical Mechanics* 7th edn (Belmont, CA: Thomson Learning)
Goldstein H 1980 *Classical Mechanics* 2nd edn (Reading, MA: Addison-Wesley)
Landau L D and Lifshitz E M 1960 *Mechanics* (Reading, MA: Addison-Wesley)
Rutherford D E 1951 *Classical Mechanics* (New York, NY: Interscience)
Stephenson R 1961 *Mechanics and Properties of Matter* 2nd edn (New York: Wiley)
Symon K 1971 *Mechanics* 3rd edn (Reading, MA: Addison-Wesley)

Chapter 5

Hamiltonian mechanics

5.1 Hamiltonian mechanics

Hamiltonian mechanics was first formulated by Irish mathematician Sir William Rowan Hamilton in 1835 CE, about 50 years after the advent of Lagrangian mechanics. Hamiltonian mechanics uses a different mathematical formalism to classical mechanics; however, the Hamiltonian formalism predicts the same outcomes as Newtonian and Lagrangian mechanics, as we will see in the course of this study.

5.1.1 Basis of Hamiltonian mechanics

Similar to the Lagrange method, the Hamiltonian method contains the *principle of least action*, which we first defined in section 4.2.2, as a basic starting point. However, it additionally contains factors that make it very useful in certain branches of physics such as statistical mechanics (which deals with a large number of particles, e.g. in a system such as a gas) or in quantum mechanics (where no particles are present and everything is expressed in terms of a wave).

For standard classical physics, say for situations were one or two particles are involved, this method is rather inconvenient and more difficult to use than the Lagrangian and the Newtonian mechanics. Hence it is not utilized much in classical physics. But it is a good idea to know about it in terms of classical mechanics before you see it in quantum mechanics and wonder where it came from.

Energy conservation

Conservation of energy arises when the Lagrangian is independent of time. The motivation for this expression for energy comes from the theory of Legendre transforms, which are explained in other textbooks, but we will not get into that here. We will just accept the definition given in the following equation to hold true:

doi:10.1088/2053-2563/aae1b9ch5

$$E = \sum_{k=1}^{N} \frac{\partial L}{\partial \dot{q}_k} \dot{q}_k - L. \tag{5.1}$$

Before talking about the Hamiltonian, let us look at the relation between E as given in equation (5.1) and the energy of the system.

Example 5.1

A particle of mass (m) is moving along the x-axis under the influence of a potential $V(x)$, where x is our standard Cartesian coordinate.

Then, the Lagrangian (L) of this motion is written as

$$\begin{aligned} L &= T - V \\ &= \frac{1}{2} m\dot{x}^2 - V(x). \end{aligned}$$

Then we use equation (5.1), knowing the generalized coordinate (q) is x in this case, to find E as

$$\begin{aligned} E &= \frac{\partial L}{\partial \dot{x}} \dot{x} - L \\ &= (m\dot{x})\dot{x} - L \\ &= m\dot{x}^2 - L \\ &= 2T - (T - V) \end{aligned}$$

$$E = T + V, \tag{5.2}$$

which is nothing other than the total energy of the system. Hence, the E in equation (5.1) represents the total energy of the system.

Example 5.2

A particle of mass (m) is moving along the xy-plane under the influence of a potential $V(x, y)$, where x and y are our standard Cartesian coordinates.

Then, the Lagrangian (L) of this motion will be written as

$$\begin{aligned} L &= T - V \\ &= \frac{1}{2} m(\dot{x}^2 + \dot{y}^2) - V(x, y). \end{aligned}$$

Then we use equation (5.1) to find E as

$$\begin{aligned} E &= \frac{\partial L}{\partial \dot{x}} \dot{x} + \frac{\partial L}{\partial \dot{y}} \dot{y} - L \\ &= [(m\dot{x})\dot{x} + (m\dot{y})\dot{y}] - L \\ &= 2T - (T - V) \end{aligned}$$

$$\boxed{E = T + V,} \tag{5.2}$$

which is nothing other than the total energy of the system, again!

Example 5.3

A particle of mass (m) moves in a horizontal plane. It is connected to the origin by a spring having a spring constant (k) and potential energy, $V = \frac{1}{2}kr^2 = \frac{1}{2}k(x^2 + y^2)$:

(a) Find L and E in Cartesian coordinates.
(b) Find L and E in polar coordinates.

Solution

(a) L and E in Cartesian coordinates.

Using the definition of the Lagrangian (L) we can write

$$L = T - V$$

$$\Rightarrow \boxed{L = \frac{1}{2}m(\dot{x}^2 + \dot{y}^2) - \frac{1}{2}k(\dot{x}^2 + \dot{y}^2)} \tag{5.4}$$

and using equation (5.1) we obtain energy (E) as

$$\begin{aligned} E &= \frac{2L}{2\dot{x}}\dot{x} + \frac{2L}{2\dot{y}}\dot{y} - L \\ &= \frac{1}{2}m(2\dot{x})\dot{x} + \frac{1}{2}m(2\dot{y})\dot{y} - L \\ &= m\dot{x}^2 + m\dot{y}^2 - \frac{1}{2}m(\dot{x}^2 + \dot{y}^2) + \frac{1}{2}k(x^2 + y^2) \end{aligned}$$

$$\Rightarrow \boxed{E = \frac{1}{2}m(\dot{x}^2 + \dot{y}^2) + \frac{1}{2}k(\dot{x}^2 + \dot{y}^2).} \tag{5.5}$$

(b) L and E in polar coordinates.

We find L in polar coordinates as

$$L = T - V$$

$$\Rightarrow \boxed{L = \frac{1}{2}m(\dot{r}^2 + r^2\dot{\theta}^2) - \frac{kr^2}{2}} \tag{5.6}$$

and to obtain E in polar coordinates

$$
\begin{aligned}
E &= \frac{2L}{2\dot{r}}\dot{r} + \frac{2L}{2\dot{\theta}}\dot{\theta} - L \\
&= (m\dot{r})\dot{r} + (mr^2\dot{\theta})\dot{\theta} - L \\
&= m\dot{r}^2 + mr^2\dot{\theta}^2 - \frac{1}{2}m(\dot{r}^2 + r^2\dot{\theta}^2) + \frac{kr^2}{2} \\
&= \frac{1}{2}m(\dot{r}^2 + r^2\dot{\theta}^2) + \frac{kr^2}{2}
\end{aligned}
$$

$$
\Rightarrow \boxed{E = \frac{1}{2}m(\dot{r}^2 + r^2\dot{\theta}^2) + \frac{kr^2}{2}.} \tag{5.7}
$$

5.2 The Hamiltonian principle

The Hamiltonian principal states:

> 'The motion of any system takes place in such a way that the integral ($\int_{t_1}^{t_2} Ldt$) always assumes an extreme value'.

In other words:

> 'Out of all possible ways a system can change in a finite time interval $[t_1 \rightarrow t_2]$ the particular motion that will occur is for which the above integral is either a maxima or a minima'

and is mathematically expressed as

$$
\delta \int_{t_1}^{t_2} Ldt = 0, \tag{5.8}
$$

where δ represents small variations in the path. This variation results from taking different paths of integration by varying the generalized coordinates and generalized velocities as functions of t.

In Hamiltonian mechanics, a classical physical system is described by a set of canonical coordinates[1] given by q_i and p_i coordinates. A typical example of canonical coordinates are Cartesian coordinates where position is given by x_i, and components of momentum are given by p_i and are referred to as *conjugate momenta.*

5.2.1 Hamiltonian function and Hamiltonian's equations

The Hamiltonian function in generalized coordinates is given mathematically as

$$
\boxed{H = \sum_k \dot{q}_k p_k - L.} \tag{5.9}
$$

[1] Canonical coordinates are defined as a set of coordinates which can be used to describe a physical system at any given point in time.

For a simple dynamic system, the kinetic energy (T) is a homogeneous quadratic function of the $\dot{q}$, whereas the potential energy (T) is a function of q. For such a system the Lagrangian is given by

$$L = T(\dot{q}_k) - V(q_k). \tag{5.10}$$

From Euler's theorem for homogeneous functions[2],

$$\begin{aligned}\sum_k \dot{q}_k p_k &= \sum_k \dot{q}_k \frac{\partial L}{\partial \dot{q}_k} \\ &= \sum_k \dot{q}_k \frac{\partial T}{\partial \dot{q}_k} \\ &= 2T.\end{aligned}$$

Put into equation (5.9), we obtain

$$\begin{aligned}H &= 2T - L \\ H &= 2T - (T - V),\end{aligned}$$

which reduces to the expression

$$\boxed{H = T + V.} \tag{5.11}$$

This function is called the Hamiltonian function or just *Hamiltonian* for short, and it equals the total energy of the system under consideration. This is a very important function and is used a lot in quantum mechanics and celestial mechanics.

5.2.2 Understanding the Hamiltonian

In order to understand the Hamiltonian formalism let us consider the following cases.

Case 1. Conservation of energy

The basic idea of physics is that energy is conserved. Classically, if an object moves from a high potential to a low potential it speeds up to conserve energy. This is mathematically represented as

$$\frac{dE}{dt} = \frac{dH}{dt} = 0. \tag{5.12}$$

The significance of this is that we can derive Newton's law from this simple requirement as shown here:

$$H = T + V \text{where } T = \frac{p^2}{2m}; \text{ and } V = V(r)$$

---

[2] Euler's theorem for homogeneous functions states that for a homogeneous function given as f of degree n we can say $[x_1\frac{\partial f}{\partial x_1} + x_2\frac{\partial f}{\partial x_2} + \cdots x_n\frac{\partial f}{\partial x_n} = nf]$ holds true.

$$H = \frac{p^2}{2m} + V(r).$$

Differentiating w.r.t. t we obtain

$$\frac{dH}{dt} = \frac{d}{dt}\left(\frac{\vec{p}\cdot\vec{p}}{2m}\right) + \frac{d}{dt}V(r).$$

Put in the value from equation (5.12) and differentiate to obtain

$$0 = \frac{1}{2m}\vec{p}\left(\frac{dp}{dt}\right) + \frac{1}{2m}\vec{p}\left(\frac{dp}{dt}\right) + \frac{dV(r)}{dt}$$

$$0 = \frac{\vec{p}}{m}\frac{d\vec{p}}{dt} + \frac{dV(r)}{dt}\frac{dr}{dr}.$$

Rearranging and simplifying, we obtain

$$\frac{\vec{p}}{m}\frac{d\vec{p}}{dt} = -\frac{dV(r)}{dr}\frac{dr}{dt}$$

$$\frac{m\vec{v}}{m}\frac{d\vec{p}}{dt} = -\frac{dV(r)}{dr}\vec{v}$$

$$\frac{dp}{dt} = -\frac{dV(r)}{dr}.$$

Using equation (2.19) we can write

$$\frac{dp}{dt} = \vec{F}. \tag{5.13}$$

This is Newton's second law, similar to equation (3.16) again!

Case 2. In free space ($V(r) = 0$)

Free space is defined as a space where there is zero potential ($V(r) = 0$) and hence there will be no force on a particle placed at that point. This means that the momentum will be conserved, $\frac{dp}{dt} = 0$, and momentum is constant, $p(t) = p(0) \equiv p_o$,

$$p_o = mv = m\frac{dr}{dt}$$

$$p_o \int_o^t dt = m \int_{r_o}^{r(t)} dr$$

$$p_o t = m[r(t) - r_o]$$

$$\Rightarrow \boxed{r(t) = r_o + \frac{p_o}{m}t.} \tag{5.14}$$

This implies that the particle moves at a constant velocity and position changes linearly with time. The motion depends on where the particle started and its initial momentum.

Case 3. On the surface of the Earth ($V(z) = mgz$)

Consider that the vertical distance from the surface of the Earth is given by z where $[z \equiv r - r_o]$, then momentum will be given as

$$p = mv = m\dot{z} = m\frac{dz}{dt}. \tag{5.15}$$

Newton's equation as given in equation (5.13) is

$$F = \dot{p} = \frac{dp}{dt}$$
$$F = -\frac{dV}{dz} = -\frac{d}{dz}(mgz) = -mg.$$

Then we can rewrite the above equations and integrate,

$$\frac{dp}{dt} = -mg$$
$$\int dp = -\int mgdt$$
$$p(t) - p_o = -mgt$$
$$p(t) = p_o - mgt,$$

which can also be written as

$$m\dot{z} = p_o - mgt$$
$$m\frac{dz}{dt} = p_o - mgt$$
$$dz = \left[\frac{p_o}{m} - gt\right]dt.$$

Integrating both sides we obtain

$$\boxed{z - z_o = \frac{p_o}{m}t - \frac{1}{2}gt^2.} \tag{5.16}$$

Note that this is the same as the second equation of motion we found in section 3.2 in equation (3.3), keeping in mind ($p_o = mv_o$).

5.2.3 From Lagrangian to Hamiltonian mechanics

The conversion from Lagrangian mechanics to Hamiltonian mechanics is basically a change of variable from velocity ($\dot{x}$) to the generalized momentum (p). In other words we can distinguish between Lagrangian and Hamiltonian mechanics by keeping the following two things in mind, at any time t, the state of a system is defined by specifying values for:

- The generalized coordinates (q) and the velocity ($\dot{q}$) in Lagrangian mechanics.
- The generalized coordinates (q) and the momentum (p) in Hamiltonian mechanics.

5.2.4 Deriving the Hamiltonian's canonical equations of motion

The Hamiltonian, defined as the sum of kinetic and potential energies, $(T + V)$, can be expressed for n equations as a function of generalized positions (q_i) and conjugate momenta (p_i), as

$$H(q_i, p_i) = \sum_i p_i \dot{q}_i(p_i, q_i) - L, \tag{5.17}$$

where L is the Lagrangian $(L = T(q_i, \dot{q}_i) - V(q_i))$ and $i = 1, 2\ldots, n$, and

$$p_i = \frac{\partial L}{\partial \dot{q}_i}$$

and

$$\dot{q}_i = \dot{q}_i(p_i, q_i).$$

Since $H = H(q_i, p_i)$, we can write the general variation equation for H as

$$\delta H(q_i, p_i) = \sum_i \left[\frac{\partial H}{\partial p_i} \delta p_i + \frac{\partial H}{\partial q_i} \delta q_i \right]. \tag{5.18}$$

We can also calculate the variation of the function H from equation (5.17) as

$$\delta H = \sum_i \left[p_i d\dot{q}_i + \dot{q}_i \delta p_i - \left(\frac{\partial L}{\partial \dot{q}_i} d\dot{q}_i + \frac{\partial L}{\partial q_i} \delta q_i \right) \right]. \tag{5.19}$$

We know by definition $p_i = \frac{\partial L}{\partial \dot{q}_i}$, which implies the first and third terms will cancel and equation (5.19) reduces to

$$\delta H = \sum_i \left[\dot{q}_i \delta p_i - \frac{\partial L}{\partial q_i} \delta q_i \right]. \tag{5.20}$$

We can simplify equations (5.20) knowing $p_i = \frac{\partial L}{\partial \dot{q}_i}$:

$$\delta H = \sum_i \left[\dot{q}_i dp_i - p_i dq_i \right]. \tag{5.21}$$

Comparing the above equations (5.18) and (5.21):

$$\boxed{\frac{\partial H}{\partial p_k} = \dot{q}_k} \tag{5.22}$$

$$\boxed{\frac{\partial H}{\partial q_k} = -\dot{p}_k.} \tag{5.23}$$

These equations (5.22) and (5.23) are known as *Hamilton's canonical equations of motion*, or just the Hamiltonian equations for short. They consist of $2n$ first order differential equations. The solution of Hamilton's equations of motion will give the path of a particle in terms of generalized positions (q_i) and conjugate momenta (p_i) as a function of time.

Hamilton's equations of motion are equivalent to Newton's equations of motion, and are similar to the Lagrangian formulation. Hamilton's equations are used to find equations of motion in any set of coordinates. As an example, consider x to be the generalized coordinate; then the above equations can be written as

$$\frac{\partial H}{\partial p_x} = \dot{x} = v, \tag{5.24}$$

where v is the corresponding velocity in the x-direction,

$$\frac{\partial H}{\partial x} = -\dot{p}_x = -F, \tag{5.25}$$

which is Newton's second law again!

Note that here we only solved/derived the Hamiltonian equation for simple conservative systems. The Hamiltonian can be shown to hold for more general systems. For example:

- Non-conservative systems.
- Systems in which the potential energy function involves $\dot{q}$.
- For systems in which Lagrangians involve time explicitly.

However, in these cases the total energy E_{total} is no longer equal to the Hamiltonian (H) alone.

5.2.5 The Hamiltonian as a conserved quantity

We can see from Hamilton's equations that the Hamiltonian (H) is conserved when the independent variable does not appear explicitly in the Hamiltonian. This can be shown as follows.

Suppose the Hamiltonian is a function of coordinate (x), momentum (p_x) and time (t). Then we can write

$$\frac{dH}{dt} = \frac{\partial H}{\partial x}\left(\frac{dx}{dt}\right) + \frac{\partial H}{\partial p_x}\left(\frac{dp_x}{dt}\right) + \frac{\partial H}{\partial t}. \tag{5.26}$$

Using Hamilton's equations (5.22) and (5.23), we have

$$\begin{aligned}\frac{dH}{dt} &= \frac{\partial H}{\partial x}\left(\frac{\partial H}{\partial p_x}\right) + \frac{\partial H}{\partial p_x}\left(-\frac{\partial H}{\partial x}\right) + \frac{\partial H}{\partial t}\\ \frac{dH}{dt} &= \frac{\partial H}{\partial t}.\end{aligned} \tag{5.27}$$

If the Hamiltonian does not depend explicitly on t then equation (5.27) becomes

$$\frac{dH}{dt} = 0, \tag{5.28}$$

which shows the Hamiltonian (H) is conserved.

5.2.6 Hamiltonian formalism extended to n particles

The Hamiltonian formalism can be extended to a system consisting of n particles:

$$H = \sum_{n=1}^{N} T_n + V, \tag{5.29}$$

where

$$T_n = \frac{\vec{p}_n \cdot \vec{p}_n}{2m_n} \tag{5.30}$$

is the kinetic energy operator of n particle and

$$V = V(\vec{r}_1, \vec{r}_2 \cdots \vec{r}_n, t) \tag{5.31}$$

is the potential energy function, now a function of the spatial configuration[3] of the system and time.

For n particles in 3D space the Hamiltonian will be written as

$$\begin{aligned} H &= \sum \frac{1}{2m_n} \nabla_n^2 + V(\vec{r}_1, \vec{r}_2 \cdots \vec{r}_n, t) \\ &= \sum \frac{1}{2m_n} \left(\frac{\partial^2}{\partial x_n^2} + \frac{\partial^2}{\partial y_n^2} + \frac{\partial^2}{\partial z_n^2} \right) + V(\vec{r}_1, \vec{r}_2 \cdots \vec{r}_n, t). \end{aligned} \tag{5.32}$$

I have just defined the Hamiltonian (H) here for the 3D case in terms of the ∇ operator but have not given details. We will discuss the ∇ operator in more detail in section 5.3.1 and also in chapter 9.

5.2.7 Route to follow to solve specific problems using the Hamiltonian equation of motion

The steps to follow to solve problems using Hamiltonian methodology are as follows:

1. Write the Lagrangian $L\ (q, \dot{q})$.
2. Find the generalized (also called conjugate) momentum ($p = p(\dot{q})$) associated with each coordinate (q) present in the system

$$p = \frac{\partial L(q_i, \dot{q}_i)}{\partial \dot{q}_i}. \tag{5.33}$$

[3] A spatial configuration is a particular set of spatial positions at some instant of time that defines a configuration.

3. From step 2, write ($\dot{q}$) as a function of q and p. In other words, find $\dot{q} = \dot{q}(q, p)$.
4. Write the Hamiltonian:

$$H(p, q) = \sum p_i \dot{q}_i - L(q_i, \dot{q}_i). \tag{5.34}$$

5. Re-express the Hamiltonian in terms of p_i, $H = H(p_i)$.
6. Substitute H into the Hamiltonian equations (5.22) and (5.23):

$$\frac{\partial H}{\partial p_i} = \dot{q}_i; \qquad \frac{\partial H}{\partial q_i} = -\dot{p}_i. \tag{5.35}$$

7. Solve the Hamiltonian equations. The usual goal is to obtain N functions $q(t)$. This generally involves eliminating the p_i in favor of $\dot{q}_i$. This will turn $2N$ first order Hamiltonian differential equations into N second order differential equations, the same as Lagrangian equations!

Example 5.4. Particle in a central force field

Central forces are very important in physics and engineering. Central forces are forces that fulfill the following conditions:

- The magnitude of the force only depends on the distance $\vec{r}$ from a fixed point O.
- The force is always directed towards or away from the fixed point O.

The fixed point is usually taken as origin (O). This means that central forces are radially outwards or radially inwards.

Examples of central forces include the gravitational force of attraction between two point masses and the Coulomb force of attraction and/or repulsion between charged particles.

Solution

In order to solve this example let us first define the coordinates for central force problems. For 2D central force problems, the coordinates are written as

$$\begin{cases} x & = r\cos\theta \\ y & = r\sin\theta \end{cases}. \tag{5.36}$$

Differentiating the above equations will give us

$$\begin{cases} \dot{x} & = \dot{r}\cos\theta - r\dot{\theta}\sin\theta \\ \dot{y} & = \dot{r}\sin\theta - r\dot{\theta}\cos\theta \end{cases}.$$

We will solve this example using both (a) the Lagrangian and (b) the Hamiltonian mechanics. We expect that both mechanics will give the same results.

(a) Lagrangian mechanics.

We will begin by writing the Lagrangian ($L = T - V$) in terms of Cartesian coordinates,

$$L = \frac{1}{2}m(\dot{x}^2 + \dot{y}^2) - V(r), \tag{5.37}$$

and then converting the Lagrangian into spherical coordinates,

$$L = \frac{1}{2}m(\dot{r}^2 + r^2\dot{\theta}^2) - V(r). \tag{5.38}$$

The Lagrangian formulation will yield two Lagrangian equations:

- The radial Lagrangian equation.
- The angular Lagrangian equation.

The *radial Lagrangian equation* will be

$$\frac{d}{dt}\left(\frac{\partial L}{\partial \dot{r}}\right) - \frac{\partial L}{\partial r} = 0$$
$$\frac{d}{dt}(m\dot{r}) - \left(mr\dot{\theta}^2 - \frac{\partial V}{\partial r}\right) = 0$$
$$m\ddot{r} - mr\dot{\theta}^2 + \frac{\partial V}{\partial r} = 0.$$

Note that the third expression in this equation is $F(r)$, since $\frac{\partial V}{\partial r} = F(r)$. From equation (2.19), we can rewrite the above equation as

$$\boxed{F(r) = m\ddot{r} - mr\dot{\theta}^2.} \tag{5.39}$$

The *angular Lagrangian equation* will be

$$\frac{d}{dt}\left(\frac{\partial L}{\partial \dot{\theta}}\right) - \frac{\partial L}{\partial \theta} = 0$$
$$\frac{d}{dt}(mr^2\dot{\theta}) - 0 = 0$$

$$\boxed{\frac{d}{dt}(mr^2\dot{\theta}) = 0.} \tag{5.40}$$

Note that from this expression, if we consider r to be a constant we will obtain that the angular acceleration $\ddot{\theta}$ is zero, which implies that the angular velocity $\dot{\theta}$ is constant.

(b) Hamiltonian mechanics.

The Hamiltonian is given as $H = T + V$,

$$H = \frac{1}{2}m(\dot{r}^2 + r^2\dot{\theta}^2) + V(r). \tag{5.41}$$

Step 1. Write the Lagrangian

$$L = \frac{1}{2}m(\dot{r}^2 + r^2\dot{\theta}^2) - V(r).$$

Step 2. Find the generalized momentum

$$p = \frac{2L}{2\dot{q}}$$

$$P_r = \frac{2L}{2r} = m\dot{r}$$

$$P_\theta = \frac{2L}{2\theta} = mr^2\dot{\theta}.$$

Step 3. Rearranging the equations in step 2,

$$\dot{r} = \frac{P_r}{m}$$

$$\dot{\theta} = \frac{P_\theta}{mr^2}.$$

Step 4. Find the Hamiltonian,

$$\begin{aligned} H &= \Sigma p_1 \dot{q}_i - L \\ &= P_r\dot{r} + P_\theta\dot{\theta} - L \\ &= P_r\left(\frac{P_r}{m}\right) + P_\theta\left(\frac{P_\theta}{mr^2}\right) - \frac{1}{2}m\left(\frac{P_r^2}{m^2} + r^2\left(\frac{P_\theta}{m^2r^2}\right)\right) - V(r) \\ &= \frac{P_r^2}{m} + \frac{P_\theta^2}{mr^2} - \frac{P_r^2}{2m} - \frac{P_\theta^2}{mr^2} + V(r) \\ \Rightarrow H &= \frac{P_r^2}{2m} + \frac{P_\theta^2}{2mr^2} + V(r), \end{aligned}$$

which is just the total energy of the system.

Step 5. The four Hamiltonian equations.

There will be four Hamiltonian equations, as follows.

Hamiltonian equation 1:

$$\begin{aligned} \dot{r} &= \frac{\partial H}{\partial P_r} \\ \dot{r} &= \frac{P_r}{m} \\ m\dot{r} &= P_r \\ mv &= P_r. \end{aligned}$$

Differentiate the above equation to obtain

$$\boxed{m\ddot{r} = \dot{P}_r.} \tag{5.42}$$

Hamiltonian equation 2:

$$\dot{P}_r = -\frac{\partial H}{\partial r}$$
$$\dot{P}_r = \frac{P_\theta^2}{mr^3} - \frac{dV(r)}{dr}.$$

Put in the value of p_r from equation (5.42),

$$m\ddot{r} = \frac{P_\theta^2}{mr^3} - \frac{dV(r)}{dr}$$

$$\boxed{m\ddot{r} = \frac{P_\theta^2}{mr^3} - \frac{dV(r)}{dr}.} \tag{5.43}$$

Note that this is the same equation as (5.39), which we found for the Lagrangian method.

Hamiltonian equation 3:

$$\dot{\theta} = \frac{\partial H}{\partial P_\theta}$$
$$\dot{\theta} = \frac{P_\theta}{mr^2}.$$

Hamiltonian equation 4:

$$\dot{P}_\theta = -\frac{\partial H}{\partial \theta}$$
$$\dot{P}_\theta = 0,$$

which implies that P_θ is a constant, just like we found for the Lagrangian case.

Example 5.5. Atwood's machine—a frictionless and massless pulley

Find the equation for the acceleration (a) for an Atwood's machine where the pulley is frictionless and massless, using (figure 5.1):

(a) Newtonian mechanics.
(b) Lagrange's mechanics.
(c) Hamiltonian's mechanics.

Solution

(a) Newtonian mechanics.

First we will solve for the acceleration using the Newtonian formalism:

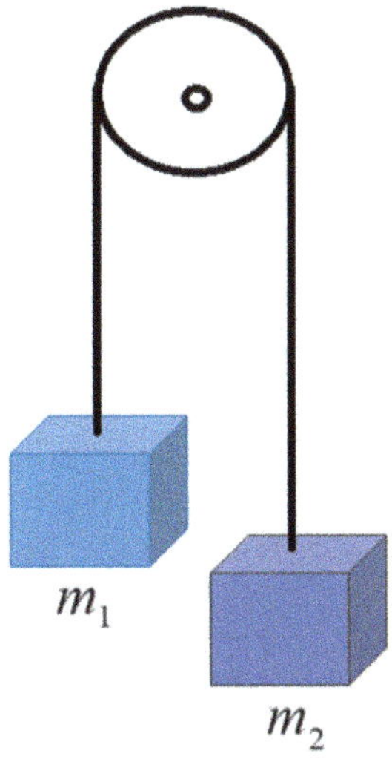

Figure 5.1. Atwood's machine.

mass 1 (m_1):

$$\begin{aligned} \Sigma F &= T_1 - m_1 g \\ m_1 a &= T_x - m_1 g; \end{aligned} \tag{5.44}$$

mass 2 (m_2):

$$\begin{aligned} \Sigma F &= - T_2 + m_2 g \\ m_2 a &= - T + m_2 g. \end{aligned} \tag{5.45}$$

Knowing that $a_1 = a_2$ and $T_1 = T_2$ we obtain

$$\begin{cases} m_1 a & = T - m_1 g \\ m_2 a & = -T + m_2 g \end{cases}.$$

Hence the acceleration will be

$$\boxed{a = \frac{(m_2 - m_1)g}{m_1 + m_2}.} \tag{5.46}$$

Putting the value of acceleration a into the equation for mass 1, equation (5.44) becomes

$$m_1\left(\frac{m_2 - m_1}{m_1 + m_2}\right)g = T - m_1 g.$$

Solve to obtain the value for tension:

$$T = \frac{2m_1 m_2}{m_1 + m_2}g. \tag{5.47}$$

(b) Lagrangian mechanics.

$$T = \frac{1}{2}m_1\dot{x}^2 + \frac{1}{2}m_2\dot{x}^2$$
$$V = -m_1gx - m_2g(\ell - x - \pi r)$$
$$V = (m_2 - m_1)gx - m_2g\ell + m_2g\pi r.$$

Using the above expressions we can write the Lagrangian (L) as

$$L = T - V = \frac{1}{2}m_1\dot{x}^2 + \frac{1}{2}m_2\dot{x}^2 - (m_2 - m_1)gx + m_2g\ell - m_2g\pi r. \tag{5.48}$$

The Lagrangian equation of motion is

$$\frac{d}{dt}\left(\frac{2L}{2\dot{x}}\right) - \frac{2L}{2x} - 0 = 0$$
$$\frac{d}{dt}[m_1\dot{x} + m_2\dot{x}] - (m_2 - m_1)g = 0$$
$$m_1\ddot{x} + m_2\ddot{x} - (m_2 - m_1)g = 0$$

$$\boxed{a = \ddot{x} = \frac{(m_2 - m_1)}{m_1 + m_2}g.} \tag{5.49}$$

(c) Hamiltonian Mechanics:

Step 1: Write Lagrangian:

$$L = \frac{1}{2}m_1\dot{x}^2 + \frac{1}{2}m_2\dot{x}^2 - m_1gx - m_2g(\ell - x - \pi r)$$

Step 2: Generalized Momentum:

$$p_x = \frac{\partial L}{\partial \dot{x}} = m_1\dot{x} + m_2\dot{x}$$

Step 3: Re-express

$$\frac{p_x}{m_1 + m_2} = \dot{x} \tag{5.50}$$

put values in the Lagrangian to express it in terms of momentum:

$$L = \frac{p_x^2}{2(m_1 + m_2)} - m_1gx - m_2g(\ell - x - \pi r) \tag{5.51}$$

Step 4: Hamiltonian Function:

$$\begin{aligned} H &= \Sigma p_i \dot{q}_i - L \\ &= p_x \dot{x} - L \end{aligned}$$

Put values of $\dot{x}$ from equation 5.50 and L from equation 5.51 to find the Hamiltonian as,

$$\begin{aligned} &= \frac{p_x^2}{m_1 + m_2} - L \\ &= \frac{p_x^2}{m_1 + m_2} - \left[\frac{p_x^2}{2(m_1 + m_2)} - m_1 g x - m_2 g(\ell - x - \pi r) \right] \\ \Rightarrow H &= \frac{p_x^2}{2(m_1 + m_2)} + m_1 g x + m_2 g(\ell - x - \pi r) \end{aligned}$$

Step 5: Hamiltonian Equations:

$$\begin{aligned} \dot{p}_x &= -\frac{\partial H}{\partial x} \\ \dot{p}_x &= -[m_1 g - m_2 g] \\ \dot{p}_x &= (m_2 - m_1) g \end{aligned}$$

and,

$$\begin{aligned} \dot{x} &= \frac{\partial H}{\partial p_x} \\ \dot{x} &= \frac{p_x}{m_1 + m_2} \end{aligned}$$

Differentiating the above equation and putting in value of $\dot{p}_x$ to find the value of acceleration (a):

$$a = \ddot{x} = \frac{(m_2 - m_1)}{m_1 + m_2} g \tag{5.52}$$

Compare equations (5.46), (5.49) and (5.52)—they are identical, which proves that no matter what methodology we use we will obtain the same values of the accelerations. As it should be.

5.3 Classical and quantum mechanics

The word *quantum* derives from the Latin, meaning ‘how much’. Quantum mechanics is a branch of physics that is essential to understanding the behavior of systems at atomic length scales. Quantum mechanics deals with the fundamental theory of nature at the subatomic level. It was developed to explain and describe the atom.

If we try to explain the physical nature of an atom by classical mechanics, we realize that electrons would eventually fall into the nucleus, since orbiting electrons emit radiation. However, atoms as we know them are stable and electrons never fall into the nucleus. Hence the classical explanation fails to explain the stability of atoms.

Without going into the details of quantum mechanics (which you can find in any quantum mechanics textbook) my intention here is for you to understand the similarities between quantum mechanics and classical mechanics.

The background that you need to appreciate quantum mechanics is to realize that quantum mechanics uses classical mechanics with one slight but essential modification, the introduction of $\hbar$. $\hbar$ is called the reduced Plank's constant, ($\hbar = h/2\pi$) with numerical value 1.05×10^{-34}J s or 6.58×10^{-16}eV s.[4] This modification changes the nature of the problem under consideration but leaves the underlying physics unchanged. As we have discussed, the Hamiltonian function is defined as the total energy of a system as

$$\begin{aligned} H &= \frac{1}{2}m(\dot{r}^2 + r^2\dot{\theta}^2) + V(r) \\ H &= \frac{p^2}{2m} + V(r). \end{aligned} \tag{5.53}$$

Then position and momentum can be found by taking partial derivatives of the above equation w.r.t. x and p, noting that p is independent of x and $V(r)$ is independent of p. We obtain

$$\frac{\partial H}{\partial x} = -\frac{\partial V(r)}{\partial x}. \tag{5.54}$$

Using the Hamiltonian equations and putting in values from equation (5.54),

$$\begin{aligned} \frac{dp}{dt} &= -\frac{\partial H}{\partial x} \\ &= -\frac{\partial V(r)}{\partial x} \\ &= F(x) \end{aligned} \tag{5.55}$$

and

$$\begin{aligned} \frac{\partial H}{\partial p} &= \frac{p}{m} \\ &= \frac{mv}{m} \\ &= \dot{x}. \end{aligned} \tag{5.56}$$

[4] eV = electron-volt is a unit of energy equal to the amount of energy gained (or lost) by the charge of one electron moving across an electric potential difference of one volt. Numerically eV is equal to approximately 1.6×10^{-19} J.

Comparing equation (5.55) to the Hamiltonian equations (5.22) and (5.23), we see that the Hamiltonian equations are just another way of looking at Newton's law specifically in situations where we are considering the momentum of the system.

Quantum mechanics is mainly taken from the Hamiltonian approach which was set up for classical mechanical problems. Keep in mind that the Hamiltonian approach emphasizes that the state of a system be determined by specifying the total energy of the system.

5.3.1 Quantum mechanical Hamiltonian

The Schrödinger Hamiltonian in quantum mechanics for a N-particle system is given as

$$\begin{aligned} \hat{H} &= \sum\nolimits_{n=1}^{N} \hat{T}_n + V \\ &= \sum\nolimits_{n=1}^{N} \frac{\vec{\hat{p}}_n \cdot \vec{\hat{p}}_n}{2m_n} + V(\vec{r}_1, \vec{r}_2 \cdots \vec{r}_N, t) \\ &= -\frac{\hbar^2}{2} \sum_{n=1}^{N} \frac{1}{m_n} \nabla_n^2 + V(\vec{r}_1, \vec{r}_2 \cdots \vec{r}_N, t), \end{aligned} \tag{5.57}$$

where ∇^2 is the Laplace operator, also called the *Laplacian*, which is a differential, scalar operator and can be applied to either the vector or scalar field. The Laplace operator is found in many fields of physics and many important equations such as Laplace's equation, Poisson's equation, the wave equation and the Schrödinger equation. For a 3D Cartesian coordinate systems it is defined as

$$\nabla^2 = \nabla \cdot \nabla = \frac{\partial^2}{\partial x^2} + \frac{\partial^2}{\partial y^2} + \frac{\partial^2}{\partial z^2}. \tag{5.58}$$

The quantum mechanical Hamiltonian is applied to quantum systems and it generates the time evolution of quantum states. If the state of the quantum system at time t is represented by a wave function as $|\psi(t)\rangle$, then the Hamiltonian operates on it as follows,

$$H\,|\psi(t)\rangle = i\hbar \frac{\partial}{\partial t}\,|\psi(t)\rangle, \tag{5.59}$$

where i is the imaginary unit and is the reduced Planck constant. This is the most general form of the time-dependent Schrödinger equation and it describes evolutions of quantum mechanical systems over time. It is named after Erwin Schrödinger, who published it in 1926 CE and was awarded the Nobel Prize in Physics in 1933 CE.

5.3.2 Expressions for the quantum Hamiltonian

Typical ways to classify the Hamiltonian are:

- The number of particles.
- The number of dimensions.

- The nature of the potential energy function.

We can generalize for one-particle systems that our Hamiltonian function will take on the following forms under various potential energy conditions:

- Unbound potential particle ($V = 0$):

$$\text{in 1D:} \quad \hat{H} = -\frac{\hbar^2}{2m}\frac{\partial^2}{\partial x^2}$$

$$\text{in 3D:} \quad \hat{H} = -\frac{\hbar^2}{2m}\nabla^2.$$

- Constant potential well ($V = V_0$):

$$\text{in 1D:} \quad \hat{H} = -\frac{\hbar^2}{2m}\frac{\partial^2}{\partial x^2} + V_0$$

$$\text{in 3D:} \quad \hat{H} = -\frac{\hbar^2}{2m}\nabla^2 + V_0.$$

- Simple harmonic oscillator ($V = \frac{1}{2}kx^2 = \frac{1}{2}(m\omega^2)x^2$):

$$\text{in 1D:} \quad \hat{H} = -\frac{\hbar^2}{2m}\frac{\partial^2}{\partial x^2} + \frac{1}{2}(m\omega^2)x^2$$

$$\text{in 3D:} \quad \hat{H} = -\frac{\hbar^2}{2m}\nabla^2 + \frac{1}{2}(m\omega^2)x^2.$$

Concept questions 5.1

1. Under what conditions is the Hamiltonian of a state of a system conserved?
2. When the Hamiltonian satisfies the equation $\frac{\partial H}{\partial q} = 0$, what can we say about the generalized momentum (p)?
3. Two equal masses (m) are connected by a spring with force constant (k). They slide across a floor. Explain why we can or cannot use ($H = T + V$) for the following cases:
   (a) When the floor is considered frictionless.
   (b) When the friction between the masses and floor is f_f.

Exercises 5.1

1. Find the equation of motion and the acceleration for a massless pulley Atwood's machine, using:
   (a) Newtonian mechanics.
   (b) Lagrangian mechanics.
   (c) Hamiltonian mechanics.
2. Find the equation for an Atwood's machine when the pulley has mass M, using:
   (a) Newtonian mechanics.
   (b) Lagrangian mechanics.
   (c) Hamiltonian mechanics.
3. Write the Hamiltonian function and Hamiltonian canonical equation for 3D motion of a projectile in a uniform gravitational field with no air resistance.

Further reading

Baierlein R 1983 *Newtonian Dynamics* (New York: McGraw-Hill)
Fowles G 1986 *Analytical Mechanics* 4th edn (New York: CBS College Publishing)
Fowles G and Cassiday G 2005 *Analytical Mechanics* 7th edn (Belmont, CA: Thomson Learning)
Goldstein H 1980 *Classical Mechanics* 2nd edn (Reading, MA: Addison-Wesley)
Landau L D and Lifshitz E M 1960 *Mechanics* (Reading, MA: Addison-Wesley)
Rutherford D E 1951 *Classical Mechanics* (New York: Interscience)
Stephenson R 1961 *Mechanics and Properties of Matter* 2nd edn (New York: Wiley)
Symon K 1971 *Mechanics* 3rd edn (Reading, MA: Addison-Wesley)

Chapter 6

Waves and oscillations

6.1 Mechanical waves

A wave is a disturbance that transfers energy from one place to another without requiring any net flow of mass. Waves can be viewed in two ways:

- As the disturbance in all of space at a single instant in time.
- As a disturbance at a single point as a function of time.

In the first case, a wave is taken as a disturbance in all of space at a single time, we would plot the magnitude of the disturbance at the given instant of time as a function of a distance coordinate, magnitude versus displacement. In the second case, where the wave is taken as a function of time at a single point in space, we plot the magnitude of the wave disturbance at a given point in space as a function of time. Note that both these pictures will give identical waves apart from the label on the x-axis.

Medium

Another really important concept in the definition of a wave is that of a *medium*. A medium is a collection of interacting particles through which the wave can propagate. The interaction of one particle with its adjacent particle allows the disturbance, a wave, to travel through the medium. Think of a wave generated in a Slinky toy. In this case, the particles of the medium are the individual coils of the Slinky. In the case of sound waves traveling in air, the particles or interacting parts of the medium are the individual molecules of air.

Waves and medium

There is a difference between what the wave does and what the particles in the medium do. As the wave travels through the medium, the particles of the medium oscillate in response to the disturbance. In a uniform medium, the wave may travel at a constant speed, however, each particle of the medium will have a constantly

doi:10.1088/2053-2563/aae1b9ch2

changing speed in response to the disturbance. The maximum speed of the medium particle is determined by the wave amplitude and its frequency. Additionally, the medium particles will return to their original condition after the wave has passed by.

6.1.1 Properties of a wave

The main properties of the waves are shown in figure 6.1 and are defined as follows.

- *Amplitude.* The height of the wave above its midline is called the amplitude of the wave. The SI unit of measurement is the meter, however, not all waves are measured in meters. It depends on the type of wave.
- *Crest.* The highest point in the wave is called the crest of the wave.
- *Trough.* The lowest point in the wave is called the trough of the wave.
- *Wavelength.* The distance between corresponding points on successive cycles is called the wavelength of a wave. It is denoted by the Greek letter lambda (λ) and is measured in units of length (meters).
- *Period.* The time it takes for one complete wave to pass a given point is called the period of a wave. The SI unit for the period is the second.
- *Frequency.* The number of complete waves that pass a particular point in one second is called the frequency of a wave. The SI unit for frequency is (1/seconds) or hertz (Hz).
- *Intensity.* The intensity of a wave or the power radiated by a source are proportional to the square of the amplitude.
- *Speed.* The speed of a point on a wave as it propagates. The SI unit for the speed of a wave is meters/second. The velocity of propagation is usually a characteristic of the medium and the type of wave, and does not depend on the parameters of the source or the detector:

$$\text{speed} = \frac{\text{distance}}{\text{time}} = \frac{\text{wavelength } (\lambda)}{\text{period } (s)} = \text{wavelength } (\lambda) \times \text{frequency } (f). \quad (6.1)$$

6.1.2 Types of waves

There are two main types of mechanical waves: transverse waves and longitudinal waves.

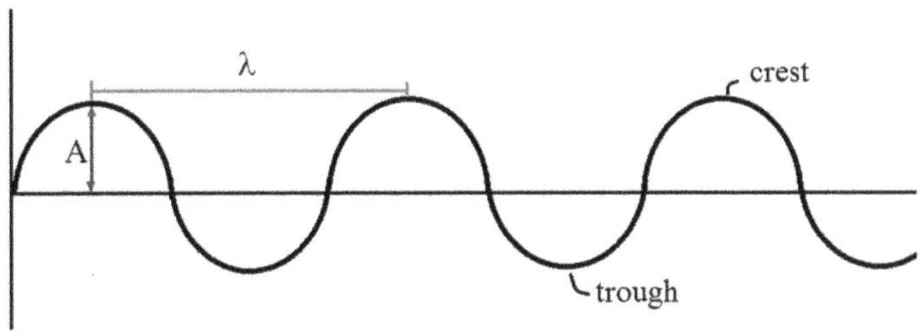

Figure 6.1. Properties of a wave.

Transverse waves
A wave is called transverse when the displacement of the medium is perpendicular to the direction of propagation of the wave (figure 6.2). Transverse waves cannot propagate in a gas or inside a liquid because there is no mechanism for driving motion perpendicular to the propagation of the wave.

A good way to visualize transverse waves is to think of surface water waves, a buoy sitting on the surface of the water goes up and down as the wave passes by and as soon as the wave is gone the buoy comes back to rest (figure 6.3).

Longitudinal waves
A wave is called longitudinal when the displacement of the medium is parallel to the direction of propagation of the wave. A good way to visualize it is to think of a compressed Slinky spring, as shown in figure 6.4. Longitudinal waves can travel through gases, liquids and solids.

Surface waves
Before moving on there is one other kind of wave that we need to discuss, *surface waves*. Surface waves, such as water waves, are generally a combination of

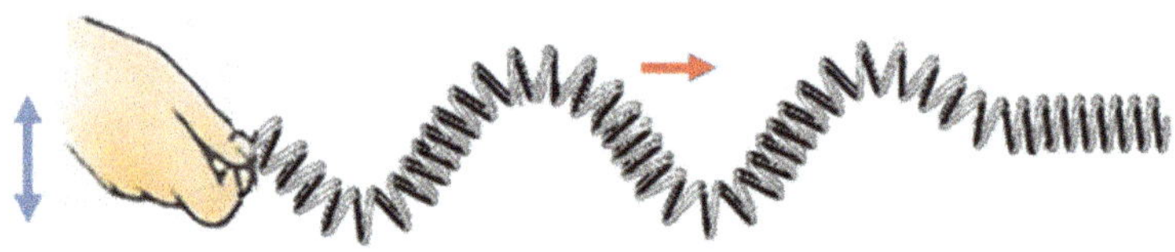

Figure 6.2. Visualizing transverse waves.

Figure 6.3. Buoy moving under transverse waves.

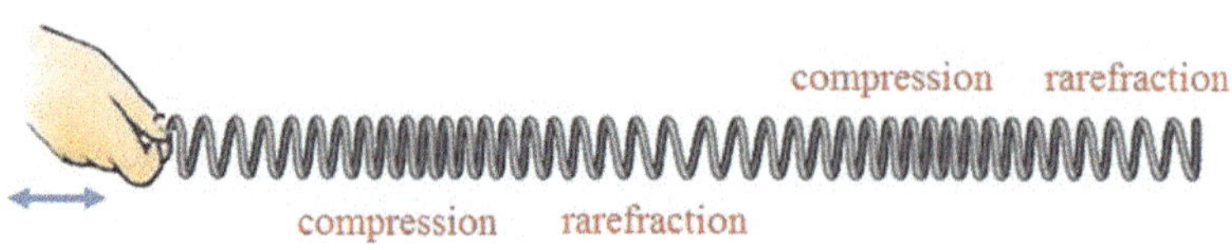

Figure 6.4. Visualizing longitudinal waves.

transverse and longitudinal waves. The particles on the surface of the water travel in circular paths as a wave moves across the surface.

Seismic waves are a result of earthquakes and are waves through the Earth's layers. Seismic waves can broadly be distinguished as *body waves*, that travel through the interior of the Earth, and surface waves, that propagate on the surface of the Earth.

Body waves are of two types, primary waves (P-waves) and secondary waves (S-waves). P-waves are pressure waves and are called primary since they are the fastest moving waves and reach the seismograph stations first. S-waves are transverse waves and travel almost 1.7 times slower that the speed of S-waves.

The difference in the arrival times of the P- and S-waves can be used to determine the distance to the event. To calculate the distance to the epicenter of an earthquake take Δt between the arrival times of the P- and S-waves in seconds and multiply by 8 km per second.

Example 6.1

A transverse wave moves along the $+x$-direction. It has a horizontal distance of 8 cm from a crest to the next trough, a frequency of 16 Hz and a vertical distance of 10 cm from a crest to the nearest trough. Calculate the following for the wave:

(a) Amplitude (A).
(b) Period (T).
(c) Wavelength (λ).
(d) Speed (v).

Solution

(a) Amplitude (A). Given the vertical distance from a crest to the nearest trough $= 10\text{ cm} = 2A$. Then amplitude $(A) = 5\text{ cm}$.
(b) Period (T). Given the frequency of the wave = 16 Hz. Then period $(T) = \frac{1}{f} = \frac{1}{16} = 0.0625\text{ s}$.
(c) Wavelength (λ). Given the horizontal distance from a crest to the next trough $(\lambda/2) = 8\text{ cm}$. Then wavelength $(\lambda) = 16\text{ cm}$.
(d) Speed (v). From equation (6.1): speed = wavelength × frequency. Then speed $= 16\text{ cm} \times 16\text{ Hz} = 256\text{ cm s}^{-1}$.

Example 6.2

Which of the following waves are longitudinal or transverse or both?

(a) Light waves. *Answer*: Light waves are electromagnetic waves, hence they are transverse.
(b) Sound waves. *Answer*: Longitudinal waves.
(c) P-type seismic waves. *Answer*: Longitudinal waves.
(d) S-type seismic waves. *Answer*: Transverse waves.
(e) Water waves. *Answer*: Both transverse and longitudinal.

6.1.3 Periodic motion

Waves can be broadly separated into either pulses or periodic waves. A pulse is a single disturbance while a periodic wave is a continually oscillating motion.

A periodic wave is a special type of oscillation which is a periodic function in both space and time. It can also be defined as a series of regularly timed disturbances in a particular medium.

6.1.4 Similarities and differences between various kinds of waves

Similarities

All waves, regardless of their origin, make or type, share the following properties:

- All waves have energy and can be reflected, refracted or diffracted, see section 6.2 for details of each.
- All waves transport energy from one place to another without the transfer of mass.

Differences

Depending on the type of wave, waves have differences that can classified as follows:

- Waves of different types propagate with widely varying speeds.
  Light/radio waves travel 3×10^8 m s^{-1} (186 000 miles/sec).
  Sound waves travel 344 m s^{-1}.
  Water waves travel a few feet/second.
- Sound waves require a medium such as a solid, liquid or gas for propagation. Light/radio waves do not need a medium to travel through. Actually, light waves also need a medium capable of supporting electromagnetic fields, and empty space can be that medium.

6.1.5 Simple harmonic motion

There is a close connection between simple harmonic motion and periodic waves. In most periodic waves, the particles in the medium experience simple harmonic motion.

Simple harmonic motion exhibits the following essential characteristics:

- It is described by a second order, linear differential equation with constant coefficients, for example for a mass–spring system it will be given by the equation

$$m\frac{d^2x(t)}{dt^2} = -kx(t). \tag{6.2}$$

- The superposition principle holds for simple harmonic waves. It states that if two particular solutions are found, their sum is also a solution of the simple harmonic equation. Solutions are generally taken to be functions of sines and cosines as

$$x(t) = A\cos(\omega t) + B\sin(\omega t), \tag{6.3}$$

where A, B and $\omega^2 = \frac{k}{m}$ are constants. For more details see section 7.1.

- The period of the motion repeats and is independent of the displacement away from equilibrium.

We will discuss simple harmonic motion in detail in chapter 7.

6.1.6 Properties of a wave

In this section let us briefly discuss some properties common to all waves.

Suppose we have a string attached to a wall on one end and being held on the other end. A wave is set up in the string by moving the hand up and down, as seen in figure 6.5.

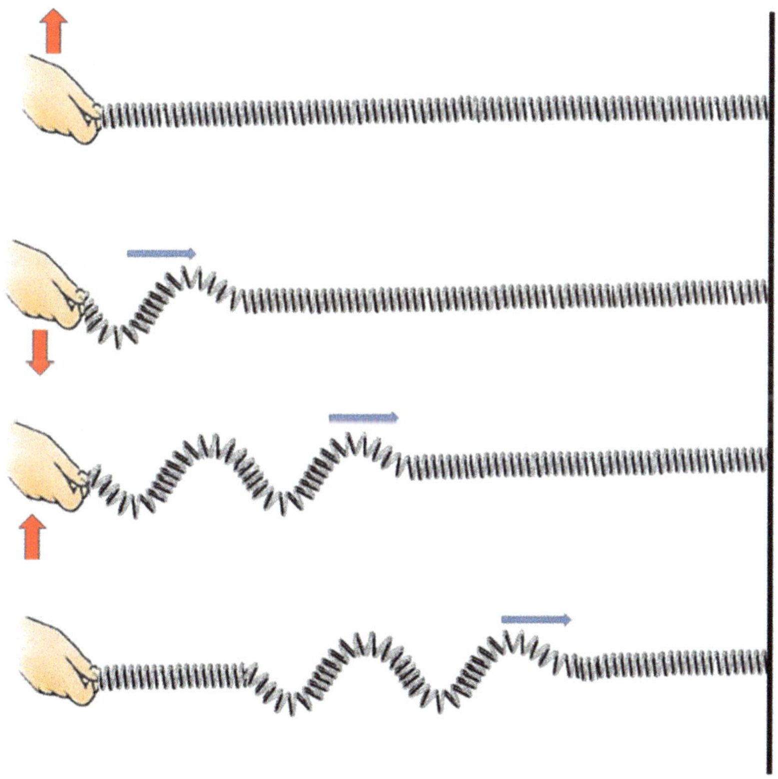

Figure 6.5. Setting up waves in a string.

6.1.6.1 Power of a wave

For a wave in one dimension, traveling from the left to the right, power is transmitted at a point, from the continuum to the left to the continuum to the right. Then in order to define the power of the wave, we begin by asking the following question: what is the rate at which the string to the left of any given point is doing work on the string to the right?

To answer this question, assume that our traveling wave is given by equation

$$\boxed{y = A\sin(kx - \omega t),} \tag{6.4}$$

where A is the amplitude (height) of the wave, k is the angular wave number measured in radians per unit distance ($\equiv\frac{2\pi}{\lambda}$) and ω is the angular frequency, measured in units of radians per second.

From equation (6.4) we can write

$$\begin{aligned}\frac{dy}{dx} &= Ak\cos(kx - \omega t)\\ \frac{dy}{dt} &= A(-\omega)\cos(kx - \omega t).\end{aligned}$$

From the definition of work (W), we know from equation (2.15) a force $\vec{F}$ does work ($=\vec{F}\cdot d\vec{s}$) on the object it is applied to. To find the power P, or rate of doing work we simply divide both sides by dt:

$$P(t) = \frac{dW}{dt} = \frac{\vec{F}\cdot d\vec{s}}{dt} = \vec{F}\cdot\vec{v}. \tag{6.5}$$

Assuming the amplitude of the wave is small means we can consider only the y-motion for the amplitude of wave. The component of the force along the vertical (y-axis) is ($-T\sin\theta$) and the velocity in the y-direction is $\partial y/\partial t$ so, using the small angle approximation for θ, we can write

$$\begin{aligned}P(t) &= -(T\sin\theta)v_y\\ &= -\left(T\frac{dy}{dx}\right)\frac{dy}{dt}\\ P(t) &= Tk\omega A^2\cos^2(kx - \omega t).\end{aligned} \tag{6.6}$$

Using $v^2 = T/\mu$, where μ is defined as mass per unit length, we substitute $v^2\mu$ for T and we obtain an expression for the power at any instant t as

$$\boxed{P = v\mu\omega^2A^2\cos^2(kx - \omega t).} \tag{6.7}$$

Over a number of cycles, the average of the $\cos^2$ function is numerically equal to $\frac{1}{2}$, so the average power transferred from the left to the right, for a wave traveling to the right, is

$$P_{\text{average}} = \frac{1}{2} v \mu \omega^2 A^2. \tag{6.8}$$

Hence, we can see from equation (6.8) that the rate of transmission of energy to the right is:

- Proportional to the speed (v) at which the wave shape is traveling to the right.
- Proportional to the mass per unit length (μ).
- Proportional to the square of the angular frequency (ω).
- Proportional to the square of the amplitude A.

6.1.6.2 Kinetic energy of a wave

The appearance of the squared terms is logical. Recall, for the traveling sine wave, each point on the wave executes simple harmonic motion with amplitude A and angular frequency (ω). So the kinetic energy is proportional to

$$\boxed{T = \mu A^2 \omega^2.} \tag{6.9}$$

6.1.6.3 Intensity of a wave (I)

For waves in three dimensions, power is transmitted through an area, which is why we introduce the term intensity (I), which we define as the *power per unit area*.

For a simple case, imagine we have a source of power P, radiating waves isotropically[1]. At a distance r, all of the radiated power is radiated through a sphere with area $4\pi r^2$. So the intensity at r from the source is

$$I = \frac{P}{4\pi r^2}. \tag{6.10}$$

However, this equation should be used with caution. Very few radiators are truly isotropic and measurements are often complicated by reflections in most cases.

6.1.6.4 Sound pressure (P)

Note that sound intensity (whose units are $\mathrm{W\,m^{-2}}$) is not the same physical quantity as sound pressure (P). Sound pressure is the local pressure change caused by the passage of a sound wave from the normal atmospheric pressure at that particular spot. The SI unit of sound pressure is the pascal ($\mathrm{Pa = N\,m^{-2}}$) and is measured by a microphone in air and with a hydrophone under water.

[1] Isotropically means the source is radiating equally in all directions. Think of it as a small source of sound held high up in the air, or a small bulb as a source of light held up so that light spreads equally on all sides. The Sun is a good example of an isotropic source of light.

6.1.6.5 Sound intensity level (SIL)

Sound intensity level (SIL) is the level of the intensity of a sound relative to a reference value. Sound intensity level is also called *acoustic intensity level* and is mathematically given as

$$L(dB) = 10 \log \frac{I}{I_0}, \tag{6.11}$$

where I is the sound intensity, I_0 is the reference sound intensity. A commonly used reference in air is 1×10^{-12} Wm^{-2}, which is the lowest sound intensity that can be heard by a human under regular room conditions.

6.1.6.6 Sound pressure level (SPL)

Sound pressure level (SPL) is a logarithmic measure of the effective pressure of a sound relative to a reference value. SPL is also called *acoustic pressure level.* It is denoted by L_p and is measured in decibels (dB). SPL is defined mathematically as

$$L_p(dB) = 20 \log \frac{P}{P_0}, \tag{6.12}$$

where P is the root-mean square sound pressure and P_0 is the reference sound pressure, which in air is 20×10^{-6}Pa.

6.1.6.7 Sound intensity level and sound pressure level

The relationship between sound intensity level (SIL) and sound pressure level (SPL) is mathematically given as

$$I \propto P^2 \tag{6.13}$$

and

$$\boxed{\frac{I}{I_0} \propto \frac{P^2}{P_0^2}.} \tag{6.14}$$

Example 6.3

A quiet room has a sound intensity of 1×10^{-8} W m^{-2}. Calculate the sound intensity in dB.

Solution

$$L(dB) = 10 \log\left(\frac{I}{I_0}\right).$$

Given that $I_0 = 10^{-12}$ Wm^{-2},

$$L = 10 \log\left(\frac{1 \times 10^{-8}}{1 \times 10^{-12}}\right) = 40 \text{ dB}.$$

Example 6.4

Find the intensity and the rms[2] values of the electric and magnetic fields at a distance of (a) 2 m and (b) 4 m from a 100 W light bulb. Assume, that there is no power loss and the bulb is an isotropic source.

Solution

Use the following equations (to find the intensity use $I = \frac{P}{A} = \frac{P}{4\pi r^2}$, where $P = 100$ W for this question):

$E_{rms} = \sqrt{\frac{I}{\varepsilon_0 c}}$, where $\varepsilon_0 = 8.85 \times 10^{-12} \frac{C^2}{N\,m^2}$

$B_{rms} = \sqrt{\frac{I\mu_0}{c}}$, where $\mu_0 = 4\pi \times 10^{-7} \frac{T\,m}{A}$.

(a) At a distance of 2 m.

Intensity = 1.9 W m^{-2}.

$E_{rms} = 27$ V m^{-1}.

$B_{rms} = 9.1 \times 10^{-8}$ T.

(b) At a distance of 4 m.

Intensity = 0.5 W m^{-2}.

$E_{rms} = 14$ V m^{-1}.

$B_{rms} = 4.6 \times 10^{-8}$ T.

Concept questions 6.1

1. Suppose you have a long, thin rod and a hammer. How must you hit one end of the rod to generate a transverse wave? A longitudinal wave?
2. A wave transmits energy. Does it also transfer momentum?
3. Why is the speed of sound generally much higher in solids than in air?

Exercises 6.1

1. A particular wave on a string has amplitude (A_1) and frequency (f_1) and it transfers energy at a rate of (P_1). If the amplitude and frequency are both doubled, at what rate will the new wave transfer energy?
2. Jack and Jill want to figure out how far the train is from the station. Jill puts her ear to the tracks and hears the train 2 s before Jack hears it through air. Can you help them by calculating how far away the train is? *Hint:* Assume that sound travels 15 times faster through the metal railway tracks than through air.
3. An intensity level change of +10 dB corresponds to what change in intensity?

[2] The rms (root-mean-square) value is the square root of the arithmetic mean of the squares of a set of numbers. If the waveform is a pure sine wave, the relationships between amplitudes and rms values are known and do not change. For a sine wave of the form $y = A\sin(\omega t)$ the rms value will be $\frac{A}{\sqrt{2}}$.

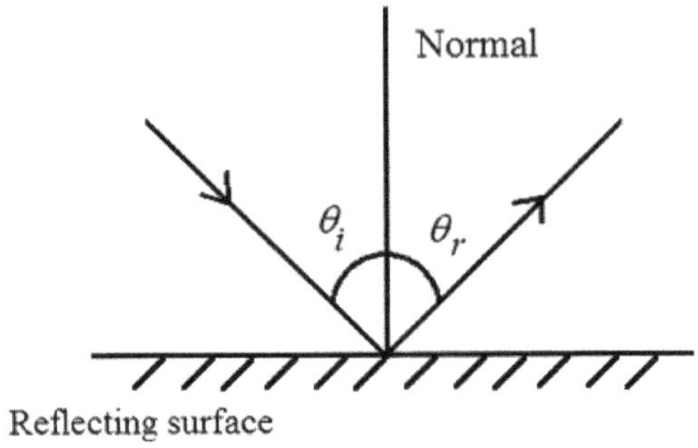

Figure 6.6. Reflection of light.

6.2 Physical properties of waves

There are some common behaviors that all waves exhibit under some standard situations. These include properties such as:

- Reflection.
- Refraction.
- Interference.
- Diffraction.
- Polarization.

We will discuss these properties very briefly in this book. For more detail see other textbooks.

6.2.1 Reflection and refraction

A wave incident on a reflective surface is reflected back, such that the angle made by the incident wave and line normal to the surface equals the angle made by the reflected wave and the same normal line, as shown in figure 6.6. This is called *reflection* of a wave.

Refraction is the name assigned to the phenomenon in which a wave changes its speed when it enters or exits from one medium to another, as shown in figure 6.7. The amount by which a wave is refracted by a material is given by the refractive index (n)[3] of the material.

The directions of incidence and refraction are related to the refractive indices of the two materials given by *Snell's law*. Snell's law states that the ratio of the sines of the angles of incidence and refraction is equivalent to the reciprocal of the ratio of the indices of refraction. Assume the refractive indices of medium 1 and medium 2 are n_1 and n_2, respectively, and θ_1 is the angle of incidence, θ_2 is the angle of refraction. Then Snell's law is mathematically given as

$$\boxed{n_1 \sin\theta_1 = n_2 \sin\theta_2.} \tag{6.15}$$

[3] In optics, the refractive index or index of refraction of a material describes how light traverses through the medium. The refractive index is a ratio and hence it is dimensionless. It is defined mathematically as $n = c/v$.

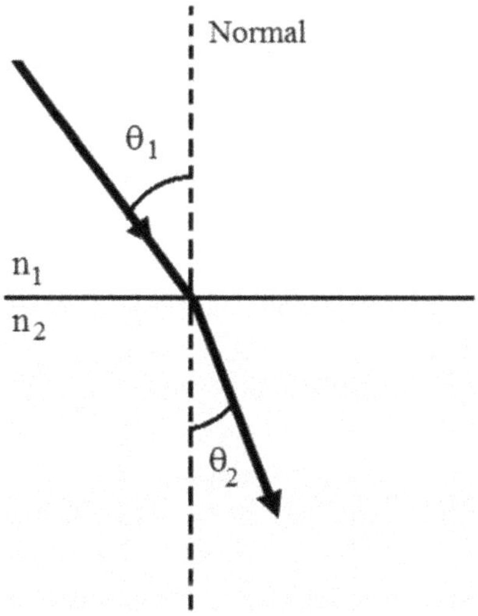

Figure 6.7. Refraction of light between two media of different refractive indices.

6.2.2 Superposition and interference

Most waves obey a principle of superposition which states that whenever two (or more) waves traveling through the same medium at the same time pass each other they superimpose each other. The net displacement of the medium will simply be the sum of the individual wave displacements. However, the waves will pass through each other without any permanent change in either of the two waves.

Note that this will be true only for non-dispersive media[4], where all frequencies travel at the same speed.

For a dispersive media, where the wave speed depends on the frequency of the wave, the waves would change their shape and higher frequencies waves will travel faster than lower frequency waves. As a result, the wave pulse spreads out and changes shape as it travels.

Interference is the interaction of two waves traveling in the same medium. Interference effects can be observed in all types of waves (for example, light waves, sound waves and water waves). Interference can occur in space or in time as explained in the following.

6.2.2.1 Interference in space

Superposition of two waves having the same frequency and moving in the same direction will result in the phenomenon of interference in space if the waves arrive at a particular point in space at the same time.

[4] In a non-dispersive medium the wave speed depends only on the physical properties of the medium, e.g. the elastic and inertial properties of a mechanical medium, and hence the wave speed is just a number and is independent of frequency. Waves traveling through this medium will maintain a constant shape.

When two or more waves are incident on the same point, the total displacement at that point is equal to the vector sum of the displacements of the individual waves.

Note the two waves will interfere *iff* they have phase coherence, i.e. they come from two coherent sources. Two sources are called coherent if they produce waves of the same amplitude, frequency, wavelength and phase.

Interference can be either constructive or destructive (figure 6.8):

- *Constructive interference.* If the two waves reach a point in phase i.e. the crest of one falls on the crest of the other wave and the trough of one wave falls on the trough of the other wave, then the two waves reinforce each other and the net wave effect increases, i.e. the net intensity is greater than the intensity of the individual waves. Such interference is called *constructive interference.*

  Maximum constructive interference of the two waves occurs when their phase difference (in radians) is $0, 2\pi, 4\pi, \ldots, 2m\pi$, where, $m = 0, 1, 2, 3, \ldots$.
- *Destructive interference.* If the two waves reach a point out of phase, i.e. the crest of one wave falls on the trough of the other, then the two waves cancel each other's effects and the net intensity of the individual wave is zero. This type of interference is called *destructive interference.*

  Complete destructive interference of two waves occurs when their phase difference (in radians) is $\pi, 3\pi, 5\pi, \ldots, (2m + 1)\pi$.
- *Intermediate interference.* Intermediate interference occurs when the two waves are added such that the crests and troughs do not line up and in this case we do not find either complete constructive or complete destructive interference. Instead, the result is between constructive and destructive interference.

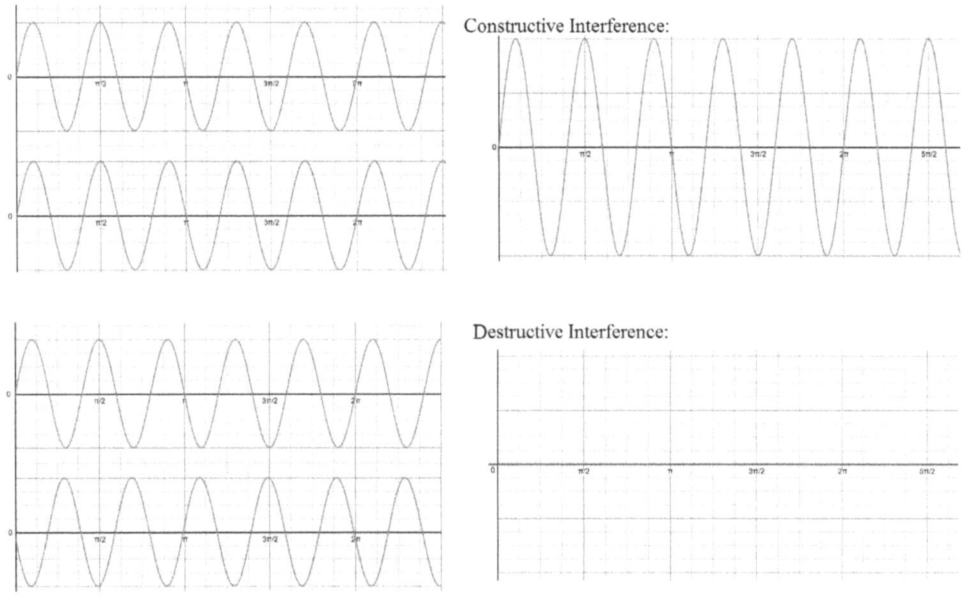

Figure 6.8. The interference of two waves. When in phase, the two waves create constructive interference (top), resulting in a wave of greater amplitude. When 180° out of phase, they create destructive interference (bottom).

Mathematically, this can be represented as

$$\begin{array}{c} 2\pi = \lambda \\ \text{phase difference} = \text{path difference}. \end{array} \tag{6.16}$$

For constructive interference, phase difference (m) is written as $2m\pi$ and path difference is written as $m\lambda$. For destructive interference, phase difference can be written as $(2m + 1)\,\pi$ and path difference as $(m + \frac{1}{2})\lambda$.

Example 6.5

Constructive interference doubles the amplitude, and hence quadruples the intensity and power. If two identical waves, each having an intensity of $1\ \mathrm{W\,m^{-2}}$, interfere perfectly constructively, the resulting wave has an amplitude of 2 m and the power from equation (6.8) will be 4 W and from equation (6.10) the intensity will be $4\ \mathrm{W\,m^{-2}}$. This means that the intensity goes up by a factor of 4 when the amplitude doubles. This appears to violate conservation of energy. Is this possible?

Think about it—in chapter 2 we said that conservation of energy can never be violated. Actually what is happening here is very interesting. From equation (6.10) we see the area over which the intensity equals $4\ \mathrm{W\,m^{-2}}$ is much smaller than the area taken-up by the two waves before they interfered. The 'extra energy' is actually coming from the destructive part of the interference where the intensity is zero and energy seems to disappear. It is actually conservation of energy at its finest. The observed pattern of interference must include both constructive and destructive interference. One cannot happen without the other and neither will happen in isolation.

6.2.2.2 Interference in time

When two sources of sound that have slightly different frequencies interfere, sometimes the crests of the waves will come together and sometimes the crest and trough will come together. If we listen to both sounds together we will hear a rise and fall of sounds, we call this phenomenon of rising and falling of the sound *beats* or, in other words, interference in time.

In music, a beat is an interference between two notes of slightly different frequencies. Musicians generally use this phenomenon to tune instruments. When the two tones are close in pitch but not identical, the difference in frequency generates beats. As the two tones gradually approach the same frequency, beating slows down and when the instruments are in tune the beating stops.

As seen in figure 6.9, two sines waves, y_1 and y_2, with slightly different frequencies, say f_1 and f_2, are traveling to the right and are given as

$$\begin{cases} y_1 &= A\cos(\omega_1 t) \\ y_2 &= A\cos(\omega_2 t) \end{cases}.$$

Suppose the two waves travel with the same speed in a non-dispersive medium and they interfere. The resulting superposition travels towards the right with the same

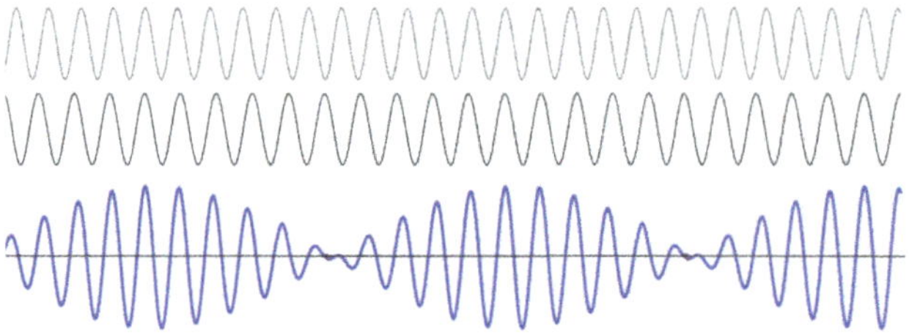

Figure 6.9. The interference of two waves with slightly different frequencies.

speed as the incident waves, but its local amplitude depends on the phase of the two individual waves:

$$
\begin{aligned}
y_{\text{net}} &= y_1 + y_2 \\
&= 2A\cos\left(\frac{\omega_1 + \omega_2}{2}t\right)\cos\left(\frac{\omega_1 - \omega_2}{2}\right).
\end{aligned}
$$

Keep in mind that $\omega = 2\,\pi f$.

The frequency of the modulation, the *beat wave*, oscillates with the average frequency of the two interfering waves and is given by

$$\text{frequency of the modulation} = \frac{f_1 + f_2}{2}. \tag{6.17}$$

The frequency of the audible beat wave, called the *amplitude envelope* of the beat, varies as the difference in frequencies:

$$f_{\text{beat}} = f_1 - f_2. \tag{6.18}$$

6.2.3 Diffraction

Diffraction is the signature phenomenon of wave motion. Bending of light around the edges of an obstacle, which results in the spread of light waves into a geometrical shadow of the obstacle, is called *diffraction* of light waves.

We can explain diffraction of light using Huygens' principle. Huygens' principle states that:

'Every point on a wavefront can be considered as a secondary source'.

When a wavefront of monochromatic light reaches an obstacle, part of the wavefront will fall on a slit of the obstacle. This part of the wavefront will now act as a secondary source. The wavelets that originate from this secondary source will now reach the shadow region where original light is not incident directly.

In the slit, there are infinitely many points which can act as point sources, and all of these are phase coherent. Hence, we can say that the diffraction of light is a superposition of infinitely many phase coherent waves! Diffraction also occurs in water waves. Figure 6.10 represents the diffraction of water waves through an opening for the cases when the opening is large and when the opening is small.

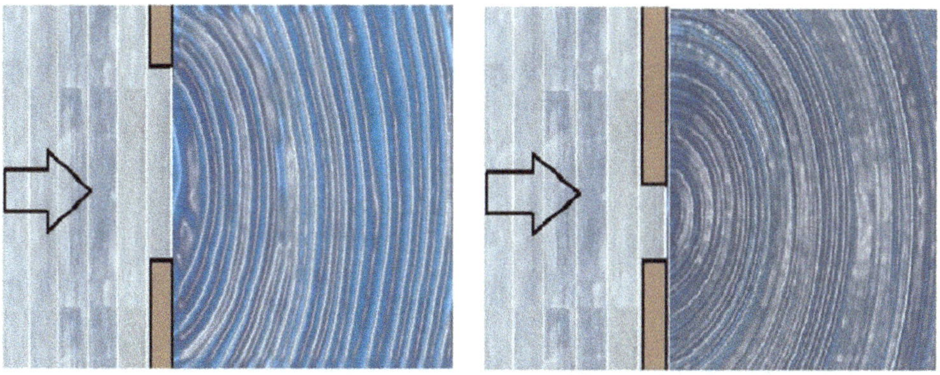

Diffraction through large opening — Diffraction through small opening

Figure 6.10. Diffraction.

6.2.4 Comparison of interference and diffraction

Similarities between interference and diffraction:

Both prove the wave nature of light.

Both occur due to superposition of phase coherent waves.

We observe minima and maxima of intensity in both processes.

Differences between interference and diffraction:

Interference occurs due to superposition of two phase coherent waves, while diffraction occurs due to superposition of an infinite number of phase coherent waves.

Maxima and minima in interference are of equal widths, however, they are not equal in size in diffraction.

Maxima have equal brightness (intensity) in interference, but in diffraction their brightness is different.

A large number of maxima and minima are observed in interference but only a few maxima and minima are observed in diffraction.

Example 6.6

A sound wave has a frequency of 540 Hz. Will this wave diffract while passing through an opening of 80 cm?

Solution

Using equation (6.1) we have $v = \lambda \cdot f \Rightarrow \lambda = \frac{v}{f}$, and the speed of sound waves at room temperature is approximately 344 m s^{-1}.

We see that $\lambda = \frac{344}{540} = 0.64$ m.

Assuming we are measuring the degree of diffraction by the departure from the geometric rays and given that the opening is 0.8 m, we can say that the sound waves will diffract a lot since 0.64 and 0.8 are close in value.

6.2.5 Polarization

Light is a transverse electromagnetic radiation wave. Light emanating from some source, the Sun or a light bulb, vibrates in all directions at right angles to the direction of propagation. The direction the light travels is determined by the direction of the vector cross product ($\vec{E} \times \vec{B}$). If $\vec{E}$ is taken along the $+x$-direction and $\vec{B}$ is along the $+y$-direction, then the propagation of the light wave will be along the $+z$-direction.

Polarization means *orientation*. It comes from the Greek word *polos*, which means the axis or pivot of a spinning globe. Wave polarization occurs for vector fields, so for light, where the vectors are electric and magnetic fields, the polarization direction is taken conventionally to be along the direction of the electric field. Polarization of an electromagnetic wave is shown in figure 6.11.

If unpolarized waves with incident intensity (I_0) pass through a polarizer, the transmitted intensity is half the incident intensity for ideal polarizers:

$$I = \frac{1}{2}I_0. \tag{6.19}$$

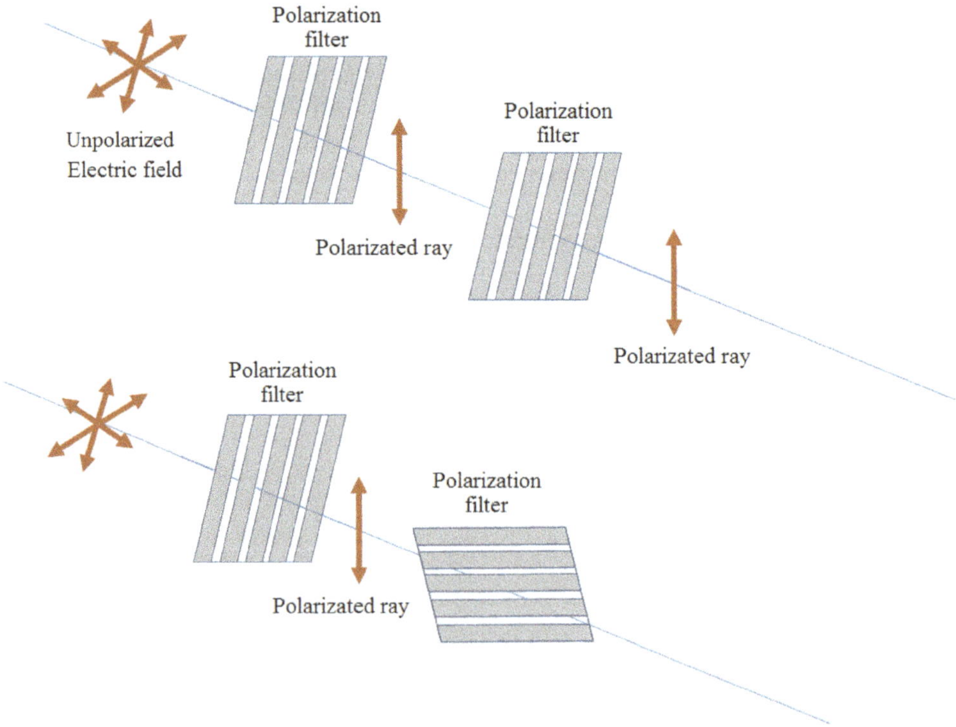

Figure 6.11. A polarizing filter has a polarization axis that acts as a slit passing through electric fields parallel to its direction.

If a linearly polarized wave is incident on a polarizer, only the component of $\vec{E}$ parallel to the transmission axis will get through. However, if there is an angle between the transmission axis and initial polarization, then

$$E = E_0 \cos\theta. \tag{6.20}$$

Intensity is proportional to the square of the amplitude, so the transmitted intensity is given by

$$\boxed{I = I_0 \cos^2\theta.} \tag{6.21}$$

Equation (6.21) is called *Malus's law* after its discoverer Etienne-Louis Malus, a French officer, engineer, physicist and mathematician.

More details of polarizations are discussed in other textbooks. We will just summarize the concept here. There are three main accepted types of polarization:

- *Plane polarization* uses a filter with a slit which provides a single vibration direction in the preferred configuration.
- *Circular polarization* uses a circular filter and has two vibration directions.
- *Elliptical polarization* uses an elliptical filter and has two vibration directions, with unequal amplitudes in the two directions.

The types of polarization are shown in figure 6.12.

Example 6.7

Suppose an unpolarized light of intensity I_1 is incident on two polarizers, one after the other. Suppose the transmission axis of the first polarizer is vertical and the transmission axis of the second polarizer is 30° with respect to the vertical. Find the polarization state of the outgoing light.

Solution

After passing the first polarizer, the light is vertically polarized. After passing the second polarizer the light will now be linearly polarized 30° with respect to the vertical.

6.3 Standing waves

Standing waves are waves that do not go anywhere. They can be produced by two identical waves traveling in opposite directions. In a bounded medium, standing waves occur when a wave with the correct wavelength meets its reflection and the two waves superimpose on each other. The interference of these two waves produces a resultant wave.

Setting up standing waves is easier than setting up traveling waves of the same amplitude. Standing waves have regions of maximum amplitude where the disturbance of the wave is largest, these points are called *antinodes*. There are also regions where the disturbance is almost zero. These points are called *nodes* of the standing waves.

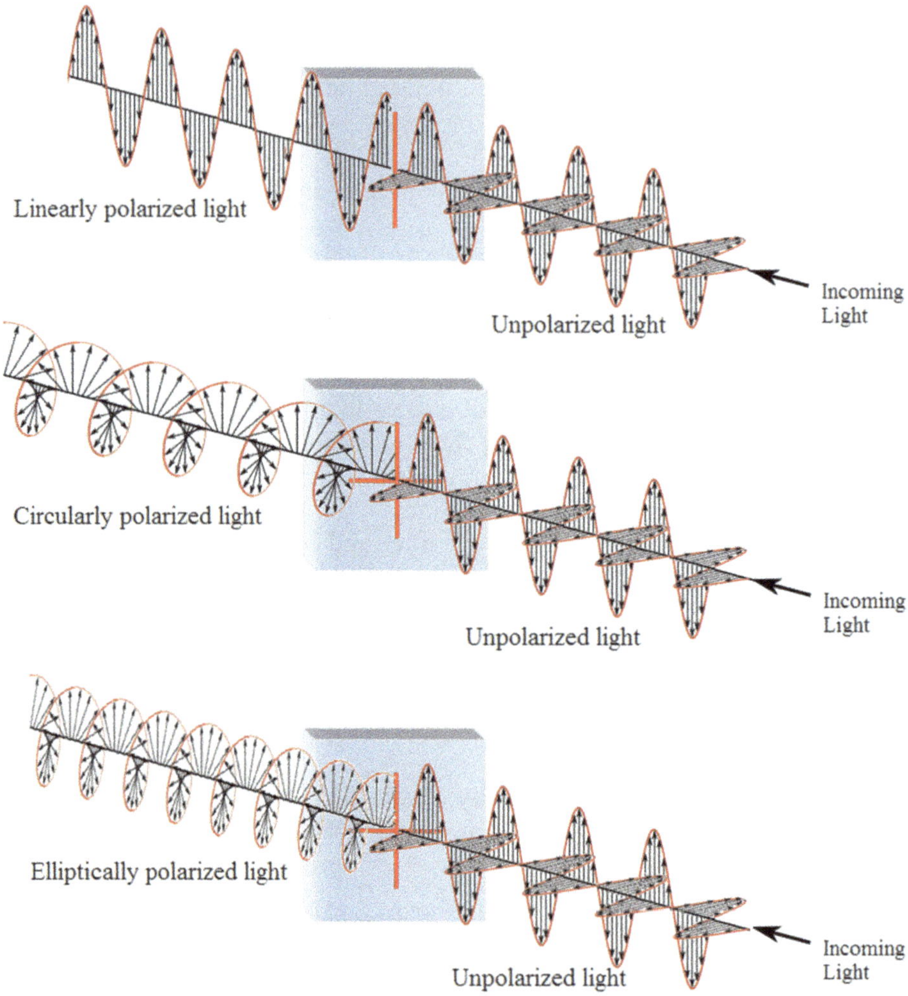

Figure 6.12. Types of polarizations.

6.3.1 Harmonics

Standing waves in any system have many natural frequencies. The set of all possible standing waves in a particular system are called the harmonics of a system. The simplest of the harmonics is called the *fundamental*. The harmonics above the fundamental are called the second harmonic, third harmonic, etc. In music theory, higher harmonics above the fundamental are called *overtones*.

6.3.2 Standing waves in strings

The velocity of a wave (v) in a stretched string is determined by the tension (T) and mass per unit length of the string ($\mu = m/L$), given as

$$v = \sqrt{\frac{T}{\mu}}, \tag{6.22}$$

and hence the frequency will be given as

$$f = \frac{v}{2L} = \frac{1}{2L}\sqrt{\frac{T}{\mu}}. \tag{6.23}$$

Ways in which standing waves can be set up in a string include:

- When the string is fixed at both ends. In this case the stationary wave must have nodes at both ends.

  Suppose the length of the string is L, then the relationship between length and wavelength is given by

$$\lambda = \frac{2L}{n}, \tag{6.24}$$

  where $n = 1, 2, 3, \ldots$.

  The frequencies of the standing waves which can be supported in this case will be

$$f = \frac{c}{\lambda} = n\frac{c}{2L}. \tag{6.25}$$

  The lowest frequency of a standing wave is called the *fundamental* frequency (f_0) and the second and third harmonics are simple multiples of this fundamental frequency (figure 6.13).

- When one end of the string is free to oscillate. The free end will be an antinode. In this case, the relationship between length and wavelength is given by

$$\lambda = \frac{4L}{(2n - 1)} \tag{6.26}$$

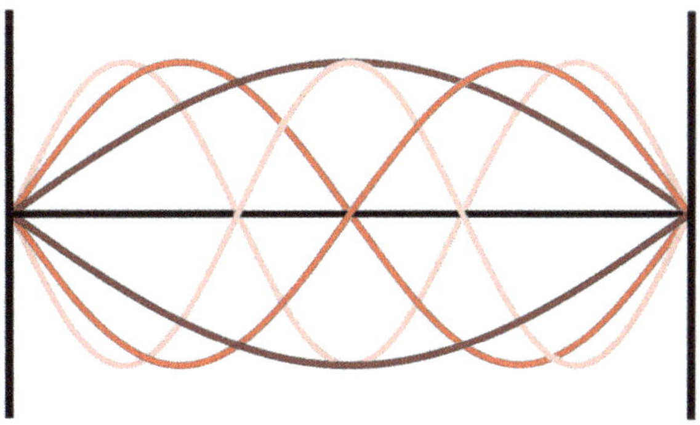

Figure 6.13. Standing waves on a string fixed at both ends.

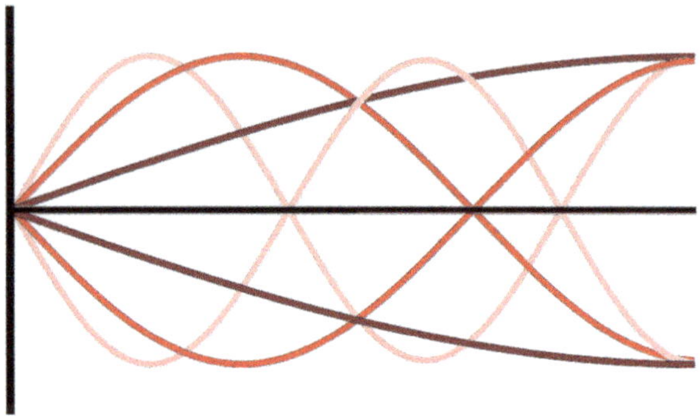

Figure 6.14. Standing waves on a string fixed at one end.

and frequency will be

$$f = \frac{c}{\lambda} = (2n - 1)\frac{c}{4L}. \tag{6.27}$$

Suppose the fundamental frequency is f_0, then the higher harmonics supported in this case will be $3f_0$, $5f_0$, $7f_0$, etc. Note that in this case there are no even harmonics, only odd harmonics occur in the string (figure 6.14).

6.4 Resonance

In order to set up standing waves in a system, we have to feed energy into the system at a particular frequency. If this energy input equals the natural frequency of the system, there is a dramatic increase in amplitude of the resultant vibrations. This condition is known as *resonance*.

Resonance is an amazing natural phenomenon where we apparently obtain a large output for very little input. A little bit of energy injected in a system at the right time accumulates and we obtain a large resulting output amplitude. Some applications include human speech, musical instruments, etc. Pushing a person on a swing is a common example of resonance.

Resonant frequency depends on the size, shape and composition of the object. In the human voice the vibrations of the vocal cords are reinforced by resonant vibrations in the oral and nasal passages. In music, resonance is used to increase the intensity (loudness) of a sound. The comparatively weak vibrations produced at the plucked string in a grand piano cause the piano to vibrate in resonance, thus greatly increasing the loudness of the sound.

Sometimes the mechanical resonances set up can produce vibrations that are so strong that they can destroy the objects they are set up in. For example, soldiers marching across bridges always break step before they step on the bridge in order to avoid a calamity of breaking the bridge by matching the resonant frequency of the bridge.

Concept questions 6.2

1. Why can you easily hear around corners but cannot see around the same corner?
2. Can sound waves be polarized? Explain.
3. Look straight up at the sky with the naked eye at the following times during the day and explain if the light you see is polarized or not. If it is, then explain the direction of polarization.
   (a) Just after sunrise.
   (b) At noon.
   (c) Just before sunset.

Exercises 6.2

1. Suppose the three-dimensional wave equation for spherical waves is given by

$$\frac{1}{r^2}\frac{\partial}{\partial r}\left(r^2\frac{\partial x}{\partial r}\right) = \frac{1}{v^2}\frac{\partial^2 x}{\partial t^2}. \tag{6.28}$$

   (a) Show that $x = \frac{A}{r}\sin(kr - \omega t)$, where A is the amplitude and k is a wave number, is a solution of this wave equation.
   (b) What are the dimensions of the constant A?
2. A particular string fixed at both ends of length 40 cm is tuned to 440 Hz. Find the three longest wavelengths of resonances of the string. What are the corresponding wavelengths that reach the listener?
3. Unpolarized light is incident on four polarizing sheets with their transmission axes oriented at 0°, 30°, 60° and 90° from the vertical, respectively. What percentage of initial intensity of the light will be transmitted?

Further reading

Arya A 1990 *Introduction to Classical Mechanics* 2nd edn (Upper Saddle River, NJ: Prentice-Hall)

Chababy R and Sherwood B 2011 *Matter and Interactions* 3rd edn (New York: Wiley)

Fowles G 1986 *Analytical Mechanics* 4th edn (New York: CBS College Publishing)

Fowles G and Cassiday G 2005 *Analytical Mechanics* 7th edn (Belmont, CA: Thomson Learning)

French A P 1971 *Vibrations and Waves* (Boca Raton, FL: Chapman and Hall)

Resnick R, Halliday D and Krane K 2002 *Physics* vol 1, 5th edn (New York: Wiley)

Rutherford D E 1951 *Classical Mechanics* (New York: Interscience)

Stephenson R 1961 *Mechanics and Properties of Matter* 2nd edn (New York: Wiley)

Symon K 1971 *Mechanics* 3rd edn (Reading, MA: Addison-Wesley)

IOP Publishing

Chapter 7

Simple harmonic oscillation

7.1 Harmonic oscillator

The simplest and most fundamental type of system undergoing a harmonic motion is a mass attached to a spring. Consider a block with mass (m) that is free to slide on a frictionless air-track. The block is attached to a fixed wall through a light spring. The downwards gravitational force (weight = mg) is canceled by the normal upwards force from the air-track. So for a mass–spring system moving along the horizontal track, the weight of the block does not play any role (figure 7.1). The only relevant force acting on the block arises from the stretching or compression of the spring. When the spring is at rest, the mass is considered to be at rest.

In general, the number of *degrees of freedom* of a system is the number of coordinates that must be specified in order to completely determine the configuration. This means that our mass–spring system has only one relevant degree of freedom, which is the position of the block. According to Hooke's law, the force from the spring is proportional to the displacement of the mass from its *equilibrium position*. Suppose the equilibrium position of the system is given by x_0 and x is the position of the block at some time (t). Then the force on the block at that instant is given by Hooke's law:

$$F_{\text{spring}} = -k(x - x_0). \tag{7.1}$$

The minus sign reminds us that spring force pulls the mass back towards the equilibrium position. The constant (k) is called the *spring constant* and has SI units of (N m^{-1}) and in terms of dimensions is given as MLT^2.

If we choose the equilibrium position to be at the origin (i.e. $x_0 = 0$) of the system and measure the position (x) of the block with respect to the origin, then the force on the block can be simplified to

$$F_s = -kx. \tag{7.2}$$

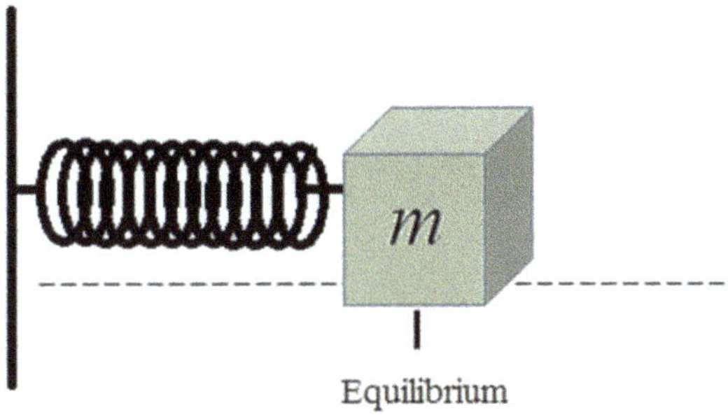

Figure 7.1. Mass–spring system.

If we let $x(t)$ be the displacement of the block as a function of time, t, then the differential form of Newton's law can be written as

$$m\frac{d^2x(t)}{dt^2} = -kx(t), \tag{7.3}$$

which is a second order differential equation. Because the system has only one degree of freedom, there is only one equation of motion.

The most general solution to the second order differential equation of motion, equation (7.3), is of the form

$$x(t) = A\cos(\omega t) + B\sin(\omega t), \tag{7.4}$$

where $\omega^2 = \frac{k}{m}$ is a constant with units of T^{-1} called the *angular frequency*[1]. Note that the ω used here is not the angular velocity of a spinning body but it instead refers to the natural motion of this oscillator.

The angular frequency is determined by the differential equation (7.3) and is numerically given by the number of radians by which the phase changes in a second. The other constants (A and B) are not determined by the differential equation (7.3), but rather by how the motion is started, i.e. by defining and knowing the initial conditions. In most cases these constants measure maximum displacement or the height of a wave and are called the *amplitudes of oscillation.*

7.1.1 Developing the equation of motion of a mass–spring system

Now let us develop the equation of motion of a mass attached to a spring undergoing a simple harmonic motion utilizing the methods we have discussed in the previous chapters.

[1] Angular frequency is a very important quantity in the study of wave phenomena. Angular frequency is a measure of the rate of rotation. A higher rate of angular frequency means the object under consideration is turning faster. In SI units, angular frequency is measured in radians per second, with dimension of T^{-1} since radians are dimensionless.

7.1.1.1. Newtonian mechanics

In the Newtonian method we start with equation (7.2):

$$\Sigma F_x = -kx$$
$$ma = -kx.$$

In 1D along the x-axis, the acceleration is given by $a = \ddot{x}$, Hence the equation of motion becomes

$$m\ddot{x} + kx = 0, \tag{7.5}$$

which is a second order differential equation.

7.1.1.2. Lagrangian mechanics

In the Lagrangian method, assume x is the displacement coordinate. Then the Lagrangian will be written as

$$L = T - V$$
$$L = \frac{1}{2}m\dot{x}^2 - \frac{1}{2}kx^2.$$

Then the Lagrangian equation will be written as

$$\frac{d}{dt}\left(\frac{\partial L}{\partial \dot{x}}\right) - \frac{\partial L}{\partial x} = 0$$
$$\frac{d}{dt}(m\dot{x}) + kx = 0.$$

Hence the equation of motion will become

$$m\ddot{x} + kx = 0, \tag{7.5}$$

which is identical to equation (7.5).

7.1.1.3. Hamiltonian mechanics

In order to develop the equation of motion we will use the methods we developed in chapter 5, section 5.2.7.

1. Write the Lagrangian $L(q, \dot{q})$, which in this case will be $L = L(x, \dot{x})$

$$L = \frac{1}{2}k\dot{x}^2 - \frac{1}{2}kx^2. \tag{7.7}$$

2. Find the generalized momentum ($p = p(\dot{q})$) associated with generalized coordinate (q) for all N coordinates present in the system. In this case the coordinate will be p_x,

$$p_x = \frac{\partial L}{\partial \dot{x}} = m\dot{x}.$$

3. From step 2, write $\dot{x}$ as a function of x and p. In other words, find, $\dot{x} = \dot{x}(x, p)$:

$$\dot{x} = \frac{p_x}{m}.$$

4. Write the Hamiltonian function:

$$H(q, p) = \sum p_i \dot{q}_i - L(q_i, \dot{q}_i)$$
$$H(x, p) = p_x \dot{x} - \left(\frac{1}{2}m\dot{x}^2 - \frac{1}{2}kx^2\right).$$

5. Re-express the Hamiltonian in terms of p_i, $H = H(p_i)$, which in this case will be p_x,$H = H(p_x)$, that we obtain from step 3,

$$H(x, p) = \frac{p_x^2}{m} - \frac{p_x^2}{2m} + \frac{1}{2}kx^2$$
$$= \frac{p_x^2}{2m} + \frac{1}{2}kx^2.$$

6. Substitute H into Hamiltonian equations, similar to equations (5.22) and (5.23):

$$\frac{\partial H}{\partial p_i} = \dot{q}_i; \qquad \frac{\partial H}{\partial q_i} = -\dot{p}_i. \tag{7.8}$$

7. Solve the Hamiltonian equations:

$$\dot{x} = \frac{\partial H}{\partial p_x}$$
$$\dot{x} = \frac{p_x}{m} \tag{7.9}$$

and

$$\dot{p}_x = -\frac{\partial H}{\partial x}$$
$$\dot{p}_x = -kx.$$

Differentiate equation (7.9) w.r.t. t,

$$\ddot{x} = \frac{\dot{p}_x}{m}$$
$$\ddot{x} = -\frac{kx}{m},$$

which we can rewrite as

$$m\ddot{x} + kx = 0, \tag{7.5}$$

which is identical to equations (7.5) and (7.6).

7.1.2 Derivation of a general solution

In order to find the general solution for the second order differential equation of motion we use the trial-and-error method. Let us pick a random solution of the form

$$x = Ae^{bt}, \tag{7.11}$$

where A and b are constants yet to be determined.

In order to prove that our chosen solution is correct, we have to show that it satisfies the second order differential equation of the form of equation (7.10):

$$m\ddot{x} + kx = 0.$$

Put values of x in the above equation:

$$m\frac{d^2(Ae^{bt})}{dt^2} + k(Ae^{bt}) = 0$$
$$mb^2Ae^{bt} + kAe^{bt} = 0.$$

Since $A \neq 0$ and $e^{bt} \neq 0$, then the only possibility is

$$mb^2 + k = 0$$
$$b^2 = -\frac{k}{m},$$

which implies the constant b is imaginary, recall $\imath^2 = -1$,

$$b = \pm\sqrt{\frac{k}{m}} = \pm\imath\omega. \tag{7.12}$$

Note, I have introduced a symbol ω, which we have already determined in section 7.1 and that we called the *angular frequency* of a system.

Thus our trial-and-error method yields two equivalent solutions:

$$x = Ae^{bt}$$
$$x = Be^{-bt}.$$

This implies that we can express the *general solution* for our second order differential equation as a linear combination by principle of superposition as

$$\boxed{x(t) = Ae^{i\omega t} + Be^{-i\omega t}} \tag{7.13}$$

or, equivalently,

$$x(t) = C\cos(\omega t) + D\sin(\omega t), \tag{7.14}$$

where constants A, B, C and D are constants and we can find the relationship between them by using Euler's equation ($e^{\pm i\theta} = \cos\theta \pm i\sin\theta$). The relationship is found to be

$$\begin{cases} A = \dfrac{C - iD}{2} \\ B = \dfrac{C + iD}{2}. \end{cases}$$

Finally, the third way of expanding the solution by using trigonometric identities is

$$x(t) = C\cos(\omega t - \phi) \tag{7.15}$$

or

$$x(t) = D\sin(\omega t - \phi), \tag{7.16}$$

where ϕ is the phase shift. The algebraic details are left as an exercise for the reader to attempt.

7.1.3 Period and frequency of a harmonic oscillation

The physical motions represented by the mathematical solutions given in section 7.1.2 are called *harmonic.*

The period is the time taken for one complete revolution or oscillation. We know the period is $T = \frac{2\pi}{\omega}$, and using the value from equation (7.12) we can rewrite the period of a mass–spring system as

$$\boxed{T = 2\pi\sqrt{\frac{m}{k}}.} \tag{7.17}$$

We define the linear frequency of this system (ω) as the number of cycles per unit time, mathematically given as $\omega = 2\pi f$ and, using the known quantities from the above equations, we find that the frequency (with SI units Hz) of the mass–spring system will be

$$\boxed{f = \frac{1}{2\pi}\sqrt{\frac{k}{m}}.} \tag{7.18}$$

7.1.4 Constants of motion and initial conditions

The solutions for the simple harmonic motion equation contains arbitrary constants and a value of phase shift (ϕ). The values of these constants are found by knowing the initial conditions of the particular problem. In this section we will calculate the values of these constants under certain initial conditions.

Case 1. The simplest example of initial conditions is when the mass is displaced from equilibrium to a particular position x_m and then released. In this case the displacement at $t = 0$ will be maximum. If we call that the amplitude (A), then

$$\text{amplitude, A} = x_m \quad \text{and} \quad \phi_0 = \frac{\pi}{2}. \tag{7.19}$$

Case 2. Another example of harmonic motion starts with the oscillator at rest at equilibrium point, $x = 0$. At time $t = 0$, the oscillator is struck, which sets the oscillator moving towards the positive x-direction with an initial velocity v_0. Suppose the initial phase is $\phi_0 = 0$. Under these initial conditions the solution gives that at $t = 0$; $x = 0$. To find the amplitude, differentiate x w.r.t. t to obtain the velocity as a function of time, then put $v = v_0$ at $t = 0$ to obtain

$$x(t) = D\sin(\omega t - \phi_o) \tag{7.20}$$

and the velocity can be found by differentiating as

$$\dot{x}(t) = D\,\omega\cos(\omega t - \phi_o). \tag{7.21}$$

At $t = 0$,

$$\dot{x}(0) = D\omega$$

$$\Longrightarrow \boxed{D = \frac{v_0}{\omega}.} \tag{7.22}$$

Case 3. (*General scenario*) for the general case, let us consider the mass to be initially displaced to some position x_0 and given an initial velocity v_0. Then the constants are found as

$$x(t) = A\sin(\omega_o t + \phi_o). \tag{7.23}$$

Taking the derivative w.r.t. time,

$$\dot{x}(t) = A\omega_o\cos(\omega_o t + \phi_o). \tag{7.24}$$

Dividing equations (7.23) and (7.24) we will obtain

$$\frac{\sin(\phi_o)}{\omega_o\cos(\phi_o)} = \frac{x(0)}{\dot{x}(0)},$$

which can be rewritten as

$$\tan(\phi_o) = \frac{\omega_o x_o}{v_o}. \tag{7.25}$$

At $t = 0$, the above equations (7.23) and (7.24) will become

$$x(0) = A \sin \phi_o \tag{7.26}$$

$$\dot{x}(0) = A\omega_o \cos \phi_o. \tag{7.27}$$

Squaring and adding equations

$$\begin{aligned} x_o^2 &= A^2 \sin^2 \phi_o \\ v_o^2 &= A^2\omega_o^2 \cos^2 \phi_o \\ \hline x_o^2 + \frac{v_o^2}{\omega_o^2} &= A^2[\sin^2 \phi_o + \cos^2 \phi_o] \end{aligned}$$

$$\Longrightarrow \boxed{A^2 = x_o^2 + \frac{v_o^2}{\omega_o^2}.} \tag{7.28}$$

This general solution reduces to either of the cases described above, which can easily seen by setting either v_o or x_o equal to zero.

7.2 Energy consideration in harmonic oscillator

The total energy in a simple harmonic oscillator is the constant sum of the potential and kinetic energies.

The kinetic energy (T) of motion of a mass is

$$T = \frac{1}{2}mv^2. \tag{7.29}$$

The potential energy (V), stored in a spring that follows Hooke's law, is

$$V = \frac{1}{2}kx^2, \tag{7.30}$$

where k is the spring constant and x is the magnitude of the displacement or deformation. In the absence of dissipative forces, the conservation of energy for these two forms is

$$\Delta T + \Delta V = 0, \tag{7.31}$$

which can be written as

$$\frac{1}{2}mv^2 + \frac{1}{2}kx^2 = \text{constant}. \tag{7.32}$$

This statement of conservation of energy is valid for all simple harmonic oscillators, including ones where the gravitational force plays a role.

Concept questions 7.1

1. Can the motion of the Moon around the Earth can be considered a simple harmonic motion? Explain your answer.
2. What will happen if washing machines were not mounted on springs?
3. Damage to buildings during an earthquake is proportional to the height of the building if the building is not properly constructed. Can you explain why?

Exercises 7.1

1. Your new AWD (all-wheel drive) car has independent suspension, in which each wheel is supported independently of the others. Your car can be considered as being mounted on four springs, called *shock absorbers* to reduce oscillations due to road conditions. If the mass of your car alone is 2000 kg and the springs are adjusted to have a frequency of 3 Hz then:
   (a) Assuming identical springs, find the force constant of the four springs.
   (b) What will be the new vibration frequency of your car if four of your friends ride along with you? Assume that the average weight of each person is 60 kg.
2. A simple harmonic motion is given by the equation

   $$y = (6m)\cos[(8 \text{ rad s}^{-1})t + 2 \text{ rad}].$$

   Find the following quantities at $t = 2$ s:
   (a) Amplitude.
   (b) Displacement.
   (c) Velocity.
   (d) Acceleration.
   (e) Frequency.
   (f) Period.
3. Suppose a wooden block of mass 10 kg is attached to a wall by means of a spring with normal length 50 cm and a spring constant 5 N m^{-1}. A 50 gm bullet is fired horizontally at the block with a speed of 150 m s^{-1} and it comes to rest in the block.
   (a) What kind of motion will be set up?
   (b) Find the amplitude of the resulting motion.
   (c) What fraction of the original KE of the bullet appears as the energy of the oscillator?

7.3 About various pendulums

The word *pendulum* comes from the ancient Latin word 'pendulus' which means 'hanging'. The concept of a swinging object has been known to man for centuries, but the first scientific investigations of the pendulum were conducted by Galileo Galilei around 1602 CE. He was the first to use it for accurate time-keeping, a technology that was used all over the globe until around the 1930s for keeping time.

A pendulum is any mass suspended in such a way that it is free to swing. The resting spot due to gravity of a pendulum is called its *equilibrium position.* When a pendulum is displaced from its equilibrium position, it will accelerate back toward the equilibrium position due to the force of gravity. We call this the *restoring force* and under the influence of the restoring force the displaced pendulum will tend towards the equilibrium position, however, the pendulum will continue past the equilibrium because of its inertia. This will cause the pendulum to oscillate about the equilibrium position, swinging back and forth.

The time for one complete cycle is called the *period* and the width of the pendulum's swing is called its *amplitude*. The period depends on the length of the pendulum and is mathematically given as

$$\boxed{T = 2\pi\sqrt{\frac{l}{g}}.} \tag{7.33}$$

7.4 Simple gravity pendulum

A simple pendulum consists of a mass (m) suspended from a pivot point by means of a string or rod of negligible mass, as seen in figure 7.2. The suspended mass is hence constrained to move on a particular path determined by the force of gravity and inertia. The position of the mass is defined by the coordinate vector $r = (x, y)$.

In the following examples, we will discuss the simple pendulum by using the various methods we have come across during the course of study of this book.

Example 7.1. The simple pendulum—Newtonian mechanics

We can treat the pendulum bob as a particle. During the swing the trajectory of a pendulum is an arc. We know in a circle, the distance along the curved line is called the arc length, mathematically given as $s = \ell\theta$, if we take the radius of the circle as ℓ and the angular displacement as θ (figure 7.3).

The restoring force is the component of the weight, mg, acting in the direction of increasing θ along the path of motion:

$$F_R = -mg \sin \theta. \tag{7.34}$$

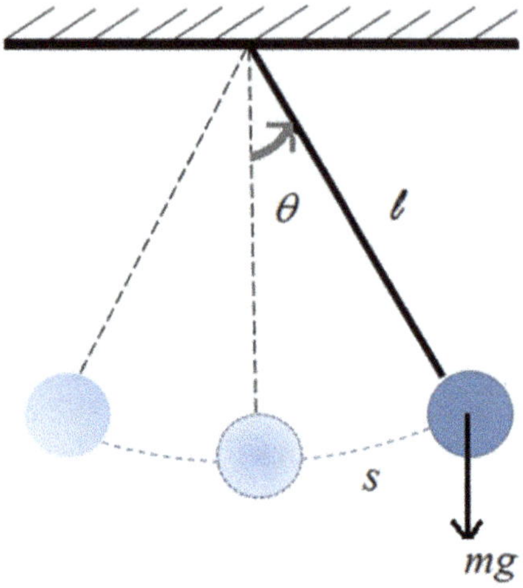

Figure 7.2. Simple pendulum.

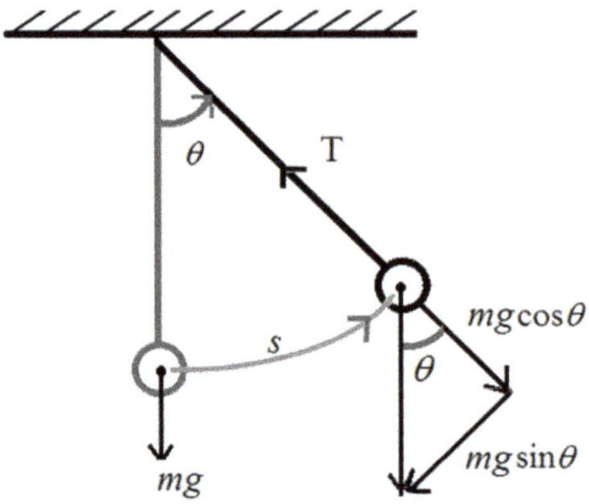

Figure 7.3. Simple pendulum.

We know for a small angle $\sin\theta \approx \theta$, so Newton's force equations along the arc will be written as

$$\Sigma F = -\, mg \sin\theta$$
$$\ddot{s} = -\, g \sin\theta$$
$$\ddot{s} + g\theta = 0$$

$$\ddot{s} + \frac{g}{\ell}s = 0, \tag{7.35}$$

or equivalently, knowing $s = \ell\theta$, as

$$\ddot{\theta} + \frac{g}{\ell}\theta = 0. \tag{7.36}$$

The solution of the equation is

$$\theta = A\sin(\omega t) + B\cos(\omega t). \tag{7.37}$$

Even though the motion is along a curved path rather than a straight line, these are both mathematically identical to the linear harmonic equations for a mass–spring system we found in section 7.1, equation (7.5), but with $(\frac{g}{\ell})$ instead of $(\frac{k}{m})$.

We can see that the motion is simple harmonic with:

- Angular frequency: $\omega = \sqrt{\frac{g}{\ell}}$.
- Period: $T = \frac{2\pi}{\omega} = 2\pi\sqrt{\frac{\ell}{g}}$.
- Frequency: $f = \frac{1}{T} = \frac{1}{2\pi}\sqrt{\frac{g}{\ell}}$.

Example 7.2. Simple pendulum—Lagrangian mechanics

Next we will develop an equation of motion using Lagrange's formalism, keeping in mind small angle approximations from trig identities, $\sin\theta \approx \theta$ and $\cos(\theta) \approx 1$:

$$\begin{cases} x = & \ell\sin\theta \\ y = & \ell\cos\theta. \end{cases} \tag{7.38}$$

We can differentiate the above equations to obtain

$$\begin{cases} \dot{x} = & \ell\cos\theta\dot{\theta} \\ \dot{y} = & -\ell\sin\theta\dot{\theta}. \end{cases}$$

Squaring and adding both sides of the above equations and knowing $\cos^2(\theta) + \sin^2(\theta) - 1$, we obtain

$$\dot{x}^2 + \dot{y}^2 = \ell^2\dot{\theta}^2.$$

This means that the kinetic and potential energies can be written as

$$T = \frac{1}{2}m(\dot{x}^2 + \dot{y}^2) = \frac{1}{2}m\ell^2\dot{\theta}^2 \tag{7.39}$$

$$V = -mg(\ell - \ell\cos\theta). \tag{7.40}$$

Hence, the Lagrangian becomes

$$\begin{aligned} L &= T - V \\ &= \frac{1}{2}m\ell^2\dot{\theta}^2 + mg\ell(1 - \cos\theta). \end{aligned} \tag{7.41}$$

Note that since r is a constant, the radial part of the Lagrangian equation will not exist and we will only be left with the angular part of the equation of motion, which will become

$$\frac{d}{dt}\left(\frac{\partial L}{\partial \theta}\right) - \frac{\partial L}{\partial \theta} = 0$$
$$\frac{d}{dt}\left(\frac{1}{2}m\ell^2\theta^2\right) + mg\ell \sin\theta = 0$$
$$\ell^2\ddot{\theta} + g\ell \sin\theta = 0$$
$$\ddot{\theta} + \frac{g}{\ell}\sin\theta = 0.$$

For small angles $\sin\theta \approx \theta$, hence the above equation becomes

$$\boxed{\ddot{\theta} + \frac{g}{\ell}\theta = 0,} \tag{7.42}$$

which is identical to equation (7.36).

Example 7.3. Simple pendulum—energy consideration

For the development of the simple pendulum equation of motion in the energy consideration method, we should begin from the conservation of energy. Keeping in mind arc length ($s = \ell\theta$),

$$\Delta E = 0$$
$$\Delta T + \Delta V = 0$$
$$\frac{1}{2}m\dot{s}^2 + mgh = \text{constant}$$
$$\frac{1}{2}\dot{s}^2 + gh = \text{constant}.$$

From figure 7.4, we can see that $h = \ell(1 - \cos\theta)$, hence the above equation becomes

$$\frac{1}{2}\dot{s}^2 + g\ell(1 - \cos\theta) = \text{constant}$$
$$\dot{s}^2 + 2g\ell(1 - \cos\theta) = \text{constant}.$$

We know that $\cos\theta \approx 1 - \frac{\theta^2}{2} + \frac{\theta^4}{4!} + \cdots$, hence the above equation can be written approximately as

$$\ell\dot{\theta}^2 + 2g - 2g\left(1 - \frac{\theta^2}{2}\right) = \text{constant}.$$

Differentiating the above equation we obtain

$$2\dot{\theta}\ddot{\theta} + 2\frac{g}{\ell}\theta\dot{\theta} = 0. \tag{7.43}$$

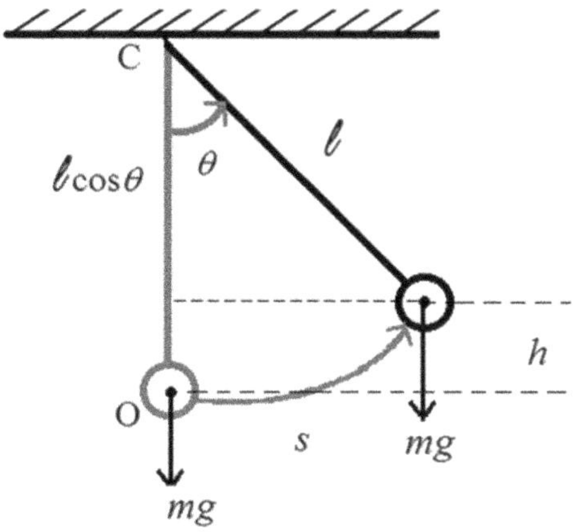

Figure 7.4. Simple pendulum.

Since $2\dot{\theta} \neq 0$ we know

$$\Longrightarrow \boxed{\ddot{\theta} + \frac{g}{\ell}\theta = 0.} \tag{7.44}$$

Example 7.4. Simple pendulum—Hamiltonian mechanics

To find the Hamiltonian canonical equation for a simple pendulum we will follow the steps we discussed in chapter 5.

Step 1. Write the Lagrangian (L):

$$\begin{aligned} L &= \frac{1}{2}m\ell^2\dot{\theta}^2 - mg\ell + mg\ell\cos\theta \\ &\approx \frac{1}{2}m\ell^2\dot{\theta}^2 - mg\ell + mg\ell\left(1 - \frac{\theta^2}{2}\right) \\ &= \frac{1}{2}m\ell^2\dot{\theta}^2 - \frac{1}{2}mg\ell\theta^2. \end{aligned}$$

Step 2. Generalized momentum:

$$P_\theta = \frac{\partial L}{\partial \dot{\theta}} = m\ell^2\dot{\theta}. \tag{7.45}$$

Step 3. Re-express:

$$\dot{\theta} = \frac{P_\theta}{m\ell^2}. \tag{7.46}$$

Step 4. Hamiltonian function:

$$\begin{aligned} H &= \Sigma p_i q_i - L \\ &= P_\theta \dot{\theta} - \left(\frac{1}{2} m\ell^2 \dot{\theta}^2 - \frac{1}{2} mg\ell\theta^2\right) \end{aligned}$$

$$H = \frac{P_\theta^2}{2m\ell^2} + \frac{1}{2} mg\ell\theta^2.$$

Step 5. Hamiltonian equations:

$$\dot{P}_\theta = \frac{\partial H}{\partial \theta}$$

$$\dot{P}_\theta = -\, mg\ell\theta$$

and

$$\dot{\theta} = \frac{\partial H}{\partial P_\theta}$$

$$\dot{\theta} = \frac{P_\theta}{m\ell^2}.$$

Differentiating w.r.t. time,

$$\ddot{\theta} = \frac{\dot{P}_\theta}{m\ell^2} = -\frac{mg\ell\theta}{m\ell^2}$$

$$\Longrightarrow \boxed{\ddot{\theta} + \frac{g}{\ell}\theta = 0.} \tag{7.44}$$

From the above examples, we clearly see from equations (7.36), (7.42), (7.44) and (7.47) that all the various methods of classical mechanics give the same equation of motion—just as we expect.

7.5 Elastic pendulum

An elastic pendulum consists of a mass (m) attached to a spring or an extendable string of stiffness (k) and unstretched length (ℓ_0). The construction of an elastic pendulum is similar to an ordinary pendulum (figure 7.5). However, since the spring and/or string is stretchable it will add additional levels of complication.

7.5.1 Elastic pendulum—Lagrangian mechanics

$$\begin{cases} x = r\cos\theta \\ y = r\sin\theta \end{cases}$$

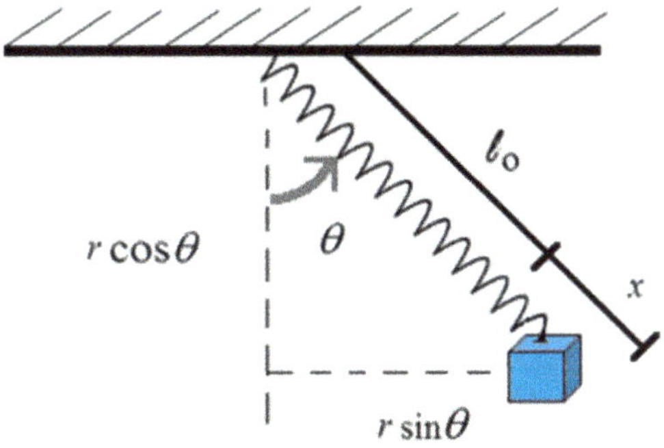

Figure 7.5. Elastic pendulum.

Differentiating the above equations we obtain

$$\begin{cases} \dot{x} = \dot{r}\cos\theta + r(-\sin\theta\dot{\theta}) \\ \dot{y} = \dot{r}\sin\theta + r(\cos\theta\dot{\theta}). \end{cases}$$

Squaring and adding these we can simplify to obtain

$$\dot{x}^2 + \dot{y}^2 = \dot{r}^2 + r^2\dot{\theta}^2, \tag{7.48}$$

hence we can write kinetic energy as

$$T = \frac{1}{2}m(\dot{x}^2 + \dot{y}^2) = \frac{1}{2}m(\dot{r}^2 + r^2\dot{\theta}^2)$$

and potential energy as

$$V = \frac{1}{2}k(r - \ell_o)^2 - mgr\cos\theta,$$

which implies that the Lagrangian will be written as

$$L = T - V = \frac{1}{2}m(\dot{r}^2 + r^2\dot{\theta}^2) - \frac{1}{2}k(r - \ell_o)^2 + mgr\cos\theta. \tag{7.49}$$

There will be two equation of motion, one for the radial component (r) and one for the angular component (θ):

- Equation of motion for radial component (r):

$$\frac{d}{dt}\left(\frac{\partial L}{\partial \dot{r}}\right) - \frac{\partial L}{\partial r} = 0$$

$$\frac{d}{dt}(m\dot{r}) - mr\dot{\theta}^2 + k(r - \ell_o) - mg\cos\theta = 0$$

$$m\ddot{r} - mr\dot{\theta}^2 + k(r - \ell_o) - mg\cos\theta = 0.$$

- Equation of motion for angular component (θ):

$$\frac{d}{dt}\left(\frac{\partial L}{\partial \dot{\theta}}\right) - \frac{\partial L}{\partial \theta} = 0$$
$$\frac{d}{dt}(mr^2\dot{\theta}) - mgr\sin\theta = 0$$
$$r^2\ddot{\theta} + gr\cos\theta = 0$$

$$\Longrightarrow \boxed{\ddot{\theta} + \frac{g}{r}\sin\theta = 0.} \tag{7.50}$$

It is worth noting here that equation (7.50) for small angles reduces to an equation of motion for a simple pendulum, as discussed in section 7.4.

7.6 Spherical pendulum

A spherical pendulum is a higher dimensional analogue of the simple pendulum. It consists of a mass m moving without friction on the surface of a sphere. The only forces acting on the mass are the reaction from the sphere and gravity. The construction of a spherical pendulum is similar to an ordinary pendulum. However, instead of swinging back and forth, the bob of a conical pendulum moves at a constant speed in a circle with the string tracing out a cone, as shown in figure 7.6:

$$\begin{cases} x = & \ell\sin\phi\cos\theta \\ y = & \ell\cos\phi\sin\theta \\ z = & \ell\cos\phi. \end{cases}$$

Differentiating the above equations:

$$\begin{cases} \dot{x} = & \ell(\cos\phi\ \dot{\phi})\cos\theta - \ell\sin\phi\sin\theta\ \dot{\theta} \\ \dot{y} = & \ell(\cos\phi\ \dot{\phi})\sin\theta - \ell\sin\phi\cos\theta\ \dot{\theta} \\ \dot{z} = & \ell\sin\phi\ \dot{\phi}. \end{cases}$$

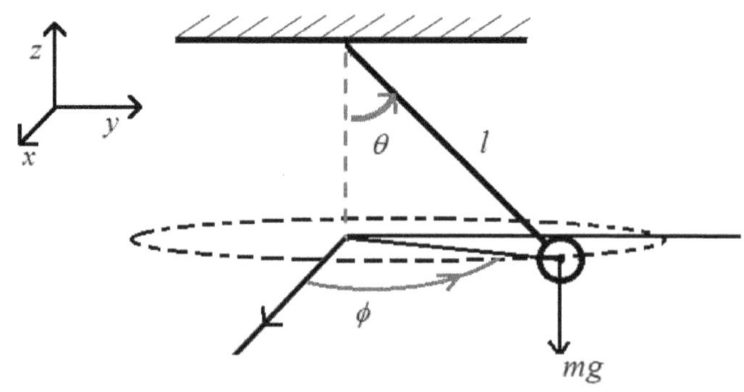

Figure 7.6. Spherical pendulum.

Squaring and adding the above equations, we can simplify to obtain

$$\dot{x}^2 + \dot{y}^2 + \dot{z}^2 = \ell^2\dot{\phi}^2 + \ell^2 \sin^2\phi\, \dot{\theta}^2. \tag{7.51}$$

Then we will obtain the kinetic energy as

$$T = \frac{1}{2}m(\dot{x}^2 + \dot{y}^2 + \dot{z}^2) = \frac{1}{2}m\ell^2(\dot{\theta}^2 + \sin^2\phi\, \dot{\theta}^2) \tag{7.52}$$

and the potential energy as

$$V = mg(-z) = -mg\ell \cos\phi. \tag{7.53}$$

Hence the Lagrangian will be given as

$$L = T - V = \frac{1}{2}m(\dot{x}^2 + \dot{y}^2 + \dot{z}^2) = \frac{1}{2}m\ell^2(\dot{\phi}^2 + \sin^2\phi\dot{\theta}^2) + mg\ell cos\phi. \tag{7.54}$$

The equations of motion for the spherical pendulum are derived below:

- *Equation of motion (radial part (ℓ)).*
  For this case the length, (ℓ) is a constant which means that the radial part does not give any major information.
- *Equation of motion* (ϕ):

$$\frac{d}{dt}\left(\frac{\partial L}{\partial \dot{\phi}}\right) - \frac{\partial L}{\partial \phi} = 0$$

$$\frac{d}{dt}\left(\frac{1}{2}mr^2(2\dot{\phi})\right) - mgr(-\sin\phi) - \frac{1}{2}mr^2[2\sin\phi\cos\phi\dot{\theta}^2] = 0$$

$$r\dot{\phi} + g\sin\phi - r\sin\phi\cos\phi\dot{\theta}^2 = 0$$

$$\dot{\phi} + \frac{g}{r}\sin\phi - \sin\phi\cos\phi\dot{\theta}^2 = 0. \tag{7.55}$$

- *Equation of motion* (θ):

$$\frac{d}{dt}\left(\frac{\partial L}{\partial \dot{\theta}}\right) - \frac{\partial L}{\partial \theta} = 0$$

$$\frac{d}{dt}\left(\frac{1}{2}mr^2\sin^2\phi(2\dot{\theta})\right) = 0$$

$$mr^2\sin^2(\phi)\dot{\theta} = 0.$$

  Since $mr^2\sin^2\phi \neq 0$, this implies that $\dot{\theta} = 0$

$$\Longrightarrow \theta = \text{ constant.} \tag{7.56}$$

Example 7.5

A 3 kg ball is suspended from the ceiling by a string of length 2 m and negligible mass. The ball is set in motion in a horizontal circle such that the string describes a cone where the angle between the string and vertical is found to be 10°. Find the angular momentum.

Solution

$$\text{Angular momentum } L \equiv I\omega, \tag{7.57}$$

where I is the moment of inertia, mathematically given as $I = mr^2$, and $\omega = \sqrt{\frac{g}{\ell \cos\theta}}$ is the angular velocity of the mass. We see from the figure that $r = \ell \sin\theta$. Then the above equation becomes

$$L = m(\ell \sin^2\theta) \times \sqrt{\frac{g}{l \cos\theta}}. \tag{7.58}$$

Solving for angular momentum we find that it numerically equals 2.3 kg m^2 s^{-1}.

Example 7.6. Double pendulum

A double pendulum is a pendulum with another pendulum attached to its end. The motion of a double pendulum is chaotic and very difficult to solve using Newtonian mechanics because it is governed by a set of coupled ordinary differential equations.

Consider a double-bob pendulum with masses m_1 and m_2 attached by rigid massless wires of lengths ℓ_1 and ℓ_2. Further, let the angles the two wires make with the vertical be denoted θ_1 and θ_2, as seen in figure 7.7. However, using Lagrangian mechanics simplifies the problem as you can see below.

- For pendulum bob 1 with mass m_1:

$$\begin{cases} x_1 = & \ell \sin\theta_1 \\ y_1 = & \ell \cos\theta_1. \end{cases}$$

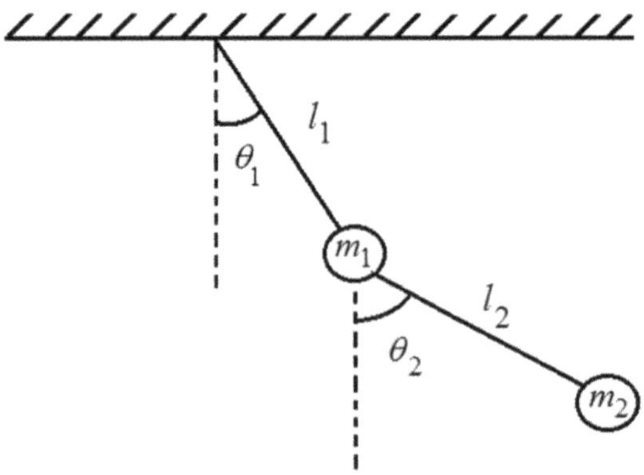

Figure 7.7. Double pendulum.

Differentiating the above equations we obtain

$$\begin{cases} \dot{x}_1 = & \ell \cos \theta_1 \dot{\theta}_1 \\ \dot{y}_1 = & -\ell \sin \theta_1 \dot{\theta}_1. \end{cases}$$

Note that $\dot{x}_1^2 + \dot{y}_1^2 = \ell^2 \dot{\theta}_1$.

- For pendulum bob 2 with mass m_2:

$$\begin{cases} x_2 = & x_1 + \ell \sin \theta_2 \\ y_2 = & y_1 + \ell \cos \theta_2. \end{cases}$$

Differentiating the above equations we obtain

$$\begin{cases} \dot{x}_2 = & \dot{x}_1 + \ell \cos \theta_2 \dot{\theta}_2 \\ \dot{y}_2 = & \dot{y}_1 - \ell \sin \theta_2 \dot{\theta}_2. \end{cases}$$

Squaring and adding the above equations we can find the kinetic and potential energy as follows.

Kinetic energy:

$$\begin{aligned} T &= \frac{1}{2} m_1 (\dot{x}_1^2 + \dot{y}_1^2) + \frac{1}{2} m_2 (\dot{x}_2^2 + \dot{y}_2^2) \\ &= \frac{1}{2} m_1 \ell^2 \dot{\theta}_1^2 + \frac{1}{2} m_2 [(\dot{x}_1 + \ell \cos \theta_2 \dot{\theta}_2)^2 \\ &\quad + (\dot{y}_1 + \ell \sin \theta_2 \dot{\theta}_2)^2] \\ &= \frac{1}{2} m_1 \ell^2 \dot{\theta}_1^2 + \frac{1}{2} m_2 [\dot{x}_1^2 + \ell^2 \cos^2 \theta_2 \dot{\theta}_2^2 + 2\dot{x}_1 \ell^2 \cos \theta_2 \dot{\theta}_2 \\ &\quad + \dot{y}_1^2 + \ell^2 \sin^2 \theta_2 \dot{\theta}_2^2 - 2\dot{y}_1 \ell^2 \sin \theta_2 \dot{\theta}_2] \\ &= \frac{1}{2} m_1 \ell^2 \dot{\theta}_1^2 + \frac{1}{2} m_2 [(\dot{x}_1^2 + \dot{y}_1^2) + \ell^2 (\cos^2 \theta_2 + \sin^2 \theta_2) \dot{\theta}_2^2 \\ &\quad + 2\dot{x}_1 \ell^2 \cos \theta_2 \dot{\theta}_2 - 2\dot{y}_1 \ell^2 \sin \theta_2 \dot{\theta}_2] \\ T &= \frac{1}{2} (m_1 + m_2) \ell^2 \dot{\theta}_1^2 + \frac{1}{2} m_2 \ell^2 \dot{\theta}_2^2 + m_2 \ell^2 \dot{\theta}_2 (\dot{x}_1 \cos \theta_2 - \dot{y}_1 \sin \theta_2). \end{aligned}$$

Potential energy:

$$\begin{aligned} V &= -m_1 g y_1 - m_2 g y_2 \\ &= -m_1 g \ell \cos \theta_1 - m_2 g (y_1 + \ell \cos \theta_2) \\ V &= -(m_1 + m_2) g \ell \cos \theta_1 - m_2 g \ell \cos \theta_2 \end{aligned}$$

and $L = T - V$. The Lagrangian equation of motion is left as an exercise for the reader to attempt.

Concept questions 7.2

1. Could you construct a true simple pendulum? Explain your answer.
2. Suppose a small child is on a swing with a large amplitude. What will happen to the frequency of the swing as its oscillations become smaller as time passes by and no one pushes the swing?
3. If a pendulum has a period of 1 s at the Equator, what will its period be at the North Pole?
4. A kindergartener sits on a swing with ropes that are 1.5 m long. Will this be considered a simple pendulum or a physical pendulum? Will it make a difference if the length of the swing is doubled to 3 m?

Exercises 7.2

1. A simple harmonic oscillator will be defined as a mass (m) attached to a spring free to move in a horizontal plane. Using the Newtonian, Lagrangian and the Hamiltonian formalisms:
   Write the equation of motion for a simple harmonic oscillator in 1D.
2. A particular simple pendulum consists of a mass attached to the string suspended from a ceiling.
   (a) Assume the initial length of the pendulum is 1 m, find the initial frequency and period of the pendulum.
   (b) If the length of the pendulum is changed to 75 cm, how does the frequency and period of the pendulum change?
   (c) How will the frequency and period of the pendulum change if the length is further reduced to 50 cm and 25 cm?
3. A particular physical pendulum consists of a meter stick pivoted by making a hole at 0 cm. If it initially oscillates freely at a frequency (f_o), find the new frequencies in terms of (f_o) and the corresponding periods if:
   (a) The meter stick is first cut at the 75 cm mark.
   (b) The meter stick is additionally cut at the 50 cm and then the 25 cm mark.
   Hint: For details of a physical pendulum please refer to chapter 15, section 15.10.
4. Compare the frequencies and periods of the two pendulums in 2 and 3.

Further reading

Arya A 1990 *Introduction to Classical Mechanics* 2nd edn (Upper Saddle River, NJ: Prentice-Hall)

Fowles G and Cassiday G 2005 *Analytical Mechanics* 7th edn (Belmont, CA: Thomson Learning)

French A P 1971 *Vibrations and Waves* (Boca Raton, FL: Chapman and Hall)

Resnick R, Halliday D and Krane K 2002 *Physics* vol 1, 5th edn (New York: Wiley)

Rutherford D E 1951 *Classical Mechanics* (New York: Interscience)

Stephenson R 1961 *Mechanics and Properties of Matter* 2nd edn (New York: Wiley)

Symon K 1971 *Mechanics* 3rd edn (Reading, MA: Addison-Wesley)

IOP Publishing

Techniques of Classical Mechanics
From Lagrangian to Newtonian mechanics
Samya Zain

Chapter 8

Gravitation and central forces

8.1 Introduction

Man has been looking up at the sky since pre-historic times. Ancient people knew about five planets. Observers were well aware that planets move in the night sky from east to west in fairly predictable paths. However, sometimes a planet turned around and moved backwards for a while. Then it appeared to do a U-turn and head westward again making a loop against the star background. It was one of the great unsolved mysteries of the ancient world.

This was the case for Mars, the god of war. The planet followed a nice, regular path, except every two years or so people observed that for a couple of months Mars moved backwards. It appeared that at these times Mars mysteriously halted, and reversed direction for as long as a few weeks before resuming its steady trek through the sky. This strange behavior was very puzzling to early sky watchers.

This issue puzzled astronomers for about 2000 years, until Nicolas Copernicus (1473–1543 CE) demonstrated it as a simple consequence of relative motion between Earth and other planets, each moving in a heliocentric orbit. Nowadays we call this behavior *retrograde motion*.

Johannes Kepler (1571–1630 CE), working to understand the orbit of Mars, developed a mathematical solution for the motion of planets, which we call Kepler's laws of planetary motion. Newton later developed his law of universal gravitation mostly from Galileo's investigations of motion of terrestrial objects and combined it with Newton's laws of motion and the world was never the same again.

8.1.1 The retrograde motion of Mars

The planets in the solar system are like race cars on an oval track that cannot leave their tracks. Venus is on the inside lane, Earth is in the middle and Mars is on the outside. The Earth moves faster than Mars and completes two orbits around the Sun in the same time as Mars completes one orbit. As a consequence, every 26 months or so, Earth comes from behind and overtakes Mars. While Earth is passing by, it

doi:10.1088/2053-2563/aae1b9ch8

appears that Mars is moving backwards. Additionally, it also appears that Mars moves up and down against the starry sky. However, this affect does not last long and as Earth moves in its orbit the illusion disappears and everything settles down once again.

This apparent backwards movement, the *retrograde motion*, explains why Mars moves backwards, but why does Mars appear to move up and down? The illusion of the up and down movement of Mars arises because Earth and Mars do not lie in the same plane. Since the planes of each planets are a bit tilted with respect to each other, it appears from our view on Earth that Mars moves up and down with respect to Earth.

8.2 Newton's law of universal gravitation

During the Great Plague, English universities closed in 1665 CE. Newton, who was an undergraduate student at Cambridge University, returned to his home in the midlands of England, where he worked on developing his ideas, and in 1686 CE he formally announced the law of universal gravitation in the *Principia* which states:

> 'Every mass in the Universe attracts every single other mass by a force pointing along the line intersecting both masses. The force is proportional to the product of the two masses and inversely proportional to the square of the distance between them'.

We can express the law vectorially by the equation

$$\boxed{\vec{F}_{ij} = G\frac{m_i m_j}{r_{ij}^2}\frac{\vec{r}_{ij}}{|r_{ij}|},} \tag{8.1}$$

where $\vec{F}_{ij}$ is the force that particle j of mass m_j exerts on particle i of mass m_i, and vector $\vec{r}_{ij}$ is directed from particle i to particle j, as shown in figure 8.1. G in this

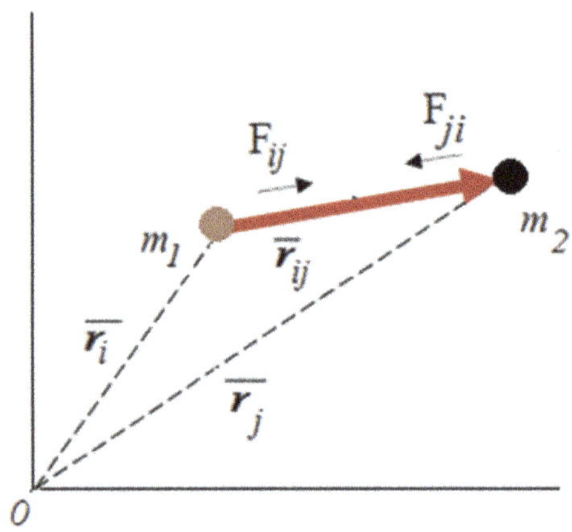

Figure 8.1. Newton's law of universal gravitation—action–reaction force pair.

equation is the constant of proportionality called the *universal constant of gravitation* or the *gravitational constant*, numerically equal to the force between two bodies of unit mass kept unit distance apart. The value of G is determined in laboratories by carefully measuring the force between bodies of known masses. The internationally accepted value of G is $(6.672\,59 \pm 0.000\,85) \times 10^{-11}\,\mathrm{N\,m^2kg^{-2}}$. Note that G is independent of the nature and sizes of the two masses and is independent of the medium in which the masses exist.

8.2.1 Central forces

We introduced central forces in chapter 5, example 5.4 and we will talk about them in detail here. The law of universal gravitation in equation (8.1) is an example of a *central force*. Central forces are defined as forces whose lines of action either emanate from or terminate on a single point or center. Think about it this way: if a point particle applies a central force, the force would always be directed toward the center, and the magnitude of this force will be the same on concentric circles drawn around it. In other words, you will not be able to tell the location of a point on a sphere by knowing the force value, you can just find the value of distance r away from the particle, where r is the radius of any concentric circle.

8.2.2 Isotropic forces

Forces are called isotropic when the magnitude of the force is independent of direction. Gravitational forces are a great example of isotropic forces.

8.2.3 Uniform gravitational field

A uniform gravitational field is one where the field lines[1] are always the same distance apart. This is almost exactly true close to the Earth's surface. However, as we move back and look at Earth from a distance, the field lines clearly radiate outwards. When viewed from an even greater distance, the complete field can be seen. Such a field is called a *radial field* and is shown in figure 8.2.

In this case the field intensity is commonly known as g, the acceleration due to gravity, and its value decreases with distance from the surface. The separation distance of the field lines gives an indication of the strength of the field. When field lines are close together the field intensity is considered to be high and when they are far apart the field intensity is taken to be low.

8.3 Gravity

Gravitation versus gravity

The force of attraction between any two neutral objects of any size and makeup is called *gravitation*. When one of these objects is a small object in the field of Earth's

[1] Field lines are useful for visualizing vector fields. They are not physically present and are merely conceptual tools.

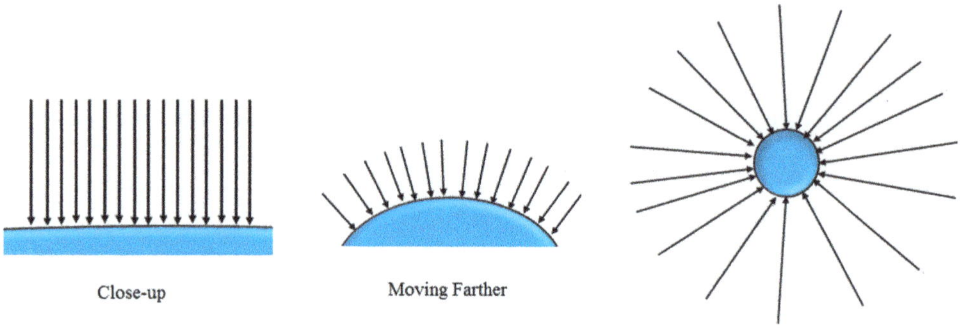

Figure 8.2. Examples of a radial field: gravitational field.

attractive force, the force of attraction on this small body due to Earth is called *gravity*.

The gravitational force that the Earth, with mass (M_E), exerts on an object of mass (m) is mathematically given by

$$F_r = -G\frac{M_E m}{r^2}, \tag{8.2}$$

where the force is directed along the line connecting the two bodies and only depends on the separation between the two bodies.

Similar to equation (8.1), equation (8.2) is also Newton's law of gravitation when objects are Earth-bound. It shows that force of gravity is a function of r, the separation between the objects and hence it is a central force.

However, Earth is not a perfect sphere but rather it bulges at the equator. Hence gravitational force varies from place to place on the surface of Earth. By definition, we know that $F = -mg$ on the surface of Earth, so the acceleration due to gravity will be given as

$$g = -G\frac{M}{R_E^2}, \tag{8.3}$$

where R_E is the radius of Earth.

Example 8.1

Referring to equation (8.2), answer the following questions:

(a) If both masses are doubled, what happens to the force?
Answer: When the masses are doubled, gravitational force will become four times larger.

(b) If the separation between masses is doubled, what happens to the force?
Answer: The gravitational force is inversely proportional to the square of the separation distance between the two objects. When separation between masses is doubled, you quarter the gravitational force.

(c) If the masses are moved ten times further what happens to the force?

Answer: When masses are moved ten times further, force goes down by a factor of 100. Hence, gravitational force is also called *inverse-square* force.

8.3.1 Variation of gravity with altitude

At height (x) above the surface of Earth acceleration due to gravity is

$$g' = -G\frac{M}{(R_E + x)^2}. \tag{8.4}$$

Comparing equations (8.3) and (8.4):

$$\frac{g'}{g} = \frac{-G\frac{M}{(R_E + x)^2}}{-G\frac{M}{R_E^2}}$$

$$= R_E^2(R_E + x)^{-2}$$

$$\frac{g'}{g} = \left(1 + \frac{x}{R_E}\right)^{-2}.$$

Expanding by using binomial theorem and ignoring higher powers of $\frac{x}{R_E}$, we obtain

$$\boxed{g' = g\left(1 - 2\frac{x}{R_E}\right).} \tag{8.5}$$

Equation (8.5) shows us that the acceleration due to gravity decreases with altitude (figure 8.3).

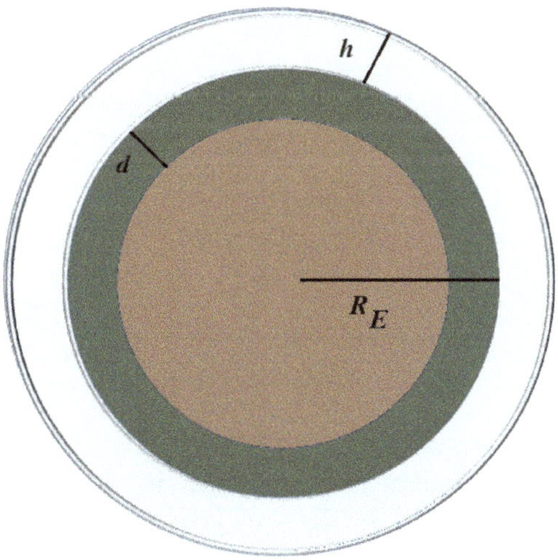

Figure 8.3. Variation of gravity with altitude and depth.

8.3.2 Variation of gravity with depth

If we now consider that the mass (m) is taken a distance d deep into the Earth, then to find the acceleration due to gravity at depth d we will assume that the Earth is a perfect sphere with mass (M_E), a uniform density (ρ_E) and radius (R_E). We can rewrite the mass in terms of the density of the Earth (ρ_E) as

$$M_E = \frac{4}{3}\pi R_E^3 \rho_E. \tag{8.6}$$

Then we can express equation (8.3) as

$$\begin{aligned} g &= -G\frac{\frac{4}{3}\pi R_E^3 \rho_E}{R_E^2} \\ &= -G\frac{4}{3}\pi R_E \rho_E. \end{aligned} \tag{8.7}$$

When we are at depth d inside the Earth, the acceleration due to gravity is only due to the mass M_d, not the entire mass of the Earth. Then the acceleration due to gravity at this point (g_d) will be given as

$$\begin{aligned} g_d &= -G\frac{M_d}{(R_E - d)^2} \\ &= -G\frac{\frac{4}{3}\pi (R_E - d)^3 \rho_E}{(R_E - d)^2} \end{aligned}$$

$$\Longrightarrow g_d = -G\frac{4}{3}\pi (R_E - d)\rho_E. \tag{8.8}$$

Dividing equations (8.7) and (8.8) we obtain

$$\begin{aligned} \frac{g_d}{g} &= \frac{R_E - d}{R_E} \\ \frac{g_d}{g} &= \left(1 - \frac{d}{R_E}\right) \end{aligned}$$

$$\Longrightarrow \boxed{g_d = g\left(1 - \frac{d}{R_E}\right).} \tag{8.9}$$

Equation (8.9) shows that acceleration due to gravity decreases as we go deeper into the Earth. At the center of the Earth ($R_E - d$) goes to zero, hence acceleration due to gravity at the center of the Earth $g'' = 0$.

Example 8.2

Find the value of acceleration due to gravity (g) on the surface of Earth.

Solution

We know that the accepted value of the mass of Earth is 6×10^{24} kg and the average radius of Earth is 6.4×10^{6} m. Then, using equation (8.3) we obtain

$$g = -G\frac{M}{R_E^2}$$

$$= -(6.67 \times 10^{-11})\frac{6 \times 10^{24}}{(6.4 \times 10^{6})^2}$$

$$\Longrightarrow g = -9.8 \text{ Nkg}^{-1}.$$

Concept questions 8.1

1. Why is the value of acceleration of gravity (g) different on the surface of the Earth and the Moon? Explain in detail.
2. Suppose the magnitude of the force of gravitation between two identical objects, with masses m_1 and m_2, is given by F. If the mass of each object and the distance between them is doubled at the same time. What will happen to the force of gravitation between two objects?
3. Write the dimensions of the universal gravitational constant (G).

Exercises 8.1

1. The mass of Mars is 6.4×10^{23} kg and its radius is 3.4×10^{6} m. What is the value of acceleration of gravity (g) on Mars?
2. How far from the Earth must a spaceship be along a line towards the Sun such that the gravitational pull from the Earth balances the gravitational pull from the Sun?

8.4 Gravitational force between a uniform sphere and a particle

We know from the law of universal gravitation that the gravitational force between two point particles is proportional to the magnitudes of their masses and is inversely proportional to the square of the distance (d) between them.

However, most objects in nature are not point particles and additionally there may be a large mass and/or volume difference between the two masses under consideration. In this case calculating the gravitational force between three-dimensional objects becomes complicated. To avoid most complications, we consider the gravitational force for highly symmetric shapes such as spheres or spherical shells only. In order to do that we will start by stating what is called the *shell theorem.*

The shell theorem

Isaac Newton proved the shell theorem, which states that:

1. A spherically symmetric object affects other objects around it as if all of its mass were concentrated at its center, called the *center of mass*. This means when calculating the force of gravity exerted on a 100 kg sphere of radius 1 m, due to Earth, the distance (d) is taken from the center of mass of the sphere to the center of mass of the Earth.
2. When the object is a spherically symmetric shell (i.e. a hollow ball) the net gravitational force on a body inside it is zero.

A corollary of the shell theorem

For an object located at a certain distance inside a uniformly filled sphere, the force exerted on it by the sphere depends only on the mass of the sphere inside the location radius. It helps if you imagine that the sphere under consideration as a collection of infinitesimally thin, concentric, spherical shells, rather like the layers of an onion. The mass outside these shells will not contribute towards the force acting on the object.

8.4.1 Gravitational force exerted on an object

When considering the gravitational force exerted on an object at a point inside or outside a uniform spherically symmetric object of radius (R), there are two simple but distinct situations that must be examined:

- A hollow spherical shell.
- A solid sphere with uniformly distributed mass.

8.4.1.1 Case 1. A hollow spherical shell

Statement 1 of shell theorem tells us the net gravitational force a spherical shell of mass (M) exerts on a body of mass (m) outside it, is the vector sum of the gravitational forces acted by each part of the shell on the body. This is similar to the situation when the mass (M) is assumed to be located completely at a point at the center of the hollow sphere.

Statement 2 of shell theorem tells us that the net gravitational force acting by a spherically symmetric shell on a point mass (m) inside it will be is zero.

Derivation

Consider a thin uniform shell of mass (M) and radius (R). Let $|r|$ be the distance from the center O of the shell to a test particle of a mass (m) placed at point (P) as seen in figure 8.4.

Assuming the particle is outside the shell (i.e. $r > R$), we can divide the total shell into circular rings, each of width = ($R\Delta\theta$) and circumference ($2\pi R \sin\theta$) where θ is

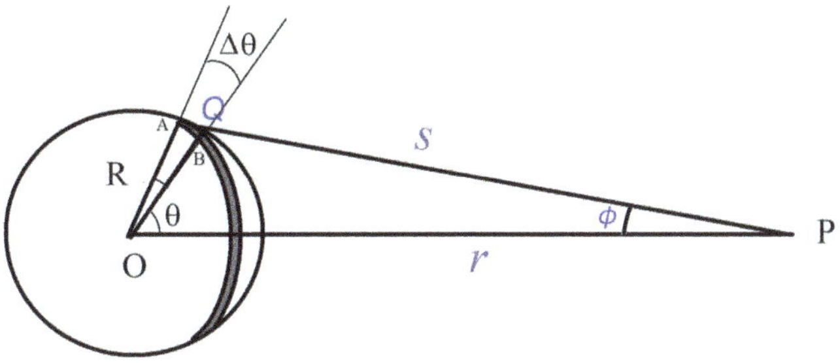

Figure 8.4. A thin hollow spherical shell.

the angle $\angle POQ$ and $\Delta\theta$ is the angle $\angle AOB$ and point Q is a point on the ring as seen in figure 8.4.

We can express the mass of the ring as $\Delta m \equiv 2\pi\rho R^2 \sin\theta\Delta\theta$, where ρ is the mass per unit area of the shell. The gravitational force exerted on P by a subsection Q of the ring is in the direction of $PQ = \Delta F$. The magnitude of the components of ΔF will be

- Along PO: $\Delta F \cos\phi$.
- $\perp$ to PO: $\Delta F \sin\phi$.

By symmetry arguments, we see that the vector sum of all the perpendicular ($\perp$) components exerted on point (P) by the whole ring cancel out. This means that the force (ΔF_{net}) is given as

$$\begin{aligned}\Delta F_{\text{net}} &= \Delta F \cos\phi \\ &= \left(\frac{Gm\Delta m}{s^2}\right)\cos\phi \\ &= \frac{Gm}{s^2}[\rho 2\pi R^2 \sin\theta\Delta\theta]\cos\phi.\end{aligned}$$

Then the total force exerted on P by the whole shell is found by integration (θ from 0 to π):

$$F = \int \Delta F_{\text{net}} = \int dF_{\text{net}} = Gm2\pi\rho R^2 \int_0^{\pi} \frac{\sin\theta\cos\phi}{s^2} d\theta. \tag{8.10}$$

The easiest way to solve this equation is by writing everything in terms of s. We know for ΔOPQ, the law of cosines will be written as

$$r^2 + R^2 - 2rR\cos\theta = s^2.$$

Differentiating this equation and knowing r and R are constants we obtain

$$\begin{aligned}rR\sin\theta d\theta &= sds \\ \sin\theta d\theta &= \frac{sds}{rR}\end{aligned}$$

and

$$\cos\phi = \frac{\text{base}}{\text{hypotenuse}} = \frac{s^2 + r^2 - R^2}{2rs}.$$

Putting values in equation (8.10) and integrating by changing the limits of integration we will obtain

$$\begin{aligned} F &= Gm(2\pi)\rho R^2 \int_{r+R}^{r-R} \frac{1}{s^2}\frac{s^2 + r^2 - R^2}{2rs}\left(\frac{sds}{rR}\right) \\ &= Gm(2\pi)\left(\frac{M}{4\pi R^2}\right)\frac{R^2}{2r^2R}\int_{r+R}^{r-R}\frac{s^2 + r^2 - R^2}{s^2}ds \\ &= \frac{GmM}{4Rr^2}\left[\int_{r+R}^{r-R} ds + \int_{r+R}^{r-R}\frac{r^2 - R^2}{s^2}ds\right], \end{aligned}$$

which can be shown by simple integration to be equal to

$$\boxed{F = \frac{GmM}{r^2}.} \tag{8.11}$$

This means that a uniform shell attracts a particle as if the mass of shell were concentrated at the center of mass.

8.4.1.2 Case 2. Solid sphere with uniformly distributed mass

The second situation is the case for a solid, uniform sphere of mass M and radius R. For the situation where the mass is outside the sphere the situation is similar to the one for the spherical shell described in section 8.4.1.1. The net gravitational force is the vector sum of the gravitational forces acted by each shell of the sphere. The resulting net gravitational force acts as if mass M is concentrated on a point at the center of the sphere.

However, for the case where the mass (m) is inside the solid sphere, say at a radius (d) from the center of-mass inside the sphere (that is, $d < R$). In this case, the contribution of all shells of the sphere outside d can be ignored. This fact can be proven by taking the shell integral again, like we did in section 8.4.1.1 with the point inside the solid sphere. I will leave that as an exercise for the reader to attempt. But from this we recognize that only the mass of the sphere within the radius $M < d$ is relevant, and can be considered to be concentrated at the center of the sphere. So, the gravitational force acting upon point mass m is

$$F = G\frac{m(M_{<d})}{d^2}, \tag{8.12}$$

where

$$M_{<d} = \frac{4}{3}\pi d^3\rho \tag{8.13}$$

and ρ is the mass density of the sphere. Combining the above two equations we obtain

$$F = \frac{4}{3}\pi G m \rho d. \tag{8.14}$$

Note that equation (8.14) clearly shows that mass (m) experiences a force that is linearly proportional to its distance, d, from the center of mass of the sphere.

8.5 Potential energy in a gravitational field: gravitational potential

Consider the work (W) required to move a test particle mass (m) along a given path in the gravitational field of another particle, say a planet of mass (M). To find the work, place the mass (M) at the origin O, so the force on mass m will be

$$\vec{F} = -\frac{GMm}{r^2}\hat{r}, \tag{8.15}$$

where $\hat{r}$ is the unit vector from mass (M) to the test mass (m). To overcome this force, an external force ($-\vec{F}$) must be applied. The work (dW) done in moving the test particle through distance $d\vec{r}$ is given by

$$dW = -\vec{F} \cdot d\vec{r} \tag{8.16}$$

$$dW = \frac{GMm\hat{r}}{r^2} \cdot d\vec{r}. \tag{8.17}$$

Next resolve $d\hat{r}$ into two components:

- The perpendicular component ($\hat{e}_{\perp}$).
- The radial component ($\hat{r}$).

Then clearly we can say

$$\hat{r} \cdot d\vec{r} = dr.$$

Then from equation (8.17), work from point $P_1(r_1)$ to $P_2(r_2)$ will be given as

$$W = GMm \int_{r_1}^{r_2} \frac{1}{r^2} dr$$

$$\Longrightarrow W = -GMm\left(\frac{1}{r_2} - \frac{1}{r_1}\right),$$

where r_1 and r_2 are the radial distances of the particle at the beginning (r_1) and end (r_2) of the path. We therefore know that:

- Work is independent of the particular path and that it only depends on the start and end points.
- The inverse-square law of force is conservative.

8.5.1 Gravitational potential energy ($V(r)$) on a particle

The general expression for gravitational potential energy comes from the law of gravity as written in equation (8.15) and is numerically equal to the work done against force of gravity to bring a mass to a given point in space.

The force of gravity is an inverse-square law—we can say that the force approaches zero for very large distances. So it makes sense to choose gravitational potential energy at an infinite distance to be numerically zero. Since the force of gravity does positive work as the mass is brought close to a planet, we take the gravitational potential energy near a planet to be negative. The general form of the gravitational potential energy of mass (m) in the presence of a large mass (M) is

$$\begin{aligned} V(r) &= -\int_{\infty}^{r} F \cdot dr \\ &= GMm\int_{\infty}^{r} \frac{1}{r^2} dr \\ &= GMm\left(\frac{1}{\infty} - \frac{1}{r}\right) \\ &= -\frac{GMm}{r}. \end{aligned}$$

Note that this is also the work done in moving a test particle from some arbitrary reference point (∞) to the point (r) in question:

$$\boxed{W = -V(r) = \frac{GMm}{r}.} \tag{8.18}$$

8.5.2 Gravitational potential (Φ)

Sometimes it is convenient to define gravitational potential (Φ), where

$$\Phi \equiv \frac{\text{PE}}{\text{mass}} = \frac{V(r)}{m}, \tag{8.19}$$

where m is the mass of the object. Then the gravitational potential in field of particle of mass M is

$$\Phi = \frac{-Gm}{r}. \tag{8.20}$$

The gravitational potential is also called the Newtonian potential and is used in the study of potential theory[2]. Potentials of the same form may also be used for solving

[2] Potential theory is the study of harmonic functions in which two fundamental forces of nature, gravity and electrostatic force, are modeled using functions called the gravitational potential and electrostatic potential. Both potentials satisfy Poisson's equation in general or Laplace's equation in vacuum. All of these are discussed in books dealing with electrostatics.

the electrostatic and magneto-static fields generated by uniformly charged or polarized ellipsoidal bodies.

8.5.3 Gravitational potential for a many particle system ($\Phi(x, y, z)$)

For a number of particles (M_1, M_2, ...) at (r_1, r_2, ...), gravitational potential at point $P(x, y, z)$ will be the sum of the gravitational potential energies of all particles as

$$\Phi(x, y, z) = \sum \Phi_i = -G\sum \frac{M_i}{r_i}. \quad (8.21)$$

8.5.4 Gravitational field intensity (I_G)

The gravitational field intensity (I_G) is the force on a unit mass at a point in the field. The units for I_G are N kg^{-1}. Gravitational field intensity I_G is mathematically given as

$$I_G = \frac{\vec{F}}{m}. \quad (8.22)$$

Also remember from Newton's law, $F = ma$, $a = F/m$ and so I_G may also be expressed in m s^{-2}, the units of acceleration. In terms of gravitational potential (Φ) we can re-express gravitational field intensity as

$$I_G = -\nabla\Phi. \quad (8.23)$$

8.5.5 Escape velocity

Escape velocity is the minimum initial speed needed for an object without propulsion to escape from the gravitational pull of a massive object. Escape velocity is a consequence of the conservation of energy in a conservative field, such as a gravitational field, and depends on the position (altitude) of the object. On the surface of the Earth, ignoring the atmospheric resistance, the escape velocity is about 11.2 km s^{-1}, which is approximately 33 times the speed of sound. The escape velocity from the surface of the Moon is about 2.4 km s^{-1}.

Note that escape velocity is actually not a velocity but rather it is a speed because an object can escape the gravitational field regardless of the direction of projection.

Derivation of escape velocity

The simplest way of deriving the formula for escape velocity is to use conservation of energy ($\Delta E = 0$). Escape velocity is the initial speed required to go from some initial point (usually taken on the surface of Earth) in a gravitational potential field of the object to infinitely far away from the gravitational pull of the planet with zero final speed with respect to the field. If we take the initial states to be given as T_i and V_i and final states as T_f and V_f, then

$$T_i + V_i = T_f + V_f. \tag{8.24}$$

Take the mass of the planet to be M, the mass of the escaping object to be m and the radius of the planet as R. Putting values in the above equation, we obtain

$$\frac{1}{2}mv_{\text{esc}}^2 + \left(-G\frac{Mm}{R}\right) = 0.$$

Solving for velocity we obtain

$$\boxed{v_{\text{esc}} = \sqrt{\frac{2GM}{R}},} \tag{8.25}$$

where G is the gravitational constant. From equation (8.25), we see that the escape velocity is independent of the mass of the escaping object. The difference that arises with respect to the mass of the escaping objects will be the amount of energy required, the more massive the object the more energy will be required.

Additionally, the escape velocity is equal to the speed that an object would have at the surface of the planet if it started at rest at an infinite distance and was pulled by gravity to that point.

Alternative expression for v_{esc}

An alternative expression for the escape velocity particularly at the surface on the astronomical object is given as

$$v_{\text{esc}} = \sqrt{2gr}, \tag{8.26}$$

where r is the distance between the center of the astronomical object and the point at which escape velocity is being calculated, and g is the gravitational acceleration at that distance.

Example 8.3

Suppose the mass of the Earth suddenly increased by five times and then by ten times, while keeping the rest of the constants the same. What will the change in escape velocity be for the new Earths?

Solution

We can use equation (8.25) to solve this problem.

(a) With one Earth mass, $v_{\text{esc}} = 11.2 \text{ km s}^{-1}$.

(b) With five Earth masses, $v_{\text{esc}} = 25.0 \text{ km s}^{-1}$.

(c) With ten Earth masses, $v_{\text{esc}} = 35.4 \text{ km s}^{-1}$.

This means that more massive objects will require higher escape velocities.

Concept questions 8.2

1. The weight of an object on the Moon is 1/6 that of its weight on Earth. How often will a pendulum clock that ticks once per second on Earth tick on the Moon?
2. How could you find the mass of the Moon?
3. Extremely high tides occur when the Moon is new. Can you explain why?
4. Does it matter what direction a rocket is pointed for its escape from Earth? Neglect air resistance and assume it has to be pointed above the horizon.

Exercises 8.2

1. Calculate the magnitude of the gravitational forces exerted on a 1 kg mass located on the equator due to:
   (a) The Earth.
   (b) The Sun.
2. A gravity train is a theoretical means of transportation intended to go between two points on the surface of a sphere, following a straight tunnel that goes directly from one point to the other through the interior of the sphere. Suppose we dig a tunnel through the center of Earth from one side to the other to form a gravity train using an elevator for transportation of goods and people. Now suppose an elevator of m is dropped in the tunnel from rest at the surface on one side. Neglecting all frictional forces and assuming Earth has a uniform density, calculate:
   (a) The force on the elevator when it is a distance (r) from the center.
   (b) The speed of the elevator when it is a distance (r) from the center.

8.6 Kepler's law of planetary motion

Kepler deduced these laws from a detailed analysis of planetary motions, mostly the motion of Mars, which is an outer planet whose orbit is elliptical and the orbit of our inner planet Venus, whose orbit is circular. Kepler used very, very carefully recorded data about the motion of Mars by Tycho Brahe (1546–1601 CE). Based on this, Kepler gave us three laws of planetary motion that we call Kepler's three laws.

Kepler's three laws

1. *Law of ellipses* (1609 CE). The orbit of each planet is an ellipse, with the Sun located at one of its focii.

2. *Law of equal areas* (1609 CE). A line drawn between the Sun and the planet sweeps out equal areas in equal times as the planet orbits the Sun.
3. *Harmonic law* (1618 CE). The square of the sidereal period[3] of a planet is directly proportional to the cube of the semi-major axis of the line of planet orbit.

8.6.1 Kepler's second law: equal areas

Let us start by asking a question: Can we consider the problem of the motion of a planet in a central field, such as gravitational field, as a 2D problem? To answer this question we start by observing that Kepler's second law is a restatement of conservation of a momentum.

Angular momentum is mathematically given as

$$\vec{L} = \vec{r} \times \vec{p}, \tag{8.27}$$

where $\vec{p}$ is the linear momentum. If we take the derivative of L with respect to t we obtain

$$\begin{aligned}\frac{dL}{dt} &= \frac{d}{dt}(\vec{r} \times \vec{p})\\ &= (\vec{r} \times \vec{p}) + \left(\vec{r} \times \frac{d\vec{P}}{dt}\right)\\ &= \vec{r} \times \vec{F}\\ &= \tau\end{aligned}$$

$$\Longrightarrow \boxed{\frac{dL}{dt} = \tau,} \tag{8.28}$$

where $\vec{F}$ is the applied force and τ is the torque (≡ moment of force). If $\vec{r}$ and $\vec{F}$ lie along a single vector line, the cross product of $\vec{r}$ and $\vec{F}$ vanishes, which means $\tau = 0$. which means $\frac{dL}{dt} = 0$, and in this case angular momentum (L) will be constant.

For a planet or a moon subject to a central force, $\vec{F_G}$:

- The line of action for the central force will be along the radius vector ($\vec{R}$) and it usually begins at a central single point.
- The vectors $\vec{R}$ and $\vec{v}$ define a plane within which the particle moves. The orientation of this plane is fixed in space and the angular momentum vector $\vec{L}$ is normal to this plane.

[3] Sidereal period is the time it takes a planet to complete one revolution about the Sun relative to the stars.

From equation (8.28) we see that the angular momentum vector is constant in both magnitude and direction. Hence the problem of motion of a particle in a central field is really a 2D problem, as seen in figure 8.5.

8.6.2 Angular momentum and areal velocity of a particle moving in a central field

I. **Magnitude of angular momentum of particle moving in a central field**

In order to calculate the angular momentum of a particle moving in a central field, we will use polar coordinates (r, ϕ) to describe the motion. The velocity of a particle in polar coordinates is

$$\vec{v} = \dot{r}\hat{r} + r\dot{\phi}\ \hat{\phi}. \tag{8.29}$$

Hence the magnitude of angular momentum is

$$\begin{aligned} L &= |\vec{r} \times m\vec{v}| \\ &= |r\hat{r} \times m(\dot{r}\hat{r} + r\dot{\phi}\ \hat{\phi})| \\ &= |r\dot{r}(\hat{r} \times \hat{r}) + r^2\dot{\phi}(\hat{r} \times \hat{\phi})|. \end{aligned}$$

From the definition of cross product we know

$$\begin{aligned} \hat{r} \times \hat{r} &= 0 \\ |\hat{r} \times \hat{\phi}| &= |\theta| = 1, \end{aligned}$$

then

$$|L| = mr^2\dot{\phi} = \text{constant}. \tag{8.30}$$

We find that for any particle moving in a central field of force (including a planet moving in F_G of the Sun) the angular momentum is constant.

II. **Areal velocity** ($\dot{A}$)

Areal velocity, also called sector velocity, is defined as the rate at which area is swept out by a particle as it moves along a curve. Areal velocity is closely linked to the angular momentum. Kepler's second law says that the *areal velocity* ($\dot{A}$) of a planet about the Sun is constant if we take the Sun to be at the origin (O). The fact that areal velocity remains constant as it travels around the Sun depends on the central nature of the gravitational force and does not depend on the strength (or variation in the strength) of the force as radius changes.

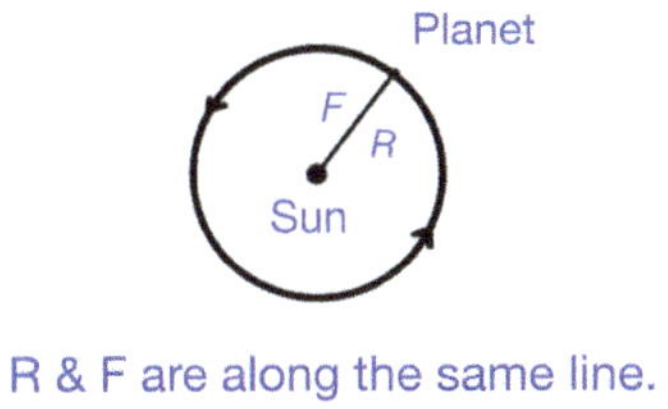

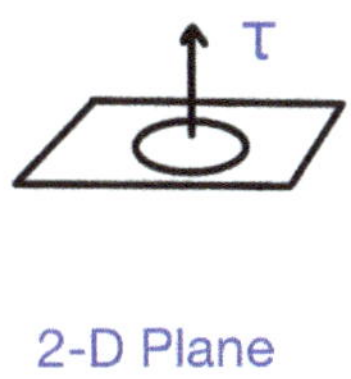

Figure 8.5. Radius vector $\vec{R}$ of the particle as it moves in a central field.

Calculate areal velocity ($\dot{A}$)
In figure 8.6 the triangular area (dA) swept out by radius vector ($\vec{r}$) as a planet moves a vector distance ($d\vec{s} = rd\phi$) in time dt along its trajectory relative to the origin of the central field (Sun) is given as

$$\begin{aligned} dA &= \frac{1}{2}|\vec{r} \times d\vec{s}| \\ &= \frac{1}{2}|r\hat{r} \times (dr \cdot \phi\hat{r} + |rd\phi|\hat{\phi})| \\ &= \frac{1}{2}|r\phi dr(\hat{r} \times \hat{r}) + r(rd\phi)(\hat{r} \times \hat{\phi})|. \end{aligned}$$

For the cross product we know $(\hat{r} \times \hat{r}) = 0$ and $(\hat{r} \times \hat{\phi} = |\theta| = 1)$, then

$$dA = \frac{1}{2}[r^2 d\phi]. \tag{8.31}$$

Differentiating w.r.t. time will give us

$$\frac{dA}{dt} = \frac{1}{2}r^2\frac{d\phi}{dt}.$$

Thus,

$$\boxed{\dot{A} = \frac{1}{2}r^2\dot{\phi}.} \tag{8.32}$$

Note that any increment of motion along radial direction ($\hat{r}$) does not:
- Add to or take away from the area (dA).
- Nor does it contribute anything to the angular momentum of the particle.

Equivalently, we note

$$\vec{v} = \frac{d\vec{r}}{dt}$$

$$\vec{v}dt = d\vec{r}.$$

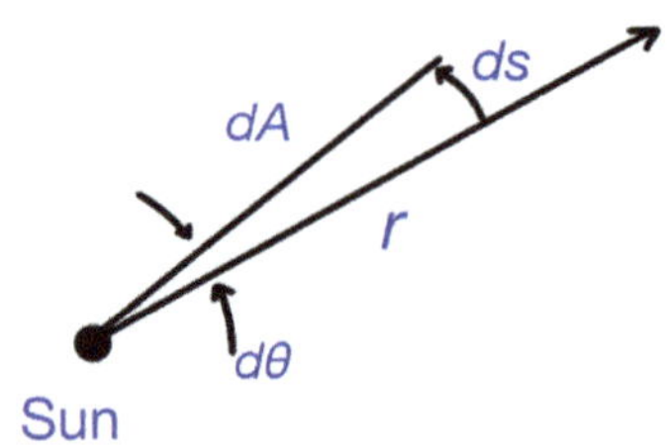

Figure 8.6. The area swept out by the radius vector $\vec{r}$ of the particle as it moves in a central field.

Then knowing angular momentum $[L = m(\vec{r} \times \vec{v})]$ we can write $[\frac{L}{m} = r^2\dot{\phi}]$ and hence equation (8.32) can be rewritten as

$$\frac{dA}{dt} = \frac{1}{2}|r^2\dot{\phi}|$$

$$\frac{dA}{dt} = \frac{1}{2}\left(\frac{L}{m}\right)$$

$$\Longrightarrow \boxed{\dot{A} = \frac{L}{2m}.} \tag{8.33}$$

This equation shows that the areal velocity is proportional to angular momentum. This is exactly what Kepler discovered for planets moving in the central gravitational field of the Sun. Summarizing, the areal velocity will be given as

$$\boxed{\dot{A} = \frac{1}{2}r^2\dot{\phi} = \frac{L}{2m} = \text{constant.}} \tag{8.34}$$

This equation shows that the time rate of change of area is constant.

Example 8.4

Suppose a planet is moving in a circular orbit under the influence of a central force $\vec{F}(r)$, where r is the radial distance between the planet and the center of the force. Find the form of $\vec{F}(r)$ in terms of areal velocities, $\dot{A}$.

Solution

For circular orbits, the transverse component of acceleration does not exist and only the radial component (centripetal acceleration) will be valid. Total acceleration in polar coordinates is given in equation (1.31) as

$$\vec{a} = (\ddot{r} - r\dot{\phi}^2)\hat{r} + (r\ddot{\phi} + 2\dot{r}\dot{\phi})\hat{\phi},$$

which reduces to

$$\vec{a}_c = (\ddot{r} - r\dot{\phi}^2)\hat{r}$$

and since r is taken to be constant, we have $\ddot{r} = 0$. Then, from Newton's force equation,

$$\Sigma F = m\vec{a}_c = -mr\dot{\phi}^2.$$

Knowing the angular momentum $L = mr^2\dot{\phi}^2$ and the areal velocity $\dot{A} = L/2m$, we can write the above equation as to find the form of $\vec{F}(r)$ in terms of areal velocity, $\dot{A}$ as

$$\boxed{\vec{F}(r) = -\frac{4m\dot{A}}{r^3}.}$$

8.6.3 Kepler's first law: the law of ellipses

To prove the first law, we develop a general differential equation for the orbit of a particle of mass (m) in the field of a central, isotropic force ($\vec{F}(r)$) due to mass (M). Then we will solve the orbital equation of the specific case of an inverse-square law of force. First, express Newton's equation of motion in 2D polar coordinates:

$$\begin{aligned}\Sigma F &= F(r)\hat{r} \\ m\ddot{r} &= -\frac{GmM}{r^2}\hat{r}.\end{aligned} \tag{8.35}$$

We can say $\vec{F}(r)$ is isotropic because it is a function of only the scalar distance (r) to the force center and we say it is central because its direction is along the radius vector.

Using the values of acceleration from equation (1.31) ($\vec{a} = (\ddot{r} - r\dot{\phi}^2)\hat{r} + (r\ddot{\phi} + 2\dot{r}\dot{\phi})\hat{\phi}$) we can write the differential equation of motion for radial and transverse components using equation (8.35) as

$$\text{radial component: } m(\ddot{r} - r\dot{\phi}^2) = -\frac{GMm}{r^2} \tag{8.36}$$

$$\text{transverse component: } m(2\dot{r}\dot{\phi} - r\ddot{\phi}) = 0. \tag{8.37}$$

Given a certain radial force function $f(r)$ we should be able to solve the differential equations (8.36) and (8.37) to obtain(r) and (ϕ) as functions of t. Usually we are interested in the path in space (the orbit) without regard to time (t).

- *Radial equation of motion.* Rearranging equation (8.36) will give us the radial equation of motion as

$$m\ddot{r} - mr\dot{\phi}^2 + \frac{GMm}{r^2} = 0 \tag{8.38}$$

- *Transverse equation of motion.* From equation (8.37) we see that since $m \neq 0$ we can write the transverse component of the equation of motion as

$$\begin{aligned}&2\dot{r}\dot{\phi} - r\ddot{\phi} = 0 \\ &\frac{d}{dt}(r^2\dot{\phi}) = 0 \\ &\Longrightarrow r^2\dot{\phi} = \text{constant (say) } l.\end{aligned}$$

From equation (10.5.1), we recall that $L = mr^2\dot{\phi} \Longrightarrow \frac{L}{m} = r^2\dot{\phi} = \text{constant}$. Then the above equation can be rewritten as

$$\frac{L}{m} = l(\text{constant}). \tag{8.39}$$

Thus l is just the angular momentum per unit mass. This is just a reinstating of the fact we developed in equation (8.30), which is that angular momentum of a particle is constant when moving under the action of a central force.

Planetary motion using Lagrangian mechanics

For a planet of mass m moving in an elliptical path about the fixed Sun of mass M_S the kinetic and potential energies in polar coordinates with origin at the Sun will be given as

$$T = \frac{1}{2}m(\dot{r}^2 + r^2\dot{\phi}^2)$$
$$V = -\frac{GMm}{r}.$$

Knowing the kinetic and potential energies we can write the Lagrangian of the motion as

$$L = T - V$$
$$L = \frac{1}{2}m\dot{r}^2 + \frac{1}{2}mr^2\dot{\phi}^2 + \frac{GMm}{r}.$$

There are two variables r and ϕ in the Lagrangian, hence we will need two Lagrangian equations, one for the radial variable (r) as

$$\begin{aligned} \frac{d}{dt}\left(\frac{\partial L}{\partial \dot{r}}\right) - \frac{\partial L}{\partial r} &= 0 \\ m\ddot{r} - mr\dot{\phi}^2 + \frac{GMm}{r^2} &= 0, \end{aligned} \tag{8.40}$$

which is identical to equation (8.38), and one for the angular component (ϕ) as

$$\begin{aligned} \frac{d}{dt}\left(\frac{\partial L}{\partial \dot{\phi}}\right) - \frac{\partial L}{\partial \phi} &= 0 \\ \frac{d}{dt}(mr^2\dot{\phi}) &= 0. \end{aligned} \tag{8.41}$$

Note that these equations are identical to equations (8.36) and (8.37).

8.6.4 Energy equation of the orbit

To find the energy equation of the orbit we first find the speed in polar coordinates and square it as $v^2 = \dot{r}^2 + r^2\dot{\theta}^2$.

Since central force is conservative $\Longrightarrow$ $E_{\text{total}} = T + V =$ constant, and is given by

$$E_{\text{total}} = \frac{1}{2}m(\dot{r}^2 + r^2\dot{\theta}^2) + V(r) = \text{constant}. \tag{8.42}$$

8.6.5 Kepler's third law: harmonic law

Let us start with asking a question, why is the Kepler's law that relates orbital period to distances from the Sun called the *harmonic law*?

The answer is very interesting, Kepler held that *the world was a beautiful place* even when it treated him badly. He believed in the Pythagorean doctrine of celestial harmony. He thought that mankind had not learned how to hear the true harmony of the worlds. In his work *Harmonic Mundi* (*Harmony of the World*), Kepler (as had Pythagoras 2000 years earlier) tried to connect the planetary motions with all fields of abstractions and harmony. Even though he failed in this attempt, it led him to the third law, hence Kepler's third law is also called the *harmonic law*.

I. **Derivation using Newton's force equations**

We can use the fact that most planets have nearly circular orbits (except Mercury, Mars and Pluto, $\epsilon = 0.206$, 0.093 and 0.249, respectively), and use the properties of circular paths (centripetal force[4], $F_c = \frac{mv^2}{R}$) to derive Kepler's third law as follows:

$$\Sigma F = F_G$$
$$F_c = F_G$$
$$\frac{mv^2}{R} = \frac{GMm}{R^2}$$
$$v^2 = \frac{GM}{R}$$
$$\left(\frac{2\pi R}{T_t}\right)^2 = \frac{GM}{R},$$

where T_t is the period of revolution, the time it takes for the planet to complete one orbit and $(2\pi R)$ is the distance traversed during one period. Rearranging the above equation and knowing $R = a$ we obtain

$$\boxed{T_t^2 = \frac{4\pi^2}{GM}a^3.} \tag{8.43}$$

This equation is Kepler's third law. If distances are taken in astronomical units (AU), where 1 AU = 1.5×10^8 km and the periods are taken in Earth years, then

$$\frac{4\pi^2}{GM} = 1 \tag{8.44}$$

[4] Centripetal force is a force that acts on a body moving in a circular path. Centripetal comes from Latin *centrum*, 'center', and *petere*, 'to seek', which means that centripetal force is always directed toward the center around which the body is moving. See chapter 10 for details on centripetal force.

and Kepler's third law takes a simple form:

$$T_t^2 = a^3. \tag{8.45}$$

II. **Derivation using areal velocity**

Alternatively, we can derive Kepler's third law if we start from the definition of areal velocity, as in equation (8.34):

$$\dot{A} = \frac{L}{2m}. \tag{8.46}$$

Integrate areal velocity over the entire orbital period (T_t):

$$\int \frac{dA}{dt} dt = \int \frac{L}{2m}$$
$$\int dA = \int \frac{L}{2m} dt.$$

For an elliptical orbit of a particle in an inverse-square field the time required to complete orbit is the period (T_t). Then the above equation can be solved to give

$$A = \frac{L}{2m} T, \tag{8.47}$$

where the area of the ellipse is given as $A = \pi ab$, where a is the *semi-major axis* and b is the *semi-minor axis* and the relation between a and b is $b = a\sqrt{1 - \epsilon^2}$ and ϵ is the eccentricity of the ellipse[5] (figure 8.7).

Then the period will be given as

$$\begin{aligned} T_t &= \frac{2mA}{L} \\ &= \frac{2m\pi ab}{L} \\ T_t &= \frac{2m\pi a(a\sqrt{1 - \epsilon^2})}{L}. \end{aligned} \tag{8.48}$$

Squaring both sides

$$\begin{aligned} T_t^2 &= \left(\frac{2\, m\pi a^2 \sqrt{1 - \epsilon^2}}{L} \right)^2 \\ &= \frac{4m^2\pi^2 a^4 (1 - \epsilon^2)}{L^2}. \end{aligned}$$

[5] The eccentricity of an ellipse is the relative distance between its center and either of its two foci. The word eccentric means 'off center'. Basically, eccentricity is a measure of how round or *not* an ellipse is. It is given by the formula eccentricity = c/a, where c is the distance from the center to a focus and a is the distance from that focus to a vertex. The eccentricity of a circle = 0 (no squishing) and the eccentricity of 1 means a parabolic orbit and the eccentricity of greater than 1 means a hyperbolic orbit.

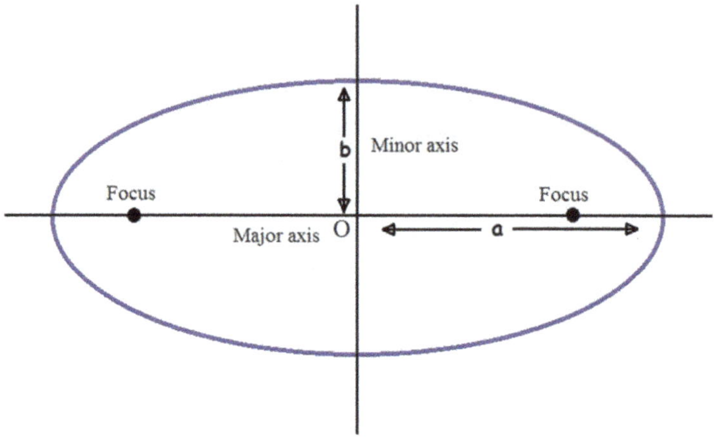

Figure 8.7. An ellipse.

Rearranging to look like equation (8.43) we can write

$$\Longrightarrow \boxed{T_t^2 = \frac{4m^2\pi^2 a(1 - \epsilon^2)}{L^2}a^3.} \tag{8.49}$$

This is also another expression of Kepler's third law, as we found in equation (8.43). However, they do not look alike so we can use dimensional analysis to compare equation (8.49) to equation (8.43).

Comparing equations **(8.43)** *and* **(8.49)**

Let us first express the units in dimensional formulas:
universal gravitational constant (G) =

$$G = \frac{Fr^2}{m_1 m_2} = \frac{[M]^1[L]^1[T]^{-2} \cdot [L]^2}{[M]^2} = [M]^{-1}[L]^3[T]^{-2}. \tag{8.50}$$

Then from equation (8.43) we see that

$$\frac{1}{GM} = [L]^{-3}[T]^2, \tag{8.51}$$

and from equation (8.49) we obtain, remembering that ϵ is unit-less,

$$\left(\frac{m}{L}\right)^2 a = \left(\frac{[M]}{[M]^1[L]^2[T]^{-1}}\right)^2 [L] = [L]^{-3}[T]^2. \tag{8.52}$$

Here we see from equations (8.51) and (8.52) that the two expressions are equivalent and hence both equation (8.43) and (8.49) can be written as

$$\boxed{T_t^2 \propto a^3.} \tag{8.53}$$

This means that square of a planet's orbital period is proportional to cube of its 'distance' from the Sun. The relative 'distance' is the semi-major axis (a) of the elliptical orbit. For a circular orbit, this distance reduces to the radius.

Example 8.5

Compare the periods of two comets, one whose semi-major axis is 3 AU and another whose semi-major axis is 12 AU.

Solution

Use equation (8.53):

Comet 1, with $a = 3$ AU, will have period $= 3^{3/2} = 5.2$ years.

Comet 2, with $a = 12$ AU, will have period $= 12^{3/2} = 41.5$ years.

Concept questions 8.3

1. How will Captain Kirk aboard Star Trek enterprise determine the value of g at the surface on an unknown planet before approaching the planet?
2. How would you determine if two objects do or do not have the same inertial mass, same gravitational mass and/or the same weight?
3. Explain in your own words what is meant by g-forces acting on a pilot in a stunt aircraft?
4. Explain the basic force law involved in rocket propulsion.

Exercises 8.3

1. Suppose scientists find a planet **B** which can sustain life. However, this planet has a much thinner atmosphere and much less gravitational force. In what way might *Homo sapiens* have evolved there to account for the difference in environment?
2. Calculate the speed of a satellite in a near circular, low Earth orbit (LEO) about the Earth.
3. Compute the mass of the Sun from the period and radius of Earth's orbit.

Further reading

Arya A 1990 *Introduction to Classical Mechanics* 2nd edn (Upper Saddle River, NJ: Prentice-Hall)

Baierlein R 1983 *Newtonian Dynamics* (New York: McGraw-Hill)

Chababy R and Sherwood B 2011 *Matter and Interactions* 3rd edn (New York: Wiley)

Fowles G 1986 *Analytical Mechanics* 4th edn (New York: CBS College Publishing)

Fowles G and Cassiday G 2005 *Analytical Mechanics* 7th edn (Belmont, CA: Thomson Learning)
Ohanian H 1985 *Physics* (New York: WW Norton and Company)
Resnick R, Halliday D and Krane K 2002 *Physics* vol 1, 5th edn (New York: Wiley)
Rutherford D E 1951 *Classical Mechanics* (New York: Interscience)
Stephenson R 1961 *Mechanics and Properties of Matter* 2nd edn (New York: Wiley)
Symon K 1971 *Mechanics* 3rd edn (Reading, MA: Addison-Wesley)

Chapter 9

Two- and three-dimensional dynamics

9.1 Introduction: general principles

In this chapter we will examine the general case of the motion of a particle in two and three dimensions. We developed a procedure/methodology for analyzing the position, velocity and acceleration for a particle moving in 1D in chapter 3. In 1D we found that when acceleration is constant, we can find the velocity at all times from equation (3.2) ($v_x = v_{ox} + a_x t$) and the position at all times from equation (3.3) ($\Delta x = x - x_o = v_{ox}t + \frac{1}{2}a_x t^2$).

In this chapter we will see how these equations can be restructured when the particle moves in two- or three-dimensional space. For motion in three dimensions the acceleration vector will be given as $\vec{a} = \ddot{x}\hat{i} + \ddot{y}\hat{j} + \ddot{z}\hat{k}$, and for motion with constant acceleration, we can obtain the general equation of motion by setting

$$a_x = \text{constant}; \quad \mathrm{a}_y = \text{constant; and} \quad \mathrm{a}_z = \text{constant}.$$

The particle starts at $t = 0$ with initial position $\vec{r}_o = x_o\hat{i} + y_o\hat{j} + z_o\hat{k}$ and initial velocity $\vec{v}_o = v_{ox}\hat{i} + v_{oy}\hat{j} + z_{oz}\hat{k}$.

Under the above initial conditions we can then write 3D vector equations of motion as:

First equation of motion:

$$\vec{v} - \vec{v}_o = \vec{a}t. \tag{9.1}$$

Second equation of motion:

$$\vec{r} - \vec{r}_o = \vec{v}_o t + \frac{1}{2}\vec{a}t^2. \tag{9.2}$$

Here, the displacement vector is given as $\Delta\vec{r} = \vec{r} - \vec{r}_o$.

Third equation of motion:

$$2\vec{a}(\vec{r} - \vec{r}_o) = \vec{v}^2 - \vec{v}_o^{\,2}. \tag{9.3}$$

When using these equations remember these two important facts:

- Each of the above equations represents three independent 1D equations, one along each axis, say the x-, y- and z-axes.
- The velocity components vary independently of one other. For example, $_{ax}$ will affect only v_x of the particle under consideration and not affect v_y or v_z, etc. This implies in a particular situation when $_{ax}$ is positive and $a_y = a_z = 0$, the velocity along the x-axis (i.e. v_x) will increase but v_y and v_z will remain unchanged.

9.1.1 Newton's law in 3D

The vector form of Newton's law in terms of linear momentum of a particle ($\vec{p} = mv$) states:

> 'The rate of change of momentum of a body is numerically equal to the total external force acting on the body',

which is mathematically written as

$$\Sigma\vec{F} = \frac{d\vec{p}}{dt}. \tag{9.4}$$

For constant mass, Newton's law in differential form in 3D will be written as

$$\boxed{\Sigma\vec{F} = m\frac{d^2\vec{r}}{dt^2} = m\frac{d\vec{v}}{dt} = m\ddot{\vec{r}}.} \tag{9.5}$$

This is a 3D vector equation and when dealing with this equation we have to consider the force components and have to be very careful while defining the factors that the force depends on. In Newtonian mechanics, force components are functions of position (x, y, z), velocity ($\dot{x}$, $\dot{y}$, $\dot{z}$) and time, t:

$$\vec{F} = \vec{F}(x, y, z, \dot{x}, \dot{y}, \dot{z}, t). \tag{9.6}$$

Under these conditions, the 3D differential equations of the form of equation (9.5) become very complicated when compared to the 1D case we dealt with in chapter 3.

Realistically, we rarely know the exact dependence of $\vec{F}$ on time, therefore we do not worry about time. Instead we focus on situations in which we know that $\vec{F}$ is an explicit function of positions and velocities.

The simplest situation most commonly encountered is where $\vec{F} = \vec{F}(x, y, z)$ only. In this case equation (9.5) can be rewritten as

$$\Sigma\vec{F} = m\ddot{x}\hat{i} + m\ddot{y}\hat{j} + m\ddot{z}\hat{k}. \tag{9.7}$$

If we consider the force ($\vec{F}$) to be a separable force[1] we can express $\Sigma\vec{F}$ in Cartesian coordinates as

$$\vec{F} = F_x\hat{i} + F_y\hat{j} + F_z\hat{k}. \tag{9.8}$$

Combining equations (9.7) and (9.8), we obtain

$$F_x\hat{i} + F_y\hat{j} + F_z\hat{k} = m\ddot{x}\hat{i} + m\ddot{y}\hat{j} + m\ddot{z}\hat{k},$$

which can be considered as being equivalent to three scalar equations in Cartesian coordinates

$$F_x = m\ddot{x};\ F_y = m\ddot{y};\ F_z = m\ddot{z}. \tag{9.9}$$

More complex problems will have $\vec{F} = \vec{F}(x, y, z, \dot{x}, \dot{y}, \dot{z})$ and so on. The more variables $\vec{F}$ depends on, the more complicated the problem becomes. Some examples include:

- Projectile motion with air resistance.
- Motion of charged particles in a static electromagnetic field.
- Fluid motion.

9.1.2 Work principle

We know that the work done on a particle causes the object to gain or lose kinetic energy. This principle can be generalized in 3D as

$$\vec{F} \cdot \vec{v} = \frac{dT}{dt}, \tag{9.10}$$

where $T = \frac{1}{2}mv^2$ represents the kinetic energy of the particle under observation. We know that $v = d\vec{r}/dt$, so the above equation can be rewritten as

$$\vec{F}d\vec{r} = dT.$$

Integrate both sides to obtain

$$\int \vec{F}d\vec{r} = \Delta T, \tag{9.11}$$

where ΔT is the change in kinetic energy of the particle from the initial to final point.

The left side of this equation is the line integral along the trajectory from some initial point (i) to some final point (f) in 3D space and represents the work done on the particle by external force ($\vec{F}$) as the particle moves from $i \to f$:

[1] A force is a separable force when the components of force of a variable depends only on that variable and does not depend on other variables.

$$\int_i^f \vec{F}_{\text{net}} \cdot d\vec{r} = W = \Delta T. \tag{9.12}$$

9.2 Some useful mathematical concepts

Before proceeding I will introduce some mathematical concepts that will be useful to know.

9.2.1 The del operator

The gradient of a function is mathematically a vector that represents the maximum spatial derivative of the function under consideration in both direction and magnitude. The gradient in 3D is represented by a vector operator called the *del operator* which is mathematically given as

$$\nabla = \frac{\partial}{\partial x}\hat{i} + \frac{\partial}{\partial y}\hat{j} + \frac{\partial}{\partial z}\hat{k}. \tag{9.13}$$

The del operator in spherical coordinates

The del operator in 3D spherical coordinates (r, θ, ϕ), where θ is the polar and ϕ is the azimuthal angle, is mathematically given as

$$\nabla = \frac{\partial}{\partial r}\hat{r} + \frac{1}{r}\frac{\partial}{\partial \theta}\hat{\theta} + \frac{1}{r \sin\theta}\frac{\partial}{\partial \phi}\hat{\phi}. \tag{9.14}$$

The del operator in cylindrical coordinates

The del operator in cylindrical coordinates (r, ϕ, z) is mathematically given as

$$\nabla = \frac{\partial}{\partial r}\hat{r} + \frac{1}{r}\frac{\partial}{\partial \phi}\hat{\phi} + \frac{\partial}{\partial z}\hat{z}. \tag{9.15}$$

Example 9.1

What force is associated with $V = \frac{1}{2}k|\vec{r}|$ in Cartesian coordinates? Name the force.

Solution

We will start the solution to this example by noting that

$$r^2 = x^2 + y^2 + z^2$$

$$\vec{F} = -\frac{\partial V}{\partial r} = \frac{1}{2}k\left(-\frac{\partial x^2}{\partial x}\hat{\imath} - \frac{\partial y^2}{\partial y}\hat{\jmath} - \frac{\partial z^2}{\partial z}\hat{k}\right)$$

$$\vec{F} = -\frac{1}{2}k(2x\hat{\imath} + 2y\hat{\jmath} + 2z\hat{k})$$

$$\vec{F} = -k\vec{r},$$

which is the formula for Hooke's law for elastic force. It is worth noting here that we saw a similar example in chapter 2, example 2.2, the formula for elastic force in 1D.

9.2.2 Curl of a vector field

Consider a general vector field $\vec{B}$, which has components that have continuous derivatives in a region of space that contains a particular surface (S), as shown in figure 9.1. Around the edge of this surface we have a curve (C) called the boundary curve.

Then according to the mathematical theorem called Stokes' theorem, we know:

> 'The line integral ($\int \vec{B} \cdot d\vec{s}$) taken over a simple, closed, smooth curve (C) that bounds an oriented, smooth surface (S) is equal to the integral of the surface unit normal dotted with the curl of $\vec{B}$ taken over the surface (S).'

This is mathematically expressed as

$$\iint_S (\text{curl of } \vec{B}) \cdot d\vec{a} = \iint_S (\vec{\nabla} \times \vec{B}) \cdot d\vec{a} = \oint_C \vec{B} \cdot d\vec{s}. \tag{9.16}$$

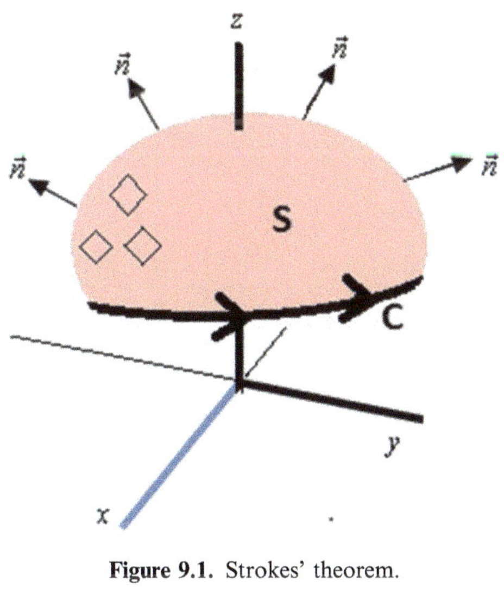

Figure 9.1. Strokes' theorem.

In this theorem, the surface S can be any surface as long as its boundary curve is given by C. Note that mathematically

$$\text{curl of } \vec{B} = (\vec{\nabla} \times \vec{B}) = \begin{vmatrix} \hat{i} & \hat{j} & \hat{k} \\ \dfrac{\partial}{\partial x} & \dfrac{\partial}{\partial y} & \dfrac{\partial}{\partial z} \\ B_x & B_y & B_z \end{vmatrix}.$$

9.2.2.1 Curl of force field ($\vec{F}$)

If a force field is a separable vector, i.e. the components of the force field can be written as

$$F = F_x\hat{i} + F_y\hat{j} + F_z\hat{k}. \tag{9.17}$$

Then we can write the curl of the force field in various coordinates as follows:

- Curl of force field ($\vec{F}$) in *Cartesian* coordinates (x, y, z):

$$\text{curl of } \vec{F} = (\vec{\nabla} \times \vec{F}) = \begin{vmatrix} \hat{i} & \hat{j} & \hat{k} \\ \dfrac{\partial}{\partial x} & \dfrac{\partial}{\partial y} & \dfrac{\partial}{\partial z} \\ F_x & F_y & F_z \end{vmatrix}.$$

- Curl of force field ($\vec{F}$) in *spherical* coordinates (r, θ, ϕ):

$$\text{curl of } \vec{F} = (\vec{\nabla} \times \vec{F}) = \frac{1}{r^2 \sin\theta} \begin{vmatrix} \hat{r} & r\hat{\theta} & r\sin\theta\hat{\phi} \\ \dfrac{\partial}{\partial r} & \dfrac{\partial}{\partial \theta} & \dfrac{\partial}{\partial \phi} \\ F_r & F_\theta & F_\phi \end{vmatrix}.$$

- Curl of force field $\vec{F}$ in *cylindrical* coordinates (r, ϕ, z):

$$\text{curl of } \vec{F} = (\vec{\nabla} \times \vec{F}) = \begin{vmatrix} \dfrac{1}{r}\hat{r} & \hat{\theta} & \dfrac{1}{r}\hat{k} \\ \dfrac{\partial}{\partial r} & \dfrac{\partial}{\partial \phi} & \dfrac{\partial}{\partial z} \\ F_r & rF_\phi & F_z \end{vmatrix}.$$

It is important to remember here that curl is a geometrical operation on a vector field that does not depend on the coordinate system.

9.3 Conservative and non-conservative forces in 3D

We talked about conservative and non-conservative forces in chapter 2, section 2.3.10, so I will provide a quick recap here for the 3D situation.

9.3.1 Conservative forces

A force is called a *conservative force* when the work done in moving a particle between two points is independent of the path followed. A conservative force depends only on the position of the object. In 3D it will be given as

$$\vec{F} = \vec{F}(\vec{r}). \tag{9.18}$$

9.3.1.1 Conditions for conservative forces

A force field, $\vec{F}$, is a conservative force if it meets these three equivalent conditions:

- The net work (W) done by the force when moving a particle through a closed path equals zero, $W \equiv \oint_C \vec{F} \cdot d\vec{r} = 0$.
- The force can be written as the negative gradient of a potential, V: $\vec{F} = -\nabla V$. The negative sign indicates that the particle is moved in the direction of decreasing potential energy.
- The curl of F is the zero vector: $\vec{\nabla} \times \vec{F} = \vec{0}$.

9.3.1.2 Conservative force and the del operator

Physically, the negative gradient of the potential energy function gives the direction and magnitude of the force that acts on a particle in a given field. We express a conservative force $\vec{F}$ as

$$\vec{F} = -\nabla V = -\frac{\partial V}{\partial x}\hat{i} - \frac{\partial V}{\partial y}\hat{j} - \frac{\partial V}{\partial z}\hat{k}. \tag{9.19}$$

The negative sign means that the particle accelerates in the direction of *decreasing* potential energy.

In terms of the del operator, the force can be written as

$$\text{curl of } \vec{F} = \vec{\nabla} \times \vec{F}. \tag{9.20}$$

This means that the curl is given as $\vec{\nabla} \times \vec{F}$, and $\vec{F}$ in terms of the scalar function V is given by the operation $\vec{F} = -\vec{\nabla} V$. Then $\vec{\nabla} \times (\vec{\nabla} V) \equiv 0$, or in other words the curl of any gradient is identically zero. Then, due to Strokes' theorem, the condition for conservative force can be summarized as

$$\boxed{\text{curl of } \vec{F} = \nabla \times \vec{F} \equiv \nabla \times (\nabla V) = 0.} \tag{9.21}$$

Example 9.2

Is the force $\vec{F} = 2xy\hat{i} + 5xz\hat{j} + 3yz\hat{k}$ a conservative or a non-conservative force?

Solution

We know from equation (9.21) that the condition for conservative force is

$$\text{curl of } \vec{F} = 0.$$

Then, in this case, we will have

$$\text{curl of } \vec{F} = (\nabla \times \vec{F}) = \begin{vmatrix} \hat{i} & \hat{j} & \hat{k} \\ \dfrac{\partial}{\partial x} & \dfrac{\partial}{\partial y} & \dfrac{\partial}{\partial z} \\ 2xy & 5xz & 3yz \end{vmatrix}.$$

Solving the matrix, we find

$$\nabla \times \vec{F} = \hat{i}\left(\frac{\partial(3yz)}{\partial y} - \frac{\partial(5xz)}{\partial z}\right) - \hat{j}\left(\frac{\partial(3yz)}{\partial x} - \frac{\partial(2xy)}{\partial z}\right) + \hat{k}\left(\frac{\partial(5xz)}{\partial x} - \frac{\partial(2xy)}{\partial y}\right)$$
$$= (3z - 5x)\hat{i} + (5z - 2x)\hat{k}.$$

From the above solution it is clear that

$$\nabla \times \vec{F} \neq 0.$$

Hence the given force is not conservative.

Example 9.3

For what of constants A and B will the force $\vec{F} = Axy\hat{i} ++ By^2\hat{j}$ be conservative?

$$\text{curl of } \vec{F} = (\nabla \times \vec{F}) = \begin{vmatrix} \hat{i} & \hat{j} & \hat{k} \\ \dfrac{\partial}{\partial x} & \dfrac{\partial}{\partial y} & \dfrac{\partial}{\partial z} \\ Axy & Bx^2 & 0 \end{vmatrix}.$$

Solving the matrix, we find,

$$\text{curl of } \vec{F} = (2Bx - Ax)\hat{k}$$
$$0 = (2Bx - Ax)\hat{k}.$$

Since $x \neq 0$ means $(2B - A) = 0$,

$\Rightarrow$ for $A = 2B$, the force will be a conservative force.

9.3.2 Non-conservative forces

If the force is not conservative, then defining a scalar potential for it is not possible. This is because taking different paths would lead to very different potential differences between initial and final points. Friction is an example of a non-conservative force. Another example of a non-conservative force can be stress, which we will talk in detail about in chapter 14, section 14.2.

9.4 Generalized conservation of energy principle in 3D

The work done by a conservative force in moving a particle from point A to point B can be written as

$$\begin{aligned}\int_{\mathrm{A}}^{\mathrm{B}} \vec{F} \cdot d\vec{r} &= -\int_{\mathrm{A}}^{\mathrm{B}} \nabla V(\vec{r}) \cdot d\vec{r} \\ &= -\int_{\mathrm{A}_x}^{\mathrm{B}_x} \frac{\partial V}{\partial x} dx - \int_{\mathrm{A}_y}^{\mathrm{B}_y} \frac{\partial V}{\partial y} dy - \int_{\mathrm{A}_z}^{\mathrm{B}_z} \frac{\partial V}{\partial z} dz \\ &= -\int_{\mathrm{A}}^{\mathrm{B}} dV(\vec{r}) \\ &= -\Delta V = -(V_{\mathrm{B}} - V_{\mathrm{A}}).\end{aligned}$$

The last step illustrates the fact that $[\vec{\nabla} V \cdot d\vec{r}]$ is an *exact* differential[2] equal to dV. Hence we see that

$$\int_{\mathrm{A}}^{\mathrm{B}} \vec{F} \cdot d\vec{r} = V_{\mathrm{A}} - V_{\mathrm{B}}. \tag{9.22}$$

The work done by any net force is always equal to the change in kinetic energy,

$$\int_{\mathrm{A}}^{\mathrm{B}} \vec{F} \cdot d\vec{r} = \Delta T = -\Delta V. \tag{9.23}$$

Therefore,

$$\Delta(T + V) = \Delta E = 0. \tag{9.24}$$

This means that total energy is constant, which is the statement of the law of conservation of energy.

9.4.1 Non-conservative forces (F_{NC}) are present

Any force that is also a function of time and/or velocity is called non-conservative. If $\vec{F}$ is a non-conservative force, it cannot be set equal to $-\nabla V$. The work increment $(\vec{F}_{\mathrm{NC}} \cdot d\vec{r})$ is not an exact differential and cannot be equated to $(-dV)$.

[2] If we have a function of the form $G = G(x_1, x_2, x_3, \ldots, x_n)$, then we can write the differential form $dG = (\partial G)/(\partial x_1)dx_1 + (\partial G)/(\partial x_2)dx_2 + \cdots + (\partial G)/(\partial x_n)dx_n$, which is called an exact or a total differential.

9.4.2 Both conservative and non-conservative forces are present

In this case, where both conservative forces F and non-conservative forces F_{NC} are present,

$$\begin{aligned}\text{total work increment} &= (\vec{F} + \vec{F}_{NC}) \cdot d\vec{r} \\ &= \vec{F} \cdot d\vec{r} + \vec{F}_{NC} \cdot d\vec{r} \\ &= -\,dV + \vec{F}_{NC} \cdot d\vec{r} \\ &= dT.\end{aligned} \tag{9.25}$$

Then the generalized form of the work–energy theorem becomes

$$\int_A^B \vec{F}_{NC} \cdot d\vec{r} = \Delta E. \tag{9.26}$$

The total energy (E) does not remain a constant throughout the motion of the particle by increasing or decreasing depending upon the nature of the non-constant force F_{NC}.

In the case of a dissipative force such as friction or air resistance, the direction of F_{NC} is always opposite to the motion. Hence $(\vec{F} \cdot d\vec{r})$ is negative and the total energy of the particle decreases as it moves through space.

9.5 The energy equation

Gravitational and spring forces have potential energies mathematically given as

$$\Delta V_{grav} = mg\Delta y = -W_{grav}$$

$$\Delta V_{spring} = \frac{1}{2}k(\Delta x)^2 = -W_{spring}.$$

The work done by conservative forces (W_{cons}) is

$$W_{cons} = -\Delta V. \tag{9.27}$$

Non-conservative forces (e.g. friction, drag) are dissipative in nature. The work they do on an object ($W_{non\text{-}cons}$) in a system converts kinetic or potential energy into heat (thermal energy, E_{th}), thus

$$W_{non\text{-}cons} = -\Delta E_{th}. \tag{9.28}$$

Note that $W_{non\text{-}cons}$ is always negative, which means when a dissipative force does work the thermal energy of a system always increases.

9.5.1 The complete energy equation

Suppose the net work done on the object by all forces internal to the system is (W_{int}) and the net work done on the object by all forces external to the system is (W_{ext}). Then the total work (ΣW) done by all forces on an object in a system can be written as

$$\Sigma W = W_{int} + W_{ext}. \tag{9.29}$$

Internal forces can be further categorized as conservative or non-conservative (dissipative). Then we can rewrite equation (9.29), knowing $W_{\text{cons}} = -\Delta V$ and $W_{\text{non-cons}} = -\Delta E_{th}$ as

$$\begin{aligned} W &= W_{\text{cons}} + W_{\text{non-cons}} + W_{\text{ext}} \\ &= (-\Delta V) + (-\Delta E_{th}) + W_{\text{ext}}. \\ &= \Delta K \end{aligned} \tag{9.30}$$

This is called the *complete energy equation.*

Concept questions 9.1

1. Considering Strokes' theorem, why can we say that $\int \vec{E} \cdot d\vec{l} = 0$ for an electric field given as $\vec{E}$ for integration around a closed path?
2. Two toy cars, one of mass m and the other of mass $4m$, are pushed on a frictionless surface from a start line to a finish line with equal constant forces. If we assume both cars to be initially at rest, which of the following statements is true when the cars reach the finish line?
   (a) The kinetic energies of both cars are equal.
   (b) The car with larger mass has the greater kinetic energy.
   (c) The car with smaller mass has the greater kinetic energy.
3. Which of the two cars in question 2 will reach the finish line first?

Exercises 9.1

1. Given the following values of potential find the force:
   (a) $V = Axyz + B$, where c A and B are constants.
   (b) $V = cr^n \sin\theta\varphi$ in spherical coordinates where C is a constant.
2. By finding the curl determine which of the following forces are conservative:
   (a) $\vec{F} = y\hat{i} - x\hat{j} + z^2\hat{k}$.
   (b) $F = kr^{-n}\hat{r}$ in spherical coordinates.
3. Find the value of the constant (a) such that the following force become conservative:

$$\vec{F} = (Ax + By^2)\,i + Cxy\,j$$

9.6 Body with variable mass

So far we have only discussed motions where the masses of the objects remain the same. However, there are situations where masses of objects change over time, such as:

- Snowballs gaining mass as they roll down a hill.
- Raindrops gathering up smaller droplets as they fall and thereby increasing their mass.

- Rockets burning fuel and ejecting gases to propel themselves will lose mass as they accelerate.
- A truck carrying a load that is not properly secured will lose mass during the trip.

In each case, mass is either continuously being added or being removed from the object in question. This change in mass affects the motion. Here we derive the general differential equation that describes the motion of such objects.

9.6.1 General case

Let us first derive the general differential equation that describes the motion in which mass is added to the body as it moves; think of a snowball rolling down a hill. In this case the rate of change of mass will be a positive quantity.

A snowball with mass (m) rolls down a hill, as it rolls it picks up more snow particles along the way. Thus, the snowball continually gathers more mass as time goes by. Assume at time t, the mass of the snowball is $m(t)$ and its velocity at that time is $v(t)$.

For the most general situation assume the small snow particles are not at rest but are also moving with a velocity $u(t)$. Then at a later time $(t + \Delta\, t)$, the snowball will have picked up an additional small amount of mass Δm. Thus, its mass at time $(t + \Delta\, t)$ will be $[m(t + \Delta t) = m(t) + \Delta m]$ and its velocity will be changed to $[v(t + \Delta t)]$. Then in time interval Δt, the total linear momentum of the system will change as

$$\Delta\vec{p} = p_{\text{total}}(t + \Delta t) + p_{\text{total}}(t). \tag{9.31}$$

Knowing $[p_{\text{total}}(t + \Delta t) = (m + \Delta m)(\vec{v} + \Delta\vec{v})]$ and $[P_{\text{total}}(t) = m\vec{v} + (\Delta m)\vec{u}]$, we can simplify equation (9.31) to obtain

$$\Delta\vec{p} = (m + \Delta m)(\Delta\vec{v}) + \Delta m\vec{v}_r, \tag{9.32}$$

where $\vec{v}_r \equiv \vec{v} - \vec{u}$ and is the velocity of Δm relative to m. Dividing the above equation by Δt and taking the limit $\Delta t \to 0$ we can rewrite

$$\frac{dp}{dt} = \vec{F}_{\text{ext}} = m\frac{d\vec{v}}{dt} - \frac{dm}{dt}\vec{v}_r. \tag{9.33}$$

The force F_{ext} in equation (9.33) represents any external force, such as gravity or air resistance, that acts on the system and equation (9.33) is the general case that we can apply to the two situations as discussed in section 9.6.

- The gaining mass case.
- The losing mass case.

When $F_{\text{ext}} = 0$, the total momentum of the system is conserved.

9.6.2 Case 1. Gaining mass

Consider that the object (say a raindrop) is falling through the atmosphere and it collects more mass as it falls by picking up water droplets suspended in air. In this scenario we can assume that the small droplets are suspended in the atmosphere such that their initial velocity is zero. Hence, $u = 0$ and in equation (9.33) $v_r = -v$ and the equation may be rewritten as

$$\frac{dp}{dt} = m\frac{d\vec{v}}{dt} + \frac{dm}{dt}\vec{v}$$

$$\vec{F}_{\text{ext}} = \frac{d(m\vec{v})}{dt}.$$

Note that this equation is valid only for situations where the initial velocity of the matter being swept up is zero.

9.6.3 Case 2. Losing mass—no external forces

Suppose the object is losing mass, such as in a rocket of mass (m), moving at velocity $\vec{v}$. The rocket loses mass in the form of ejected fuel, at a rate equal to (dm/dt), hence the sign of (dm/dt) will be negative. Here the term ($\frac{dm}{dt}v_r$) will be called the *thrust* of the rocket. The direction of thrust is opposite to the direction of the relative velocity ($\vec{v}_r$) of the exhaust products and can be interpreted as an additional force on the rocket due to the gas expulsion.

First we solve the equation of motion for the simplest case in which the external force on it is zero during rocket motion, i.e. $\vec{F}_{\text{ext}} = 0$. This implies the rocket is not subject to any force such as the force of gravity, air drag, etc, so equation (9.33) reduces to

$$m\frac{d\vec{v}}{dt} = \frac{dm}{dt}\vec{v}_r$$

and solving this equation we obtain

$$\vec{v} = \vec{v}_o + \vec{v}_r \ln\left(\frac{m}{m_o}\right), \tag{9.34}$$

where m_o is the initial mass of the rocket plus unburned fuel, say when it is still grounded on Earth, m is the mass at any given time and v_r is the speed of the ejected fuel relative to the rocket. Note that in equation (9.34) we have a logarithmic function, which means the rocket initially must have a large fuel-to-payload ratio to attain the high speeds needed for launching satellites into space.

Equation (9.34) is also sometimes written as

$$\boxed{\Delta\vec{v} = \vec{v}_r \ln\left(\frac{m}{m_o}\right).} \tag{9.35}$$

Using, $\vec{v} - \vec{v}_o = \Delta\vec{v}$.

9.6.3.1 Rocket equation

Alternatively, equation (9.34) can also be written as

$$\boxed{m(t) = m_0 e^{-\Delta v/v_r},} \tag{9.36}$$

which is the well known *rocket equation*, and it gives the mass of the rocket at any time t, as a function of the initial mass m_0, $\Delta\vec{v}$ and v_r, which is the velocity of the rocket exhaust with respect to the rocket.

From the above equations, we see that for any given $\Delta\vec{v}$ and m_0, increasing v_r increases m (payload plus rocket mass) and decreases m_p. Current technology allows v_r ranging from about 2500–4500 m s^{-1} for chemical rockets and about 105 m s^{-1} for rockets with ion engines. However, in ion engines fewer particles are needed but the thrust is quite low, whereas chemical rockets have high mass ejections and also have high thrust.

Note that if $dm/dt = \dot{m}$ is constant, we can write $m(t) = m_0 + \dot{m}t$, and substitute into equation (9.35) to find $\vec{v}$ as a function of t,

$$v = v_0 + v_r \ln\left(1 + \frac{\dot{m}}{m_0}t\right). \tag{9.37}$$

Recall that $\frac{dm}{dt}$ is negative as the mass decreases with time in the rocket motion.

9.6.4 Case 3. Losing mass—where $\vec{F}_{\text{ext}} \neq 0$

In section 9.6.3 we dealt with the losing mass case where $\vec{F}_{\text{ext}} = 0$, which is definitely not true on the surface of Earth. Equation (9.34) shows that even while ignoring external forces, such as gravity and air drag, we need a large amount of fuel to put a small payload into orbit around the Earth.

9.6.4.1 When F_{ext} is gravity: $F_{\text{ext}} = mg$

We may be able to ignore air resistance, or at least minimize its effect by careful shaping and construction of the rocket. However, we cannot ignore gravity as long as we are on Earth. Hence, the general case cannot hold for real-life problems. In this situation equation (9.33) will become

$$\frac{dp}{dt} = \vec{F}_{\text{ext}} = mg = m\frac{d\vec{v}}{dt} + \frac{dm}{dt}v_r. \tag{9.38}$$

Note that in this case the second term in the above equation ($\frac{dm}{dt}V$) has a positive sign. This is because for the losing mass case ($\frac{dm}{dt}$) is negative. Choosing the upward direction as positive and rearranging we obtain

$$-g = \frac{d\vec{v}}{dt} + \frac{dm}{dt}\frac{v_r}{m}$$

$$-\frac{g}{v_r}dt = \frac{1}{V}dv + \frac{1}{dm}$$

$$-\frac{g}{v_r}dt - \frac{1}{dm} = \frac{1}{v_r}dv. \tag{9.39}$$

Velocity as a function of time

For the rocket to *lift-off* from the surface of Earth, the fraction ($\frac{dm}{m}$) must be bigger than ($\frac{g}{v_r}dt$). That means that the rocket must use a lot of fuel to achieve lift-off or it must eject a lot of matter (dm) at a very high exhaust velocity (v_r). Equation (9.39) can be integrated to give

$$\vec{v} = \vec{v}_o - v_r \ln\left(\frac{m}{m_o}\right) - gt$$

$$\vec{v} = \vec{v}_o - v_r \ln\left(\frac{m}{m_o}\right) - g\frac{m - m_o}{\dot{m}}.$$

Parameter of goodness

The reciprocal of the constant $\frac{g}{V}$ is called the *parameter of goodness* for any given type of rocket. This term is also called the *specific impulse* (τ_s) of the rocket engine:

$$\frac{\tau_s}{1} = \frac{v_r}{g}. \tag{9.40}$$

The value of the specific impulse is proportional to the exhaust velocity of the rocket and has dimensions of time. The exhaust velocity of the rocket depends on the chemical combinatorics in the rocket's combustion chamber and the shape of the rocket nozzle. A well-designed chemical rocket typically has an exhaust velocity of about 3000 m s^{-1}, where the average molecular weight of the combustibles is about 30. Which in turn shows that

$$\tau_s = \frac{v_r}{g} \approx 300 \text{ s} \approx 5 \text{ min}. \tag{9.41}$$

Final velocity attained by the rocket

To find the final velocity attained by the rocket, we have to integrate equation (9.39) from initial time t_o where the rocket is on the ground with $v_o = 0$ up to the time of burnout (τ_B),

$$\frac{dv}{v_r} = -\frac{dm}{m} - \frac{g}{v_r}dt$$

$$\frac{1}{v_r}\int_0^{v_f} dv = -\int_{m_0}^{m_f} \frac{1}{m}dm - \frac{1}{\tau_s}\int_0^{\tau_B} dt,$$

which simplifies to

$$\frac{v_f}{v_r} = \ln\left(\frac{m_0}{m_f}\right) - \frac{\tau_B}{\tau_s}.$$

Hence the final velocity attained by the rocket will be given as

$$v_f = v_r \ln\left(\frac{m_0}{m_f}\right) - v_r \frac{\tau_B}{\tau_s}. \tag{9.42}$$

Take initial mass $m_0 = m_R + m_p + m_F$; where m_R is the mass of the rocket, m_p is the mass of the payload and m_F is the mass of the fuel. Assume that at the final velocity where the mass is m_f, the mass of the fuel goes to zero then $m_f(=m_R + m_p)$, so from equation (9.42),

$$\frac{v_f}{v_r} = \ln\left(\frac{m_R + m_p + m_F}{m_R + m_p}\right) - \frac{\tau_B}{\tau_s}$$

$$e^{\left(\frac{v_f}{v_r} + \frac{\tau_B}{\tau_s}\right)} = \frac{m_R + m_p + m_F}{m_R + m_p},$$

which can be shown with a bit of algebra to be simplified to

$$\frac{\text{mass of fuel}}{\text{mass of rocket + payload}} = \frac{m_F}{m_R + m_p} = e^{\left(\frac{v_f}{v_r} + \frac{\tau_B}{\tau_s}\right)} - 1. \tag{9.43}$$

This means it takes about 105 kg of fuel to place 1 kg of stuff into orbit. That is a lot! However, this ratio is larger than that which is typically required. For example, the lift-off weight of Saturn V was about 3.2 million kg and it could put 100 000 kg into orbit. That is a ratio of about 32 kg of fuel for every kilogram of orbital stuff. Why is our result a factor of 3 larger? Answer: use more efficient two-stage rockets.

Two-stage rockets are the more efficient way to launch payloads into low Earth orbits (LEO). The tanks holding fuel for the first stage are jettisoned after the first stage burn is completed, as they are not needed any more. This reduces the overall mass and hence reduces the fuel requirement for the rest of the flight.

9.6.4.2 Launch strategy

The Earth rotates towards the east with a period of about 23.9 h. This results in a tangential velocity of $v_t = 466\,\mathrm{m\,s^{-1}}$ at the equator. The corresponding value for higher latitudes is $v_t = 466\cos(\theta)$, where θ is the latitude of position on the Earth. Mostly we want to use v_t to add to the energy available when placing a satellite in orbit. Therefore, most launch sites are located as close to the equator as possible and the launch occurs with a velocity component directed towards the east.

Table 9.1 shows the latitudes of some the international launch sites. A major criterion for choosing a launch site is public safety; both in terms of the possibility of a launch failure and exhaust fumes.

9.6.5 Case 4. Gravity plus drag: $F_{\text{ext}} = -mg - D$

The effect of the drag force, D, is harder to quantify. It turns out that for many important applications drag effects are very small. The drag force is characterized in terms of a drag coefficient, C_D,

Table 9.1. Latitudes of some international launch sites.

Kennedy Space station, Cape Canaveral, FL	28.5°
Baikonur Cosmodrome, Russia	45.6°
Vandenberg, CA	34.4°
Xichang Space Center, Sichuan, China	28.3°

$$D = \frac{1}{2}\rho v^2 A C_D, \tag{9.44}$$

where A is the cross-sectional area of the rocket and ρ is the density of air. The air density changes with altitude z, and may be approximated as

$$\rho = \rho_o e^{-z/H}, \tag{9.45}$$

where H is taken to be at the *scale height*[3] of the atmosphere and is usually taken to be approximately 8000 m (the height of Mount Everest) and ρ_o is the air density at sea level.

When we add drag to our system, the differential equation that results for the velocity cannot be integrated explicitly and, in practice, is integrated numerically. It is interesting to note, however, that the effect of drag losses is usually quite small, and it is often reasonable to ignore it in a first calculation.

To see the importance of drag we can estimate the value of the ratio (D/mg). At conditions typical for maximum drag, $\rho \approx 0.25\ \text{kgm}^{-3}$ and $v = 700\ \text{m s}^{-1}$. Considering a rocket of mass = 12 000 kg and cross-sectional area = 1 m^2 and $C_D = 0.2$, we have

$$\frac{D}{g} = \frac{\frac{1}{2}\rho v^2 A C_D}{mg} = \frac{0.25 \times 1 \times 0.2 \times 700^2}{2 \times 12\,000 \times 9.8},$$

which gives us the ratio $D/mg = 0.02$. Which implies that the drag force is about 2% of the gravitational force and hence can be ignored for most purposes.

Example 9.4. Conveyor belt problem

Suppose sand from a stationary hopper falls at the rate of 5 kg s^{-1} onto a conveyor belt moving at a constant speed of 0.5 m s^{-1} under the action of a constant horizontal force F_{ext}, as shown in figure 9.2. Find:

(a) The rate of change of momentum of sand in the horizontal direction.
(b) The force of friction exerted by the belt on the sand.
(c) The work done by F_{ext} in 2 s.
(d) The kinetic energy gained by the falling sand each second due its horizontal motion.

[3] Scale height, $H \equiv \frac{k_B T}{mg}$ where k_B is the Boltzmann constant and T is the temperature.

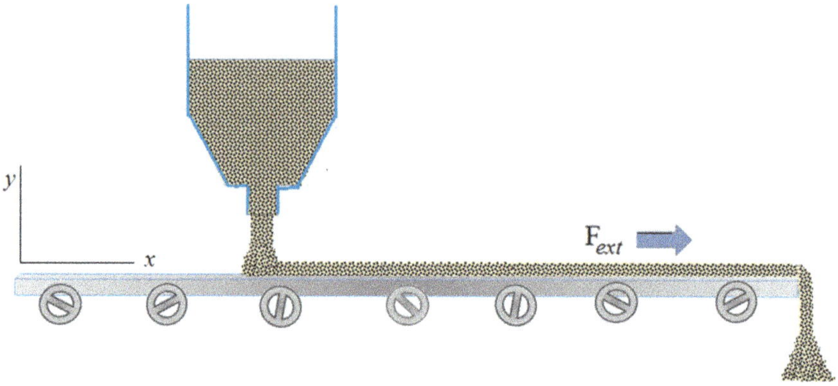

Figure 9.2. Conveyor belt.

Solution

(a) The mass of sand falling on the belt in time (dt) is dm. Sand falls vertically on the belt which means that along the horizontal direction the initial momentum of falling sand is zero. Once on the belt we can consider the motion along the x-direction only. Then the change in total momentum of the sand is

$$dp_x = v_x dm + m dv_x,$$

where v_x is the speed of the conveyor belt and since v_x is constant, we have $dv_x = 0$ and we obtain

$$dp_x = v_x dm.$$

Hence the rate of change in the total momentum of the sand is

$$\frac{dp_x}{dt} = v_x \frac{dm}{dt} = (0.5\ \text{ms}^{-1})(5\ \text{kgs}^{-1}) = 2.5\ \text{N}.$$

(b) The net force is equal to the frictional force. From Newton's force law

$$\Sigma F = F_f = 2.5\ \text{N}.$$

(c) The work can be found by knowing the force exerted on the belt:

$$\Delta W_{\text{ext}} = \int_a^b F_{\text{ext}} \cdot dx = F_{\text{ext}} \cdot v_x \Delta t = (5)(0.5)(2) = 5\ \text{J}.$$

(d) The rate of change in the total kinetic energy of the sand is

$$\frac{dK_{\text{tot}}}{dt} = \frac{(dm \cdot v_x^2 - 0)/2}{dt} = \frac{1}{2}\left(\frac{dm}{dt}\right) \cdot v_x^2 = \frac{1}{2}(5\ \text{kgs}^{-1})0.5^2 = 0.625\ \text{Js}^{-1}.$$

Hence total kinetic energy is

$$\Delta K_{\text{tot}} = \frac{dK_{\text{tot}}}{dt}\Delta t = 1.25\ \text{J}.$$

Concept questions 9.2

1. Can you think of any other variable-mass systems than the ones mentioned in section 9.6?
2. What is the concept behind the balloon-powered car? Think about the similarities and the differences between the operations of a balloon-powered car and a propulsion rocket?
3. Can a rocket reach a speed greater than the speed of the exhaust gases that propel it? Explain your answer.

Exercises 9.2

1. To keep a conveyor belt moving when it transports luggage requires a greater driving force than for an empty belt. What additional driving force is needed if the belt moves at a constant speed of $1\ \text{m s}^{-1}$ and the rate at which luggage is placed on one end of the belt and removed at the other end is $30\ \text{kg s}^{-1}$? Assume that the luggage is dropped vertically onto the belt; persons removing luggage grab hold of it and bring it to rest relative to themselves before lifting it off the belt.
2. A 6000 kg rocket is set for vertical firing. The exhaust speed is $2\ \text{km s}^{-1}$. How much gas must be ejected each second to supply the thrust needed:
   (a) To overcome the weight of the rocket?
   (b) To give the rocket an initial upward acceleration of $18.3\ \text{m s}^{-2}$?
   Note that in this case gravity is present as an external force.

Further reading

Arya A 1990 *Introduction to Classical Mechanics* 2nd edn (Upper Saddle River, NJ: Prentice-Hall)

Baierlein R 1983 *Newtonian Dynamics* (New York: McGraw-Hill)

Chababy R and Sherwood B 2011 *Matter and Interactions* 3rd edn (New York: Wiley)

Fowles G and Cassiday G 2005 *Analytical Mechanics* 7th edn (Belmont, CA: Thomson Learning)

Goldstein H 1980 *Classical Mechanics* 2nd edn (Reading, MA: Addison-Wesley)

Landau L D and Lifshitz E M 1960 *Mechanics* (Reading, Ms: Addison-Wesley)

Ohanian H 1985 *Physics* (New York: WW Norton and Company)
Resnick R, Halliday D and Krane K 2002 *Physics* vol 1, 5th edn (New York: Wiley)
Rutherford D E 1951 *Classical Mechanics* (New York: Interscience)
Stephenson R 1961 *Mechanics and Properties of Matter* 2nd edn (New York: Wiley)
Symon K 1971 *Mechanics* 3rd edn (Reading, MA: Addison-Wesley)

Chapter 10

Circular and projectile motion

10.1 Motion in higher dimensions

In this chapter we will look at the case of motion in which the particle moves at a *constant speed* in a circular path. This motion is called *uniform circular motion*. This kind of motion is set up when we swing a ball attached to a string over our heads and is similar to the planetary motion we saw in chapter 8. In order to look at this motion let us consider the following examples.

Example 10.1

What kind of motion does the position vector of a particle given by

$$\vec{r} = A\sin(\omega t)\hat{\imath} + A\cos(\omega t)\hat{\imath} \tag{10.1}$$

represent in 2D?

Solution

1. Let us find the magnitude of the position vector

$$|\vec{r}| = \sqrt{A^2\sin^2(\omega t) + A^2\cos^2(\omega t)} = \sqrt{A^2} = A.$$

   This implies that the magnitude of $\vec{r}$ remains constant, which is only true for a circular path. This implies this is a circular path.
2. The velocity is

$$\dot{r} = \vec{v} = \omega A\cos^2(\omega t)\hat{\imath} - \omega A\sin^2(\omega t)\hat{r}.$$

3. The speed is

$$|\dot{r}| = \sqrt{A^2\omega^2\cos^2(\omega t) + A^2\omega^2\sin^2(\omega t)} = \sqrt{A^2\omega^2} = A\omega.$$

doi:10.1088/2053-2563/aae1b9ch10

4. The acceleration is

$$\ddot{r} = \vec{a} = -\omega^2 A \sin(\omega t)\hat{i} - \omega^2 A \cos^2(\omega t)\hat{j} \,. \qquad (10.2)$$

Here v is tangent to the circular path and a always points towards the center of the circle.

Comparing equations (10.1) and (10.2), we obtain

$$\vec{a} = -\omega \vec{r}, \qquad (10.3)$$

hence $\vec{a}$ and $\vec{r}$ are opposite in direction.

10.2 Uniform circular motion

Uniform circular motion describes a motion of an object moving in a circular path at a constant speed, where the distance of the body from the center of the circle remains constant at all times, as we saw in example 10.1.

10.2.1 Centripetal acceleration

Since the body is constantly changing direction as it travels around the circle, the velocity of the object, being a vector, changes at all times as well. This varying velocity results in an acceleration called the *centripetal acceleration.* Centripetal acceleration has a constant magnitude and is always directed towards the center of the circle at all times and is perpendicular to the velocity at any given time.

The general result for any particle of mass (m) that moves in a circle at a constant speed (v) is that the magnitude of the centripetal acceleration ($|a_c|$) is

$$|a_c| = \frac{v^2}{r}. \qquad (10.4)$$

Centripetal acceleration always points towards the center of the circle. The units of centripetal acceleration are the same as those of linear acceleration, which we can see dimensionally as

$$|a_c| = \frac{[v]^2}{[r]} = \frac{[L]^2[T]^{-2}}{[L]} = [L][T]^{-2}. \qquad (10.5)$$

10.2.2 Centripetal force

Centripetal acceleration is produced by a *centripetal force*, which is a force that keeps the object moving in a circular path. Centripetal force has a constant magnitude and is always directed towards the center. Figure 10.1 illustrates the velocity and acceleration vectors for motion at four points in the orbit. The velocity (v) is tangent to the circular path, hence even when the magnitudes of the velocities are identical, no two velocities point in the same direction.

The magnitude of the centripetal force ($|\vec{F}_c|$) on an object of mass (m) moving at tangential speed (v) along a path with radius of curvature (r) is

$$\Sigma F = |F_c| = m\vec{a}_c = m\frac{v^2}{r}. \tag{10.6}$$

Keep in mind that centripetal force is not a new force. When the particle moves in a circle there may be many forces that act on it. The resultant sum of all those forces *must* point towards the center of the circle and we call that *centripetal force.*

10.3 Rotational motion

Displacement around a circular path is often given in terms of an angle (ϕ). This angle is the angle between a straight line drawn from the center of the circle to the object's starting position on the circle, and a straight line drawn from the object's ending position on the circle to center of the circle. See figure 10.2 for a visual

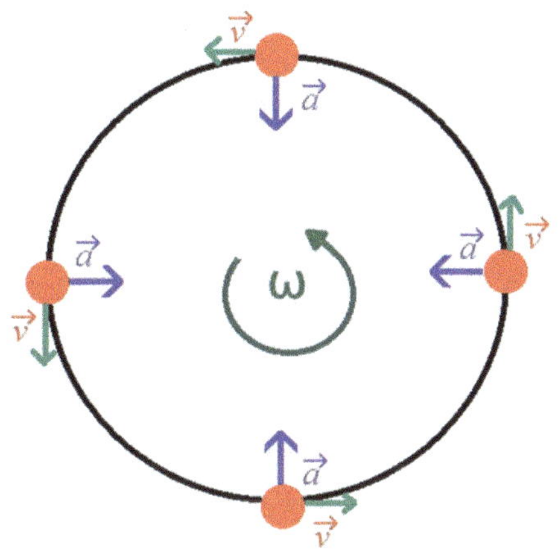

Figure 10.1. Top view of a ball traveling at constant speed around a circle.

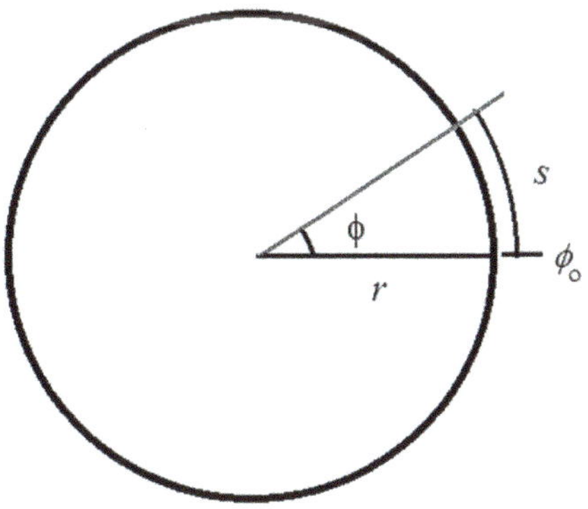

Figure 10.2. Top view—circle arc length $s = r\phi$.

representation of the angle where the point (P) started on the x-axis and moved to its present position. The angle (ϕ) describes how far it moved.

Consider a point (P) moving in a circular motion. Let r be the radius (the distance from pivot to point P), then as point P moves through an arc of length (s), as shown in figure 10.2, the *arc length* and the radius determine the angle through which the reference line rotates as

$$s = r\phi. \tag{10.7}$$

Since the circumference (C) of a circle of radius (r) is mathematically given as ($C = 2\pi r$), we can take angular measures as

$$1 \text{ revolution} = 360° = 2\pi \text{ radians}$$
$$1 \text{ radian} = 57.3° = 0.159 \text{ revolutions}.$$

Before we begin a formal discussion let us define a few important concepts that we will use:

- *Angular displacement.*
  The angular displacement of point P is ($\Delta\phi = \phi_2 - \phi_1$) during time interval ($\Delta t = t_2 - t_1$).
- *Angular velocity.*
  Average angular velocity (ω_{av}) of a particle is defined as

$$\omega_{av} = \frac{\Delta\phi}{\Delta t}. \tag{10.8}$$

  The SI units for angular velocity are radians per second (rad s^{-1}) and it has dimensions of $[T]^{-1}$. Angular velocity is also expressed in revolutions per second. Angular velocity is positive when ϕ is increasing and is negative when ϕ is decreasing.
- *Instantaneous angular velocity.*
  Instantaneous angular velocity is defined when the time interval is very small:

$$\omega = \frac{d\phi}{dt}. \tag{10.9}$$

- *Angular acceleration.*
  Average angular acceleration (α_{av}) of a particle is defined as

$$\alpha_{av} = \frac{\Delta\omega}{\Delta t}. \tag{10.10}$$

  The SI units for angular acceleration are radians per second squared (rad s^{-2}) and it has dimensions of $[T]^{-2}$. Angular acceleration is positive when ω increases and will be negative when ω decreases.

- *Instantaneous angular acceleration.*

  Instantaneous angular acceleration is defined when the time interval is very small:

$$\alpha = \frac{d\omega}{dt}. \tag{10.11}$$

10.3.1 Rotation with constant acceleration

To derive the equations of motions for rotating objects with constant acceleration, we need to specify initial conditions of the motion. At $t = 0$, let the initial position be given as ($\phi = \phi_o$) and initial velocity be given as ($\omega = \omega_o$). Also assume that the acceleration around the z-axis is a_z, a constant. Then we can derive equation of motion for circular paths as follows.

First equation of motion:
We will start from the definition of acceleration and solve as follows:

$$\frac{d\omega}{dt} = \alpha$$

$$d\omega = \alpha dt$$

$$\int_{\omega_o}^{\omega} d\omega = \int_0^t \alpha dt$$

$$\boxed{\omega - \omega_o = \alpha t.} \tag{10.12}$$

Second equation of motion:
Starting from the first equation of motion (10.12):

$$\omega - \omega_o = \alpha t$$

$$\frac{d\phi}{dt} = \omega_o + \alpha t$$

$$\int_{\phi_o}^{\phi} d\phi = \int_0^t (\omega_o + \alpha t)dt$$

$$\phi - \phi_o = \omega_o t + \frac{1}{2}\alpha t^2$$

$$\boxed{\Delta\phi = \omega_o t + \frac{1}{2}\alpha t^2.} \tag{10.13}$$

Third equation of motion:

$$\alpha = \frac{d\omega}{dt}$$
$$\alpha = \frac{d\omega}{dt}\frac{d\phi}{d\phi}$$
$$\alpha = \omega\frac{d\omega}{d\phi}$$
$$\alpha d\phi = \omega d\omega$$
$$\int_{\phi_o}^{\phi} \alpha d\phi = \int_{\omega_o}^{\omega} \omega d\omega$$
$$\alpha(\phi - \phi_o) = \frac{\omega^2 - \omega_o^2}{2}$$

$$\boxed{2\alpha\Delta\phi = \omega^2 - \omega_o^2.} \tag{10.14}$$

Note the similarities between these equations and the ones we derived for linear motion in chapter 3, section 3.2.

10.3.2 Relations between linear and angular quantities

To find the relationship between linear and angular quantities we start from equation (10.7) and differentiate it w.r.t. time:

$$\begin{aligned} s &= r\phi \\ \frac{ds}{dt} &= r\frac{d\phi}{dt} \\ v_T &= r\omega. \end{aligned} \tag{10.15}$$

Here, v_T is the linear (tangential) speed at point P and ω is the corresponding angular speed.

Differentiating equation (10.15) w.r.t. time we obtain

$$\frac{dv_T}{dt} = r\frac{d\omega}{dt}$$
$$a_T = r\alpha.$$

Here, a_T is the magnitude of the tangential component of linear acceleration at point P and α is the corresponding angular acceleration.

We also know that the radial (or centripetal) component (a_R) of acceleration is

$$a_R = \frac{v_T^2}{r} = r\omega^2. \tag{10.17}$$

Example 10.2

A wheel starts from rest and spins counterclockwise with a constant angular acceleration of 3 rad s^{-2}. Calculate the angular displacement and the angular speed of the wheel after 5 s.

Solution

Since the wheel starts from rest we can take, at $t = 0$, the initial displacement ϕ_o and initial angular speed ω_o to be zero. Then we can find the angular displacement using equation (10.13) as

$$\begin{aligned} \phi &= \phi_o + \omega_o t + \frac{1}{2}\alpha t^2 \\ &= 0 + 0 + \frac{1}{2}(3\text{rads}^{-2})(5^2) \\ \Rightarrow \phi &= 37.5 \text{ rad} \end{aligned}$$

and the angular speed of the wheel by using equation (10.14):

$$\begin{aligned} \omega^2 - \omega_o^2 &= 2\alpha\Delta\phi \\ \omega^2 - 0 &= 2(3 \text{ rads}^{-2})(37.5\text{rad}) \\ \Rightarrow \omega &= 15 \text{ rads}^{-1}. \end{aligned}$$

Example 10.3

If the radius of the wheel in equation (10.2) is 0.5 m, at time 5 s, calculate the following:

(a) The tangential speed of a point (P) on the rim.

Answer:

$$v_T = r\omega = (0.5 \text{ m})(15 \text{ rads}^{-1}) = 7.5 \text{ ms}^{-1}.$$

(b) The tangential acceleration of point (P):

Answer:

$$a_T = r\alpha = (0.5 \text{ m})(3 \text{ rads}^{-2}) = 1.5 \text{ ms}^{-2}.$$

(c) Radial acceleration of point (P):

Answer:

$$a_r = r\omega^2 = (0.5 \text{ m})(15 \text{ rads}^{-2}) = 112.5 \text{ ms}^{-2}.$$

Example 10.4

A car enters a curve of radius of 30 m. If your car can tolerate a maximum acceleration of 1g without skidding, what is the maximum possible speed you can drive at?

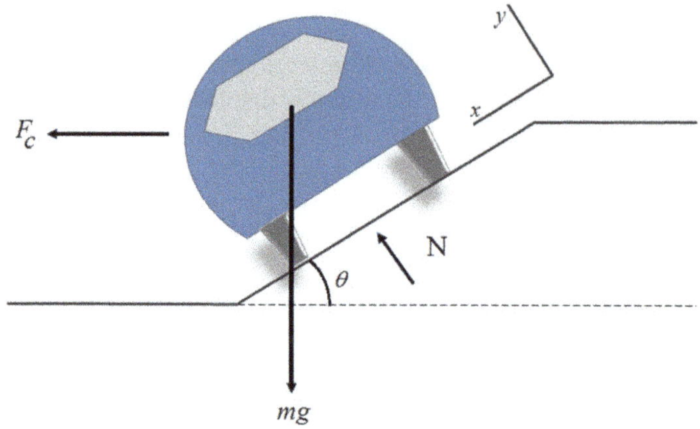

Figure 10.3. Banked curved.

Solution

$$a_c = \frac{v^2}{a}$$
$$v_{\max} = \sqrt{ra}$$
$$v_{\max} = \sqrt{30g} = 17\ \text{ms}^{-1}.$$

This means that the car could safely drive at 17 m s^{-1} or 38 miles/h around the bend.

10.3.3 Banked curves

While driving you will inevitably encounter a curved section of a road. On a completely flat curve with no banking, the car will slide off the road if you go fast. On the other hand, if the banking is too steep, the car will slide towards the center unless you drive very fast around the curve, which is dangerous.

10.3.3.1. Maximum speed and the banking angle

The horizontal components of the friction and normal force provide the centripetal acceleration that keeps the car moving in a circle. Writing Newton's force equations at maximum speed (v) on the incline, referring to figure 10.3, we obtain

$$\begin{cases} \Sigma F_x &= N\sin\theta + \mu N\cos\theta \\ \Sigma F_y &= N\cos\theta - \mu N\sin\theta - mg \end{cases} \tag{10.18}$$

where N is the normal contact force[1] and μ is the coefficient of friction. Which can be rewritten as

$$\begin{cases} m\dfrac{v^2}{r} &= N\sin\theta + \mu N\cos\theta \\ 0 &= N\cos\theta - \mu N\sin\theta - mg. \end{cases} \tag{10.19}$$

[1] Normal force is the contact force between the surface and the object. Normal force acts perpendicular to the surface.

Solving the above equations for maximum velocity (v) we obtain

$$v_{\max} = \sqrt{\frac{rg(\sin\theta + \mu\cos\theta)}{\cos\theta - \mu\sin\theta}}. \tag{10.20}$$

There are two limiting cases to consider when we talk about banking angles:

- *The frictionless case.* When $\mu = 0$, the equation (10.20) gives: $v_{\max} = \sqrt{rg\tan\theta}$.
- *A flat road.* When $\theta = 0$; the equation (10.20) gives: $v_{\max} = \sqrt{rg\mu}$.

Example 10.5

Engineers bank curves on roads such that a vehicle going around the curve does not have to rely on friction between its tires and the road surface in order to round the curve. Suppose that the radius of curvature of a given curve is $r = 60$ m, and we wish to make the recommended speed $= v = 40$ km h^{-1}. At what angle (θ) should the curve be banked?

Solution

Using the frictionless case equation for $\tan\theta$,

$$\tan\theta = \frac{v_{\max}^2}{rg},$$

will give banking angle (θ) = 11.8°.

10.4 Rectilinear motion and rotation about a fixed axis

We summarize the various quantities in rectilinear motion and rotational motion about a fixed axis in the table below.

	Rectilinear motion	Rotational motion
Mass	mass m	moment of inertia (I)
Position	$\vec{r} = (x, y, z)$	ϕ
Velocity	$\vec{v} = \dot{r}$	$\omega = \dot{\phi}$
Acceleration	$\vec{a} = \ddot{r}$	$\alpha = \ddot{\phi}$
Newton's second law	net force $\Sigma\vec{F} = m\vec{a} = m\ddot{r}$	net torque $\Sigma\tau = I\alpha$
Momentum	linear momentum $\vec{p} = m\vec{v}$	angular momentum $L = I\omega = I_z\dot{\phi}$
Potential energy	$V(x) = -\int F(x)dx$	$V(\theta) = -\int \tau d\phi$
Kinetic energy	$T = \frac{1}{2}m\dot{x}^2$	$T = \frac{1}{2}I_z\,\dot{\phi}^2$
Work	$W = \int F \cot dx$	$W = \int \tau \cdot d\phi$
Power	$P = F \cdot v$	$P = \tau\omega$
Equations of motion	$v = v_o + at$	$\omega = \omega_o + \alpha t$
	$\Delta x = v_o t + \frac{1}{2}at^2$	$\Delta\omega = \omega_o t + \frac{1}{2}\alpha t^2$
	$v^2 = v_o^2 + 2a\Delta x$	$\omega^2 = \omega_o^2 + 2\alpha\Delta\omega$

Concept questions 10.1

1. When you stir a cup of tea, the floating tea leaves collect at the center of the cup rather than on the outer rim. Why do you think this happens?
2. While flying a plane at a constant altitude a pilot wishes to make a 90° turn. Why does he bank the plane in order to do so?
3. If we take the average speed of the Earth around the Sun to be 3×10^4 ms^{-1}, what is the force required to maintain the motion of Earth around the Sun?

Exercises 10.1

1. What is the centripetal acceleration of a point on the rim of a tire, with a diameter of 25 inches, when the car is doing 60 miles per hour?
2. Assume that the road that we are traveling on is completely flat and the coefficient of friction (μ) is 0.8. What is the maximum speed with which a car can round a curve of radius 100 m?
3. A particular speedway with a curve of radius 500 m is banked at 22°. If the driver does not want to rely on friction, at what speed in miles per hour should he take this curve to be safe?

10.5 Harmonic oscillator in higher dimensions

Consider the motion of a particle under a linear restoring force, directed towards the equilibrium point. We can put our origin at the equilibrium point. Then Hooke's law says

$$\vec{F} = -k\vec{r}, \tag{10.21}$$

which results in Newton's law expressed as

$$m\ddot{\vec{r}} = -k\vec{r}. \tag{10.22}$$

This is the three-dimensional (3D) generalization of the linear oscillator defined by a differential equation, called the *linear isotropic oscillator*. This can be represented by a particle attached to a set of elastic springs as shown in figure 10.4.

10.5.1 SHM as the projection of a rotating vector

A point (P) moving on a circular path with a constant angular velocity undergoes uniform circular motion. The projection of this movement on the x-axis results in simple harmonic motion. Figure 10.5. shows that the velocity of this point around the circle is given as $v_{\max}$, and the velocity of its projection is v. Note that these velocities form a similar triangle to the displacement triangle.

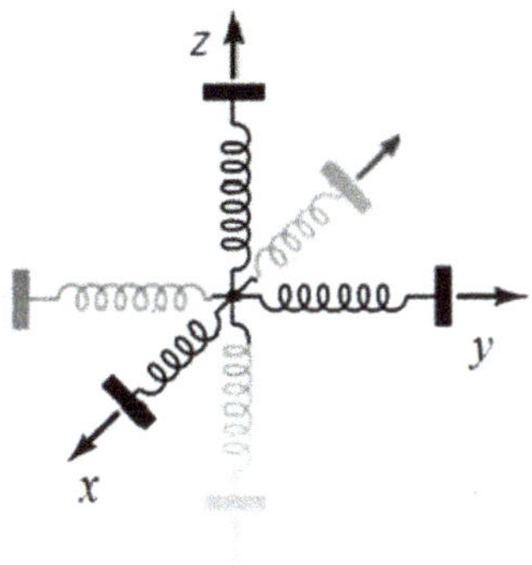

Figure 10.4. 3D isotropic oscillator.

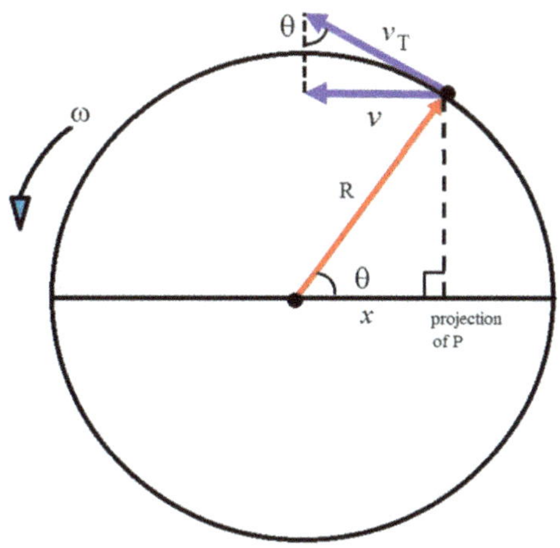

Figure 10.5. Projection of a circular path.

We can now define some quantities that will be useful in our discussion:

- *Speed.*

  The speed of a point around the orbit is

$$v = r\frac{d\theta}{dt} = r\omega, \tag{10.23}$$

  where the angular rate of rotation is ω. Thus v is a constant, and the velocity vector $\vec{v}$ also rotates with constant magnitude v, at the same angular rate ω.

- *Acceleration.*

  The acceleration in uniform circular motion is always directed inward and is given by

$$a = v\frac{d\theta}{dt} = v\omega = \frac{v^2}{r}. \tag{10.24}$$

  This acceleration acts to change the direction of v, but not the magnitude of velocity, the speed.

10.5.2 2D isotropic oscillator

For motion in a single plane, the above equation (10.22) is equivalent to the following equations:

$$m\ddot{x} = -kx \xrightarrow{\text{with sol.}} x = A\cos(\omega t + \alpha) \tag{10.25}$$

$$m\ddot{y} = -ky \xrightarrow{\text{with sol.}} y = B\cos(\omega t + \beta), \tag{10.26}$$

where $\omega = \sqrt{\frac{k}{m}}$ and constants of integration A, B and α, β are determined from initial conditions in any given case.

10.5.2.1 Equation of path

In order to find the equation of the path we must eliminate t between the two equations. To do that we will write $y = B\cos(\omega t + \beta)$ as

$$\begin{aligned} y &= B\cos(\omega t + \beta + \alpha - \alpha) \\ y &= B\cos(\omega t + \alpha + \Delta), \end{aligned} \tag{10.27}$$

where $\Delta = \beta - \alpha$. Using the trigonometic identity for the cosine we can write

$$\frac{y}{B} = \cos(\omega t + \alpha)\cos\Delta - \sin(\omega t + \alpha)\sin\Delta.$$

From equation (10.25) use $\cos(\omega t + \alpha) = \frac{x}{A}$ and $\cos^2(\omega t + \alpha) + \sin^2(\omega t + \alpha) = 1$ to obtain $\sin(\omega t + \alpha) = \sqrt{1 - (\frac{x}{A})^2}$

$$\frac{y}{B} = \frac{x}{A}\cos\Delta - \sin(\omega t + \alpha)\sin\Delta$$

$$\frac{y}{B} = \frac{x}{A}\cos\Delta - \sqrt{1 - \frac{x^2}{A^2}}\sin\Delta$$

$$\frac{y}{B} - \frac{x}{A}\cos\Delta = -\sqrt{1 - \frac{x^2}{A^2}}\sin\Delta.$$

Squaring and simplifying, we obtain

$$\frac{y^2}{B^2} + \frac{x^2}{A^2} - 2\frac{xy}{AB}\cos\Delta = \sin^2\Delta. \tag{10.28}$$

This equation (10.28) is a quadratic equation in x and y. We know that the general quadratic equation of the form $ax^2 + bxy + cy^2 + dx + ey = f$ will represent an ellipse, a parabola or a hyperbola, when its discriminant $(b^2 - 4ac)$ is negative, zero, or positive, respectively.

In this case the constants are $a = \frac{1}{A^2}$, $b = \frac{2}{AB}\cos\Delta$, $c = \frac{1}{B^2}$, etc. So we see that the discriminant of the quadratic equation is given as $(b^2 - 4ac) = -(\frac{2\sin\Delta}{AB})^2$; which is negative and hence we obtain an *elliptical path*, as seen in figure 10.6.

10.5.3 Particular cases

Case 1. If the phase difference is $\Delta = \frac{\pi}{2}$, then $\sin\frac{\pi}{2} = 1$ and $\cos\frac{\pi}{2} = 0$, which implies that the equation of path (10.28) reduces to

$$\frac{y^2}{B^2} + \frac{x^2}{A^2} = 1 \tag{10.29}$$

Equation (10.29) is that of an ellipse whose axis coincides with the coordinate axis.

Case 2. When the phase difference is $\Delta = 0$, then $\cos 0 = 1$ and $\sin 0 = 0$, which implies the equation of path (10.28) becomes

$$y = \frac{B}{A}x, \tag{10.30}$$

and when the phase difference is $\Delta = \pi$, then $\cos\pi = -1$ and $\sin 0 = 0$, which implies the equation of path (10.28) becomes

$$y = -\frac{B}{A}x. \tag{10.31}$$

Recall that $y = mx + c$ is the equation of a straight line. Then these equations are equations of a straight line motion.

Case 3. In the general case it is possible to show that the axis of the elliptical path is inclined to the x-axis by angle ψ, where

$$\tan 2\psi = \frac{2AB\cos\Delta}{A^2 - B^2}. \tag{10.32}$$

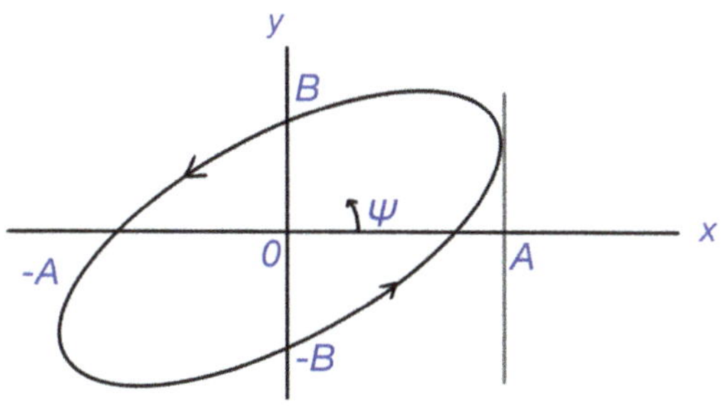

Figure 10.6. Elliptical path of a 2D isotropic oscillator.

10.5.4 Energy

We know $V(x) = \frac{1}{2}kx^2$ for the 1D harmonic oscillator. For a 2D harmonic oscillator, with springs of spring constants k_1 and k_2, the potential energy will be given as

$$V(x, y) = \frac{1}{2}k_1x^2 + \frac{1}{2}k_2y^2.$$

This value of $V(x, y)$ gives the correct result:

$$F_x = \frac{\partial V}{\partial x} = -k_1x - k_2y.$$

For the isotropic case we know that spring constants $k_1 = k_2 = k$, and we can write the potential energy as

$$V(x, y) = \frac{1}{2}k(x^2 + y^2)$$
$$V(x, y) = \frac{1}{2}k(r^2).$$

The total energy can therefore be expressed as

$$\boxed{E_T = T + V = \frac{1}{2}mv^2 + \frac{1}{2}kr^2.} \tag{10.33}$$

Exactly the same as for the 1D case!

Note that this may also be extended to the 3D case without major complications by the addition of the z-coordinate. It is left as an exercise for the reader to attempt.

Concept questions 10.2

1. What changes would you need to make in a harmonic oscillator in order to double the maximum speed of the oscillating object?
2. Answer the following questions for a hanging mass–spring system undergoing simple harmonic motion:
   (a) Where is the velocity of the mass maximum? Zero?
   (b) Where is the spring potential energy of the max? Zero?
   (c) Where is the amplitude of the mass maximum? Zero?
   (d) Where is the net force on the mass zero?
   (e) Where is the gravitational potential energy of the mass zero? Maximum?

Exercises 10.2

1. A massless spring with spring constant (k) is attached at one end of a block of mass M resting on a frictionless horizontal table. The other end of the spring is fixed to a wall. A bullet of mass m_b is fired into the block from the left with a speed v_0 and comes instantaneously to rest in the block.
   (a) How fast is the block moving immediately after the bullet comes to rest?
   (b) What is the amplitude of the resulting simple harmonic motion?
   (c) How long does it take the block to first return to the initial, equilibrium position x_0?
2. A log boat made of a single layer of wood logs may be approximated as a rectangular block of mass M, cross-sectional area A, height H and density ρ_1. It floats partially submerged in water with density ρ_2. Assume x is the depth of the boat submerged in water. If a boy of mass m jumps on it, the boat will oscillate. *Hint*: density is defined in section 11.3 as ρ = mass/Volume (V), where V is the volume, and for a boat assume $V = AH$.
   (a) Is the resulting motion of the boat simple harmonic?
   (b) Find the period of its motion.

10.6 Motion of a projectile in a uniform gravitational field

Projectile motion is a special type of motion in which the object (called a projectile) is thrown on Earth and it moves under the effect of gravity alone. In this case there is no engine or self-propulsion on the projectile.

In projectile motion, the horizontal and the vertical motion are independent of each other. This is called the *principle of compound motion* and was used by Galileo in 1638 CE.

Example 10.6

Suppose you have a motion represented by the equation $\vec{r}(t) = at\hat{\imath} + (bt - \frac{1}{2}gt^2)\hat{k}$, where a, b are constants and g represents the acceleration due to gravity. What is the equation of the path of motion?

Solution

The position vector of a particle as a function of time is given by the equation

$$\vec{r}(t) = at\hat{\imath} + (bt - \frac{1}{2}gt^2)\hat{k}.$$

Its velocity as a function of time will be

$$\vec{v}(t) = \dot{r} = a\hat{\imath} + (b - gt)\hat{k}.$$

Its acceleration as a function of time will be

$$\vec{a}(t) = \ddot{r} = -g\hat{k}.$$

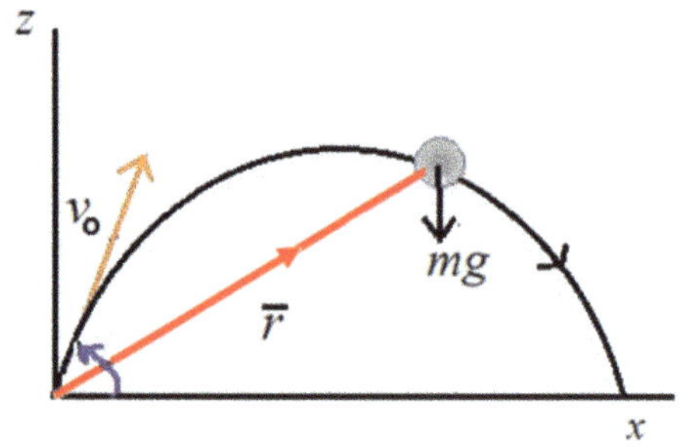

Figure 10.7. Position of projectile in the *xz*-plane.

The acceleration is in the negative z-direction with magnitude g, which is the acceleration due to gravity. The path of motion is a parabola, similar to the path seen in figure 10.7.

10.7 Projectile motion: no air resistance

For simplicity, we first consider the case of projectile motion without air resistance. The only significant force acting on the projectile is gravity, which acts down towards the center of the Earth and causes a downward acceleration. The inertia of the projectile ensures that no external horizontal force is needed to maintain the horizontal velocity of the object.

The simplest case is for projectiles that do not rise too high or travel too far. Here we can take the acceleration of gravity, g, to be constant and choose the force function to be separable and conservative.

10.7.1 Projectile motion: no air resistance—Newtonian method

For our case, let us choose the vertical direction to be the z-axis and set the coordinate system as shown in figure 10.7, where the origin is taken at the initial position of the projectile. Also let v_0 be the initial speed of the projectile. Then the components of the initial velocity will be given as

$$\begin{cases} v_{ox} &= v_o \cos\theta \\ v_{oz} &= v_o \sin\theta \end{cases} \tag{10.34}$$

The position vector of the projectile at any time will be given as

$$\vec{r} = x\hat{\imath} + z\hat{k} = (v_o \cos\theta)\hat{\imath} + (v_o \sin\theta)\hat{k}. \tag{10.35}$$

Then the equation of motion using the Newtonian method can be written as

$$\begin{cases} \Sigma F_x &= 0 \\ \Sigma F_z &= -mg \end{cases} \tag{10.36}$$

Generalizing to 3D, the motion of the projectile will be given by

$$m\ddot{r} = -mg\hat{k}. \tag{10.37}$$

10.7.2 Projectile motion: no air resistance—Lagrangian method

Let the motion again be defined in the xz-plane. The means that the generalized coordinates are (x, z). Hence we can write the Lagrangian as

$$L = \frac{1}{2}m(\dot{x}^2 + \dot{z}^2) - mgz. \tag{10.38}$$

The equations of motion along the x- and z-axes will be given as

$$\frac{d}{dt}\left(\frac{\partial L}{\partial \dot{x}}\right) - \frac{\partial L}{\partial x} = 0 \tag{10.39a}$$

$$\frac{d}{dt}\left(\frac{\partial L}{\partial \dot{z}}\right) - \frac{\partial L}{\partial z} = 0. \tag{10.39b}$$

Finding the partials using equation (10.38) and putting in equations (10.39*a*) and (10.39*b*), we obtain

$$\frac{d}{dt}(m\dot{x}) = 0 \tag{10.40a}$$

$$\frac{d}{dt}(m\dot{z}) + mg = 0. \tag{10.40b}$$

We see from equation (10.40*a*) that the horizontal momentum is constant and the combination of equations (10.40*a*) and (10.40*b*) gives

$$m\ddot{r} = -mg\hat{k}, \tag{10.41}$$

which is identical to equation (10.37).

10.7.3 Projectile motion: no air resistance—Hamiltonian method

We have derived the Lagrangian for a projectile in section 10.7.2 as

$$L = \frac{1}{2}m(\dot{x}^2 + \dot{z}^2) - mgz. \tag{10.38}$$

The Hamiltonian function for 3D motion of a projectile in a uniform gravitational field with no air resistance will be

$$H = \frac{1}{2}m(\dot{x}^2 + \dot{z}^2) + mgz \tag{10.43}$$

and the Hamiltonian canonical equations will be as follows.

For the x-axis:

$$\frac{\partial H}{\partial p_x} = \dot{x}$$
$$\frac{\partial H}{\partial x} = 0$$

and

$$\begin{aligned} p_x &= \frac{\partial H}{\partial \dot{x}} \\ &= m\dot{x}. \end{aligned}$$

Differentiating, we obtain

$$\begin{aligned} \dot{p}_x &= m\ddot{x} = 0 \\ \Rightarrow p_x &= \text{constant}. \end{aligned} \tag{10.44}$$

For the z-axis:

$$\frac{\partial H}{\partial p_z} = \dot{z}$$
$$\frac{\partial H}{\partial z} = mg$$

and

$$\begin{aligned} p_z &= \frac{\partial H}{\partial \dot{z}} \\ &= m\dot{z}. \end{aligned}$$

Differentiating, we obtain

$$\dot{p}_z = m\ddot{z} = -mg. \tag{10.45}$$

We can combine equations (10.44) and (10.45) to obtain

$$m\ddot{r} = -mg\hat{k}, \tag{10.41}$$

which is identical to equations (10.37) and (10.41)—just as we expect!

10.7.4 Important properties of projectile motion

Some important properties of projectile motion are as follows.

10.7.4.1 Acceleration of the projectile

The velocity in the horizontal direction ($v_0 \cos\theta$) is constant, hence $a_x = 0$, whereas the velocity in the vertical direction ($v_0 \sin\theta$) increases as the object falls, hence $a_y = g$. The components of the acceleration are

$$a_x = 0; \quad a_y = g. \tag{10.47}$$

10.7.4.2 Velocity of the projectile

The horizontal component of the velocity of the object remains unchanged throughout the motion. The downward vertical component of the velocity increases linearly, because the acceleration due to gravity is constant. Then using the definition of velocity and by integrating equation (10.37) we obtain velocity in 3D as

$$\dot{r} = -\hat{k}gt + v_0, \tag{10.48}$$

which can be expanded in terms of components as

$$\begin{aligned}\vec{v} &= v_{ox}\hat{\imath} + v_{oy}\hat{k} \\ &= (v_0 \cos\theta)\hat{\imath} + (v_0 \sin\theta - gt)\hat{k}.\end{aligned} \tag{10.49}$$

10.7.4.3 Displacement as a function of time

Integrating equation (10.48) we obtain displacement as a function of time in 3D as

$$\vec{r}(t) = \frac{1}{2}gt^2\hat{k} + v_0 t + r_0, \tag{10.50}$$

where r_0 is the constant of integration and represents the initial position of the object.

In terms of unit vectors equation (10.50) can be written as

$$\vec{r} = (v_0 \cos\theta)t\hat{\imath} + \left[(v_0 \sin\theta)t - \frac{1}{2}gt^2\right]\hat{k}. \tag{10.51}$$

Equation (10.51) is a vector equation in 3D. The position of the particle at any time, t can be re-expressed in components as

$$x = v_{ox}t = v_o\cos(\theta)t \tag{10.52a}$$

$$y = v_{oy}t = 0 \tag{10.52b}$$

$$z = v_{oz}t - \frac{1}{2}gt^2 = v_o\sin(\theta)t - \frac{1}{2}gt^2. \tag{10.52c}$$

So now we can show as Galileo did in 1609 CE (400 years ago) that the path of the projectile is a *parabola*. From equation (10.52*a*) we see that $t = \frac{x}{v_0 \cos\theta}$, and putting it into equation (10.52*c*) we obtain

$$z = (\tan\theta)x - \left(\frac{g}{2v_0^2 \cos^2\theta}\right)x^2, \tag{10.53}$$

which is clearly like ($z = \alpha x + \beta x^2$), the equation of a parabola, where α is given by ($\tan\theta$) and β is ($\frac{g}{2v_0^2 \cos^2\theta}$).

10.7.4.4 Maximum height (H) of the projectile

When an object is projected, its height increases but its vertical component of velocity (v_z) decreases until it reaches zero. The greatest height that the object will reach is called the *max height* or *peak of motion.*

We use the horizontal component of velocity $v_x = v_0 \cos\theta$ and from the third equation of motion $v_z^2 - v_0^2 = 2gz$ and, rearranging, $\frac{v_0^2}{2g} = z$ to obtain

$$z_{\max} = H = \frac{v_0^2 \sin^2\theta}{2g}. \tag{10.54}$$

10.7.4.5 Time to reach max height (H)

We calculated the velocity in equation (10.49). To find the time to reach max height, use the fact that at max H: $v_z = 0$. Then the z-component of equation (10.49) becomes

$$v_0 \sin\theta - gt_{\max} = 0,$$

and rearranging it we can obtain

$$t_{\max} = \frac{v_0 \sin\theta}{g}. \tag{10.55}$$

10.7.4.6 Total time of flight

The total time for which the projectile remains in the air is called the time of flight. Total time of flight: $T_{\text{total}} = T_{\text{up}} + T_{\text{down}}$. Since we are ignoring air resistance, the time taken by the projectile upwards is the same as time taken downwards; we say that time is symmetric:

$$T_{\text{total}} = \frac{2v_0 \sin\theta}{g}. \tag{10.56}$$

10.7.4.7 Horizontal range of the projectile

From equation (10.52*a*),

$$\begin{aligned} R &= v_{0x} T_{\text{total}} \\ &= (v_0 \cos\theta)\left(\frac{2v_0 \sin\theta}{g}\right), \end{aligned} \tag{10.57}$$

which using trigonometric identity $2\sin\theta\cos\theta = \sin(2\theta)$ can also be written as

$$R = \frac{v_0^2 \sin 2\theta}{g}. \tag{10.58}$$

From the above equation, we see that maximum range will occur when $R = \frac{v_c^2}{g}$, which will happen when $\sin 2\theta = 1$; this means R is maximum for $\theta = 45°$.

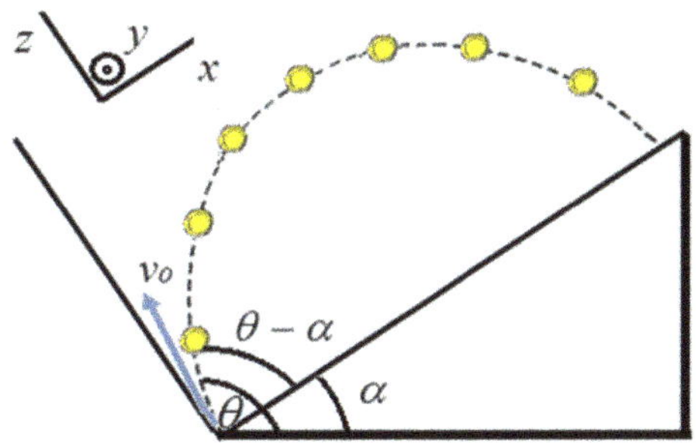

Figure 10.8. Projectile on a slope.

It is important to note that the range and the maximum height of the projectile do not depend upon its mass. Hence range and maximum height are equal for all bodies that are thrown with the same velocity and direction.

10.7.5 Projectile on an inclined plane

Sometimes you have to throw a projectile upwards or downwards while you are standing on a slope. What happens in this case? Do the equations we wrote above still hold or do we need to change our theory?

In order to address the new situation we could either analyze the system with our normal coordinate system with respect to the vertical and horizontal planes of the Earth, which I will call the *normal coordinate system*, or we could change the orientation to match the incline, which I will call the *incline coordinate system*. In this approach we will take the x-axis to be parallel to the incline and the z-axis to be normal to the surface, as seen in figure 10.8.

Taking the coordinate system to be along the slope makes the mathematics a bit easier, so here I will take that approach. However, in this approach the angle of projection needs to be handled very carefully. I will keep the angle of projection as θ and the angle of incline as α when measured w.r.t. the horizontal in the normal coordinate system. This means that the initial projectile angle w.r.t. the incline coordinate system will be $(\theta - \alpha)$ and hence the initial velocity components will be

$$\begin{cases} \Sigma v_x &= v_0 \cos(\theta - \alpha) \\ \Sigma v_z &= v_0 \sin(\theta - \alpha) \end{cases} \tag{10.59}$$

and the components of acceleration will be

$$\Sigma a_x = -\, g \sin \alpha \tag{10.60a}$$

$$\Sigma a_z = -\, g \cos \alpha . \tag{10.60b}$$

The negative signs mean the two components are in opposite directions to the positive directions of the coordinates.

We can also find the velocities by integrating equations (10.60) to obtain

$$\begin{cases} \Sigma v_x = v_{0x}\cos(\theta - \alpha) - g \sin \alpha t \\ \Sigma v_z = v_{0z}\sin(\theta - \alpha) - g \cos \alpha t. \end{cases} \tag{10.61}$$

10.7.5.1 Time of flight (T_{total})

The net displacement in the z-direction is not zero in the normal coordinate system; however, in the incline coordinate system net displacement in the z-direction is still zero. Then we can write

$$\begin{aligned} y &= v_z T_{\text{total}} - \frac{1}{2} a_z T_{\text{total}}^2 \\ 0 &= v_0 \sin(\theta - \alpha) T_{\text{total}} - \frac{1}{2}(-g \cos \alpha) T_{\text{total}}^2 \\ T_{\text{total}} &= \frac{2 v_0 \sin(\theta - \alpha)}{g \cos \alpha}. \end{aligned} \tag{10.62}$$

10.7.5.2 Range (R)

Range is the maximum distance traveled by the projectile. We can calculate the range of flight by considering motion in both the x- and z-directions. The important thing to understand here is that the motion along the x-direction (along the incline) is not constant, as is clear from equation (10.60*a*). Recall that in the normal coordinate system a_x was taken to be zero. Then in this case

$$\begin{aligned} x &= v_x T_{\text{total}} - \frac{1}{2} a_x T_{\text{total}}^2 \\ R &= (v_0 \cos(\theta - \alpha))\left(\frac{2 v_0 \sin(\theta - \alpha)}{g \cos \alpha}\right) - \frac{1}{2}(-g \sin \alpha)\left(\frac{2 v_0 \sin(\theta - \alpha)}{g \cos \alpha}\right)^2. \end{aligned}$$

Using the trigonometric identity, $2\sin^2(\theta - \alpha) = 1 - \cos^2(\theta - \alpha)$, we can simplify the above equation to

$$R = \frac{v_0^2}{g \cos^2 \alpha}(\sin 2(\theta - \alpha)\cos \alpha - \sin \alpha + \sin \alpha \cos 2(\theta - \alpha))$$

and then using another trigonometric identity, $\sin(A + B) = \sin A \cos B + \cos A \sin B$,

$$R = \frac{v_0^2}{g \cos^2 \alpha}(\sin(2\theta - \alpha) - \sin \alpha), \tag{10.63}$$

which gives us the expression for the range (R) of a projectile on an incline. Note that when $\alpha = 0$, this equation reduces to the normal range calculated in equation (10.58).

Example 10.7

A soccer player hits a ball up from the foot of an incline of angle 45° with a velocity 30 m s^{-1}. The angle of projection as measured from the horizontal is 60°.

(a) What is the speed of the ball when the projectile is parallel to the incline?

(b) How far away will the ball land?

Solution

(a) Using the incline coordinate system makes this question easier to solve. According to given values, $\theta = 60°$, $\alpha = 45°$, then $(\theta - \alpha) = 15°$. We know that

$$\text{speed} = |v_{\text{total}}| = \sqrt{v_x^2 + v_z^2}. \tag{10.64}$$

The component of velocity in the z-direction is zero. For motion in the z-direction, we have from equation (10.61)

$$0 = 30 \text{ m s}^{-1} \sin 15° - g \cos 45° t$$
$$\Rightarrow t = 1.1 \text{ s}.$$

The motion in the x-direction is

$$v_x = (30 \text{ ms}^{-1})\cos 15° - g(\sin 45°)(1.1 \text{ s}) = 21 \text{ m s}^{-1}$$
$$\text{speed of ball} = |v_{\text{total}}| = \sqrt{v_x^2 + 0} = 21 \text{ m s}^{-1}.$$

(b) The range of the ball can be calculated using equation (10.63) as

$$R = \frac{v_0^2}{g\cos^2\alpha}(\sin(2\theta - \alpha) - \sin\alpha)$$
$$R = 45 \text{ m}.$$

Concept question 10.3

1. A small cart fires a ball straight up into the air as it moves with constant speed. Where will the ball land? Where will the ball land if the cart is accelerating?
2. A missile is launched at an angle of 45° from a battleship. At what point in its trajectory does this missile have the least speed?
3. What path do all projectiles follow?
4. What remains constant in projectile motion?

Exercises 10.3

1. Bob wants to throw a ball to Sally, who is 100 feet away from him. Bob's best throwing speed is 80 mph.
   (a) If he throws the ball horizontally from 3 feet above the ground, will it reach Sally?
   (b) From the same initial height, at what upward angle must Bob throw the ball at his maximum capacity if Sally is to catch it 3 feet above the ground?
   (c) What will be the time of flight in case 2?
2. A swimmer is swimming towards a ship at 2 m s^{-1}. As he approaches the ship the shipmates need to get a survival kit to him. This kit is thrown at 10 m s^{-1} at 60° above the deck which is 10 m above the sea-level. For this kit to land at the front of the swimmer, at what distance from the ship should the swimmer be when the kit is thrown? Ignore air resistance.
3. From a 10 m platform, a ball of mass 1 kg and a bullet of 20 g are simultaneously thrown horizontally. The ball is projected at 5 m s^{-1}, whereas the speed of the bullet is 500 m s^{-1}. Compare the vertical displacements of the ball and the bullet at the end of 1 s.

Further reading

Arya A 1990 *Introduction to Classical Mechanics* 2nd edn (Upper Saddle River, NJ: Prentice-Hall)

Baierlein R 1983 *Newtonian Dynamics* (New York: McGraw-Hill)

Chababy R and Sherwood B 2011 *Matter and Interactions* 3rd edn (New York: Wiley)

Fowles G 1986 *Analytical Mechanics* 4th edn (New York: CBS College Publishing)

Fowles G and Cassiday G 2005 *Analytical Mechanics* 7th edn (Belmont, CA: Thomson Learning)

Goldstein H 1980 *Classical Mechanics* 2nd edn (Reading, MA: Addison-Wesley)

Landau L D and Lifshitz E M 1960 *Mechanics* (Reading, MA: Addison-Wesley)

Ohanian H 1985 *Physics* (New York: WW Norton)

Resnick R, Halliday D and Krane K 2002 *Physics* vol 1, 5th edn (New York: Wiley)

Rutherford D E 1951 *Classical Mechanics* (New York: Interscience)

Stephenson R 1961 *Mechanics and Properties of Matter* 2nd edn (New York: Wiley)

Symon K 1971 *Mechanics* 3rd edn (Reading, MA: Addison-Wesley)

IOP Publishing

Techniques of Classical Mechanics
From Lagrangian to Newtonian mechanics
Samya Zain

Chapter 11

Fluid-statics

11.1 Types of materials

There are five known states of matter: solids, liquids, gases, plasma and Bose–Einstein condensates. The first four states are shown in figure 11.1. The main difference between the structures of the states is in the densities of the particles.

11.1.1 Solids

Solids have a definite shape and maintain their volume. Atoms that make up a solid are tightly packed together and are unable to move about.

The atoms are fixed in place but they are free to vibrate in place. Electrons in metal are free to move about and are responsible for conductors. The particles of a solid are already so tightly packed together that increasing pressure will not compress the solid further to a smaller volume on a macroscopic scale.

11.1.2 Liquids

The liquid state of matter is an intermediate phase between a solid and gas. The liquid particles are not fixed in a regular arrangement, but are close enough to each other to have a definite volume. Particles in a liquid flow around each other. Liquid particles flow and fill the lowest portion of a container, taking on the shape of the container. Liquids have very limited compressibility. Particles of a liquid tend to be held by weak intermolecular attraction. This cohesive force pulls the particles together to form drops or streams. Liquids, like solids, cannot be compressed.

Some properties of liquids include:

- *Surface tension.* Cohesion or stickiness is the tendency for the same kind of particles to be attracted to one another. This cohesive force accounts for the surface tension of a liquid. Surface tension arises because particles in a liquid are attracted to each other. Particles on the surface do not have as many neighbor particles and hence will have higher energy. This gives the surface

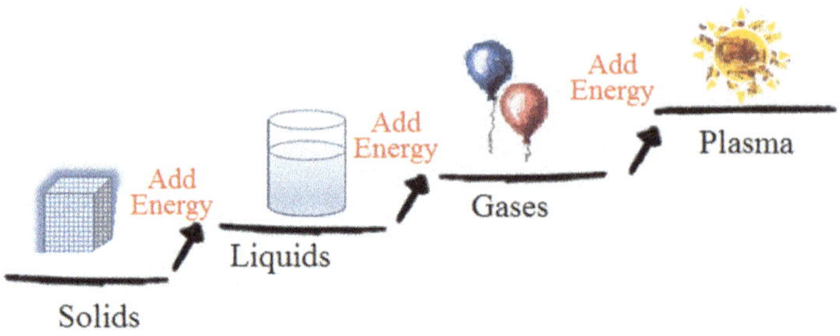

Figure 11.1. States of matter.

an energy proportional to its area. Hence the surface wants to shrink, resulting in surface tension. For example, the surface tension of water is great enough to support the weight of an insect such as a water skipper.

Surface tension (γ) is defined as the ratio of the surface force F to the length L along which the force acts and is measured in SI units of N m^{-1}:

$$\gamma = \frac{F}{L}. \tag{11.1}$$

Note that surface tension is not a force but a force per unit length.

- *Adhesion*. Forces of attraction between different types of particles is called *adhesion*. Particles of a liquid adhere to the walls of the container holding the liquid to various degrees. The combination of cohesive and adhesive forces results in a slight upward concave curve that exists at the surface of most liquids, called the 'meniscus'. The accurate measurement of the volume of a liquid in a graduated cylinder is done from the bottom of the meniscus.
- *Viscosity*. Viscosity is a measure of the resistance of a liquid to flow. A liquid that flows very slowly is more viscous (thicker) than a liquid that flows easily and quickly (thinner). For example, honey is more viscous than water. Viscosity can usually be reduced by heating the liquid. When heated, the particles of the liquid move faster, allowing the liquid to flow more easily.
- *Evaporation*. When surface particles in a liquid gain enough kinetic energy, either by absorbing heat or with collisions with each other, to escape the system evaporation occurs. As particles with higher kinetic energies (hence faster) escape, the remaining particles in the liquid have lower average kinetic energy, and the temperature of the liquid cools. This phenomenon is known as *evaporative cooling*.

11.1.3 Gases

Gas is a state of matter with no fixed shape and no fixed volume. Gases have lower density than other states of matter. There are empty spaces between particles that allow particles to move freely and collide with one another. This causes them to diffuse, or spread out, until they are evenly distributed throughout the volume of the container.

The particles in a gas are spread wide apart, which means gases can be compressed to a smaller volume by applying pressure. Compressed particles exert more force (or pressure) on the interior volume of the container. Similarly, when more gas particles enter a container, there is less space for the particles to spread out, and they become more compressed.

11.1.4 Plasma

Plasma is not common on Earth, but it may be the most common state of matter in the Universe. Stars are essentially superheated balls of plasma. Plasma is made up of highly charged particles with extremely high kinetic energy (figure 11.1).

11.1.5 Bose–Einstein condensates

Bose–Einstein condensate is a state of matter that is cooled to extremely low temperatures, very close to absolute zero (−273.15 °C). At this extremely low temperature, molecular motion almost comes to rest and kinetic energy goes to almost zero, and the atoms begin to clump together to form a *super atom*.

This state was first predicted theoretically in the 1920s, but was first created in 1995 by Eric Cornell and Carl Weiman. They used a combination of lasers and magnets to cool a sample of rubidium to within a few degrees of absolute zero and gave us the first sample of a Bose–Einstein condensate. Some applications of Bose–Einstein condensate include:

- The formation of a superfluid that flows without friction.
- To study quantum mechanics on a macroscopic level.
- To simulate conditions on Earth that might apply in black holes in order to study black holes.

11.2 Fluid-statics

In this chapter we will primarily study the properties of liquids.

Fluid-statics is the branch of mechanics that deals with the properties of liquids and the study of incompressible fluids at rest or in equilibrium. Incompressible fluids are also called perfect liquids.

A fluid is called a perfect fluid when it can be completely characterized by its rest frame and has no shear stresses, viscosity or heat conduction through it. Real fluids on the other hand are sticky, contain heat and also conduct heat.

11.2.1 Types of fluids

Fluids may be classified into the following five main types:

- Ideal or perfect fluid.
- Real fluid.
- Newtonian fluid.
- Non-Newtonian fluid.
- Ideal plastic fluid.

Let us discuss each one very briefly here. See other textbooks for more details.

- *Ideal fluid.*
  An ideal fluid is a fluid that is incompressible, irrotational and non-viscous.
  - It is incompressible because the density of an ideal fluid is constant.
  - It is irrotational, as the flow of an ideal fluid is smooth with no turbulence, with no tangential stresses between two neighboring layers.
  - It is non-viscous as it has no internal friction (viscosity: $\eta = 0$).

  Ideal fluids do not actually exist in nature, but are used when solving fluid flow problems to make the situation simpler.
- *Real fluid.*
  All the fluids in actual practice are real fluids. Their motion is known as viscous flow and they have viscosity ($\eta > 0$), which will be discussed further in chapter 13, section 13.2.
- *Newtonian fluid.*
  A Newtonian fluid is a real fluid in which the shear stress, discussed in detail in chapter 14, section 14.2, is directly proportional to rate of shear strain (or velocity gradient) as defined in chapter 14, section 14.3.
- *Non-Newtonian fluid.*
  A non-Newtonian fluid is a real fluid in which the shear stress is not proportional to the rate of shear strain.
- *Ideal plastic fluid.*
  An ideal plastic fluid is a fluid in which the shear stress is more than the yield value and the shear stress is proportional to the rate of shear strain (or velocity gradient).

11.3 Pressure and density in fluid-statistics

11.3.1 Pressure

Pressure is force per unit area, and is formally defined as

> 'Pressure is the force applied perpendicular to the surface of an object per unit area over which that force is distributed'.

The traditional unit used for pressure is pound-force per square inch (psi). But the standard SI unit for pressure is the *pascal*, which is newtons per square meter. Pressure may also be expressed in terms of a standard established constant called atmospheric pressure (atm), which is approximately equal to typical air pressure at mean sea-level on Earth, and is numerically equal to 1 atm = 101 325 Pa = 1.01×10^5 Pa. Weather forecasters use another unit, called a bar, to represent atmospheric pressure; 1 bar $\approx$ 1 atm.

For an object sitting on a table, the force pressing on the table is equal to the weight of the object regardless of the surface in contact. However, pressure will depend on the surface area in contact. Different orientations will have a different area in contact with the surface and therefore exert a different pressure. More surface contact will give lower pressure, whereas less surface contact will result in higher pressure.

11.3.2 Density

The density of a substance is its mass per unit volume. The most commonly used symbol for density is ρ and is expressed as

$$\rho = \frac{m}{V}, \tag{11.2}$$

where m is the mass and V is the volume of the substance. Density is expressed in kilograms per cubic meter ($\mathrm{kg\,m^{-3}}$).

Density (ρ) is an intensive property of the object. Increasing just the mass will not change ρ as long as the volume is allowed to change as well. However, if the volume is kept constant but the mass is increased, ρ will increase. However, ρ varies with temperature and pressure. This variation is typically small for solids and liquids but is much greater for gases. Increasing the pressure on an object decreases the volume of the object and thus increases its density.

Example 11.1

While walking on a frozen lake you hear ice cracking beneath you. What should you do to get off the lake safely?

(a) Lie flat on the ice and belly crawl toward shore.
(b) Jump up and down to reduce contact time with ice.
(c) Run as fast as you can towards the shore.
(d) Walk slowly towards the shore.

Solution

Answer: (a). Lie flat on the ice and belly crawl toward shore. The net force on the ice due to your weight will remain the same in all situations, however, the pressure will be different in each case. By lying flat, you distribute your weight over the widest possible area, thus reducing the pressure (force per unit area) and increasing your chances of survival.

11.4 Pressure in fluid-statistics

Suppose we have a perfect fluid at rest in a container. Assume we put a small object with area (ΔA) inside the fluid. Fluid exerts a force (ΔF) on one side of this area but an equal and opposite force ($-\Delta F$) is exerted by the fluid on the other side of this area. The pressure (P) at this area will be

$$P = \frac{\Delta \vec{F}}{\Delta A} = \lim_{\Delta A \to 0} \frac{\Delta \vec{F}}{\Delta A} = \frac{d\vec{F}}{dA}, \tag{11.3}$$

which can be rewritten as

$$P \cdot dA\hat{n} = \vec{F}, \tag{11.4}$$

where $\hat{n}$ is the unit vector along the outward drawn normal. Please note that P in equation (11.4) seems like it is a scalar quantity, it is not. Pressure is actually the scalar part of a stress tensor, although for simple problems it is usually considered a scalar, just like temperature. The force exerted by the fluid at rest on any area must always be at right angles to the area. When the force is not perpendicular, then it has a component along the surface, which results in an equal and opposite force being exerted on the fluid that causes the fluid to move. This negates the fluid-static condition since the fluid starts to move and we will have the condition of fluid-dynamics, which we will talk about in chapter 13.

Pascal's principle states that pressure applied to a fluid in a closed container is transmitted equally to every point of the fluid and to the walls of the container (figure 11.2).

11.4.1 Variation of pressure with depth

People who scuba dive know that the pressure increases as we dive to greater depths. The deeper you go, the higher the pressure. The increasing water pressure with depth limits how deep a submarine can go.

To calculate the variation of pressure with depth, consider an imaginary right circular cylinder of mass m and cross sectional area (A) and height $y_2 - y_1 = \Delta y$, as shown in figure 11.3.

The cylinder is immersed in water and is held stationary at a certain depth, as shown in figure 11.4. The cylinder is in equilibrium, i.e. there is no acceleration in the system, under the following conditions:

- The vertical downwards force due the weight of the fluid in the cylinder.
- The force on the bottom and top of the object due to pressure.
- Horizontal forces on the object equal zero as they are only due to the pressure of the fluid and cancel out on both sides, $\Sigma F_x = 0$.
- $\Sigma F_y = 0$ because there is no acceleration in the system.

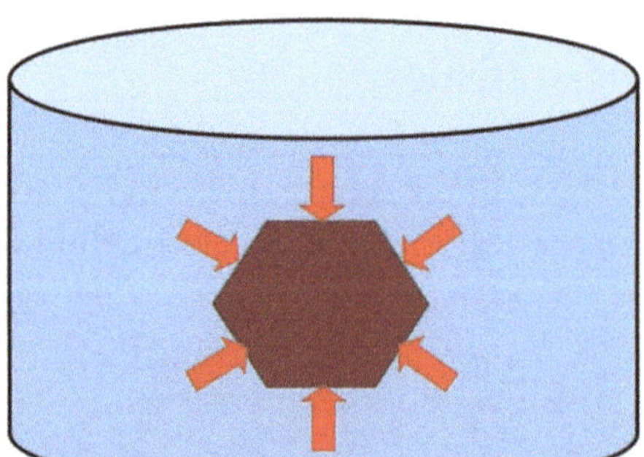

Figure 11.2. Pressure is always perpendicular to the surface of an object.

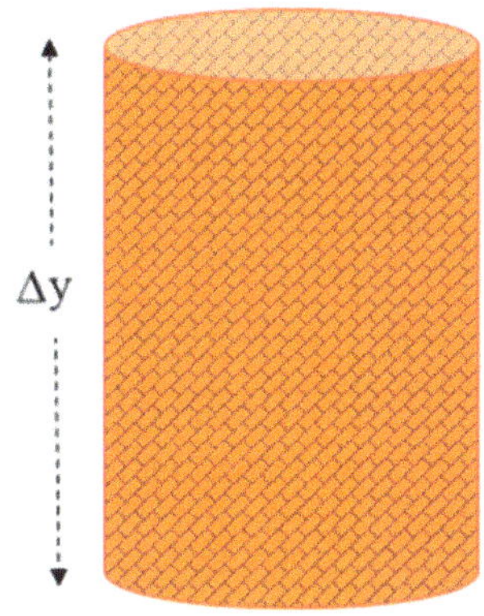

Figure 11.3. Cylinder height.

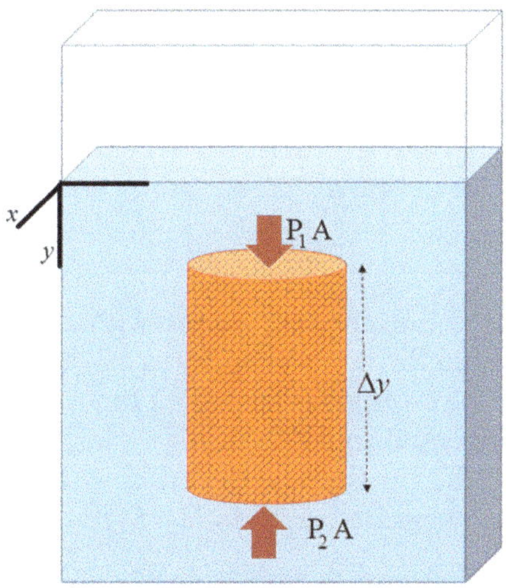

Figure 11.4. Solid cylinder immersed in liquid.

The mass of this object is given as

$$dm = \rho dV = \rho(dA)(dy)$$

and its weight is given as

$$(dm)g = \rho dVg = \rho(dA)(dy)g.$$

Writing Newton's force equation for this situation and using figure 11.5 we will obtain

$$\begin{aligned}\Sigma F_y &= F_{\text{up}} - F_{\text{down}} - mg \\ 0 &= PA - (P + dP)A - (\rho A dy)g,\end{aligned}$$

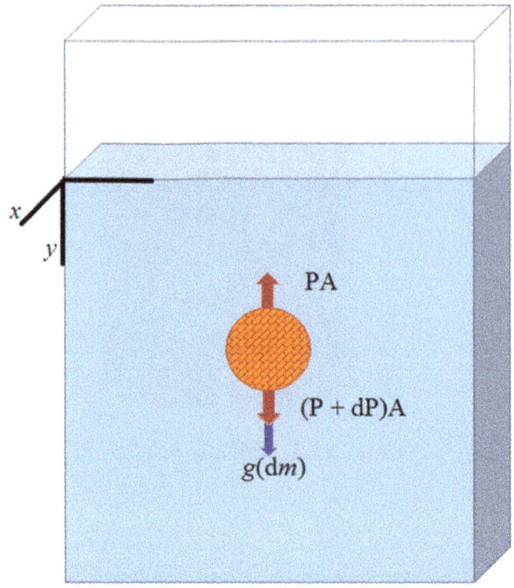

Figure 11.5. Forces acting on the object in terms of pressures.

which simplifies to

$$\boxed{\frac{dP}{dy} = -\rho g.} \tag{11.5}$$

This equation tells us how the pressure varies with elevation above a particular reference level in static equilibrium. When *dy* is positive, elevation increases and pressure decreases. The factor (ρg) is often called the *weight density* of the fluid and for water it is numerically equal to 9800 N m^{-3}.

If we assume P_1 to be the pressure at elevation y_1 and P_2 to be the pressure at elevation y_2, then we can integrate equation (11.5) to obtain

$$P_2 - P_1 = -\rho g(y_2 - y_1). \tag{11.6}$$

Assuming that in this case the density of the fluid and g remain constant and do not change, equation (11.6) gives the pressure change for a homogeneous, incompressible liquid.

11.4.2 Atmospheric pressure

In Earth's atmosphere, we take sea-level as a reference point and call the pressure there the atmospheric pressure (P_o). This is the maximum pressure due to Earth's entire atmosphere. The atmospheric pressure at sea-level is called the *mean sea-level pressure*.

In this case, we can replace P_o by P_2 that corresponds to y_2 and a general P for p_1 corresponds to y_1 and taking ($y_2 - y_1$) as the depth (h) below the surface, we would like to calculate P. Then equation (11.6) can be rewritten as

$$P_o - P = -\rho g(h)$$

$$\boxed{P = P_o + \rho g h,} \tag{11.7}$$

which gives us the variation of pressure with height.

11.4.3 Earth's atmospheric pressure as a function of height

As the elevation increases, there is less and less atmospheric mass, so the atmospheric pressure decreases. Assume this pressure variation occurs smoothly from the Earth's surface to the top of the mesosphere[1].

Similarly, we know that the density (ρ), acceleration due to gravity (g) and the temperature (T) all vary with altitude. If we assume that the variation in g and T is negligible, we can treat them as being constant over the elevation range. We can also assume that density is proportional to the pressure, i.e. as pressure decreases so does density. We can then write

$$\rho = \frac{P}{P_o}\rho_o$$

and put this value in equation (11.5):

$$\frac{dP}{dy} = -\rho g$$

$$\frac{dP}{dy} = -\frac{P}{P_o}\rho_o g.$$

Rearrange and integrate, taking sea-level as reference ($y = 0$; $P = P_o$):

$$\int_{P_o}^{P} \frac{1}{P} = -\frac{\rho_o g}{P_o}\int_0^h dy$$

$$\ln\frac{P}{P_o} = -\frac{\rho_o g}{P_o}h$$

$$\boxed{P = P_o e^{-\frac{\rho_o g}{P_o}h},} \tag{11.8}$$

which gives the expression for the variation of pressure with increasing elevation. It can be seen if we use the values of constants $P_o = 1.01 \times 10^5$ Pa, $g = 9.8$ m s^{-2}, ρ_o at

[1] The layer of the Earth's atmosphere directly above the stratosphere is called the mesosphere. The lower boundary of the mesosphere usually starts from about 50 to 65 km and the upper boundary (mesopause) is usually around 85–100 km above sea-level. In the mesosphere, the temperature decreases as height increases and the top of the mesosphere is the coldest part of Earth's atmosphere, with temperatures as low as −100 °C.

$20° = 1.21$ kg m^{-3}, and we calculate the factor ($\frac{\rho_0 g}{P_0} = \frac{1}{8550\text{m}}$), which is the height at which the atmospheric pressure reduces by a factor of e. Using equation (11.8) we can calculate that at height 20 km (65 616.8 feet) from sea-level the atmospheric pressure will be 0.1 atm and at a height of 40 km the atmospheric pressure will be 0.01 atm.

Example 11.2

A barometer, as shown in figure 11.6, is an instrument used to measure atmospheric pressure. The weight of the mercury in the column creates a vacuum at the top of the tube. Mercury in the column will always balance the atmospheric force exerted on the reservoir. High atmospheric pressure exerts more force on the reservoir, which forces the mercury higher in the column, whereas low atmospheric pressure will exert less force on the reservoir, which forces the mercury lower in the column. Compute the atmospheric pressure on a day when the height of a mercury barometer is 76 cm. Assume the density of mercury is 13.6 g cm^{-3}.

Solution

The weight of mercury in the column is $W = mg$, where $m = \rho\ V = \rho\ (AH)$, from which we can write force per unit area due to mercury:

$$\frac{W}{A} = \rho g H. \tag{11.9}$$

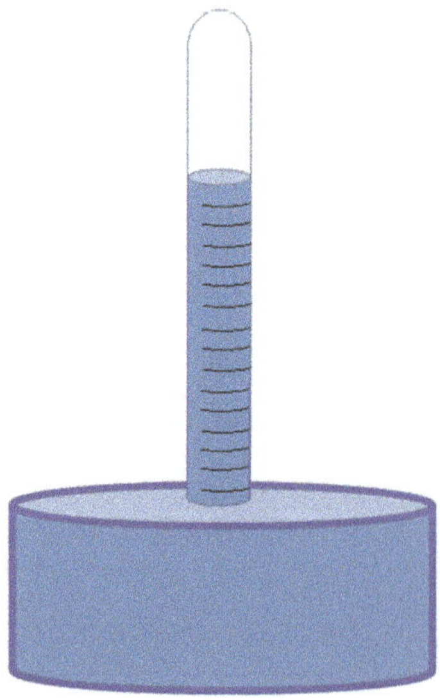

Figure 11.6. Simple mercury barometer.

For a barometer, we know the weight of mercury balances the atmospheric pressure, hence we can write

$$\begin{aligned} P &= \rho g H \\ &= (13\,600 \text{ kg m}^{-3})(9.8 \text{ m s}^{-2})(0.76 \text{ m}) \\ &= 101292.8 \text{ Pa}. \end{aligned}$$

Example 11.3

The pressure in a static water pipe in the basement of an apartment house is 3 atm, but three floors up it reduces to 1.5 atm. What is the height between the basement and the third floor? Assume the density of water is 1000 kg m^{-3}.

Solution

We can use equation (11.6) to solve this problem, given $p_1 = 3$ atm $= 3.0 \times 10^5$ Pa and $p_2 = 1.5$ atm $= 1.5 \times 10^5$ Pa,

$$p_2 - p_1 = -\rho g(y_2 - y_1)$$

$$\frac{p_2 - p_1}{\rho g} = h.$$

Putting in values we calculate the height to be $h = 15.3$ m or about 50 feet.

Concept questions 11.1

1. Estimate your average density using concepts learned in this book. How can you confirm your answer?
2. In circuses, people are amazed when they see mystics lying on a bed of nails. It seems an impossible task and it is assumed only the very brave are able to accomplish this task. On the basis of what we have discussed can you explain what is going on? Would you as an amateur be able to perform this seemingly painful task?
3. Why do airbeds or water beds feel more comfortable than a normal mattress?

Exercises 11.1

1. Find the force acting on the bottom of an aquarium having a base 1 foot by 2 feet containing water to a depth of one foot.
2. Find the height of the atmosphere, assuming the density at sea-level to be 1.21 kg m^{-3}, under the following conditions:
   (a) When the air density is assumed to be constant.
   (b) When the air density is assumed to decrease linearly from maximum to zero.
3. What is the hydrostatic pressure difference in blood pressure between a 6 foot tall man's head and feet?

11.5 Archimedes' principle

Archimedes' principle is named after Archimedes of Syracuse, who first discovered this law in 212 BCE. It describes the magnitude of a buoyant force.

Buoyant force
Buoyant force is an upward force exerted by a fluid that opposes the weight of an immersed object. In a fluid, pressure increases with depth as a result of the weight of the overlying fluid. Thus the pressure at the bottom of a column of fluid is greater than at the top of the column, as seen in figure 11.7.

Archimedes' principle
Archimedes' principle states that,

> 'The buoyant force exerted by a fluid on an object in the fluid is equal in magnitude to the weight of fluid displaced by the object',

$$F_B = F_g \text{ (displaced fluid)}. \tag{11.10}$$

Example 11.4

You put an ice-cube into your glass and note that the ice-cube floats on the surface. What percentage of the volume of the ice-cube will be above the level of water? Take the density of ice as 917 kg m^{-3}.

Solution
According to Archimedes' principle, the apparent loss in weight of an immersed object is equal to the weight of the liquid displaced. The volume of water displaced by the ice-cube will be equal to the mass of the ice-cube. Fresh, liquid water has a density of 1 g cm^{-3} and we can convert the density of ice to obtain 0.917 g cm^{-3}.

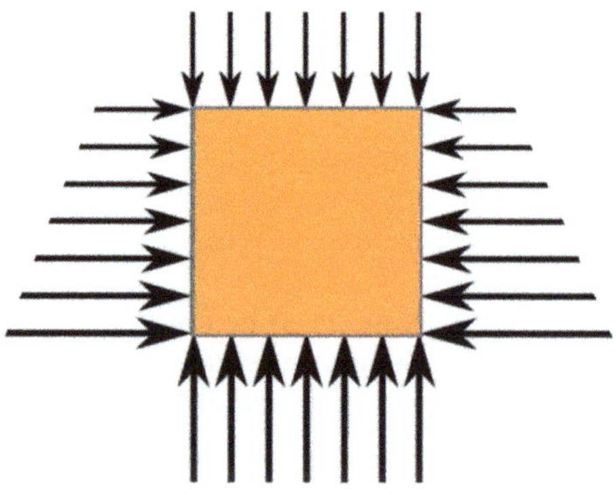

Figure 11.7. Pressure on all sides of a cube. The magnitude of the force is represented by the size of the arrows.

The ice-cube has weight $W_{ice} = M_{ice}g$ and the buoyant force is equal to the weight of the displaced water, $W_{water} = M_{water}g$. Additionally, since the ice-cube floats on water, its weight exactly balances the buoyant force:

$$W_{ice} = W_{water}$$
$$M_{ice}g = M_{water}g$$

$$V_{ice}\rho_{ice}g = V_{water}\rho_{water}g$$
$$V_{ice} = \frac{\rho_{water}}{\rho_{ice}}V_{water}.$$

The fraction of ice above the water level is

$$\frac{\text{volume of ice}}{\text{volume of water}} = \frac{\rho_{water}}{\rho_{ice}} = \frac{0.917}{1}.$$

Hence, the percentage of ice submerged in water = 91.7% and only 8.3% is above the water level.

11.5.1 Proof of Archimedes' principle

To prove Archimedes' principle, consider a cube of volume $dV = (dx\ dy\ dz)$ immersed in a fluid with density ρ as shown in figure 11.8. Horizontal forces act on the four vertical faces of the cube. Horizontal forces are equal and opposite, so they cancel each other's effect. Forces acting on the horizontal faces are called vertical forces and they act on the top and bottom faces.

The difference between the upwards force on the lower face and the downwards force on the upper face will be

$$\begin{aligned} F_{diff} &= Pdxdz - \left(P + \frac{dP}{dy}dy\right)dxdz \\ F_y &= -\frac{dP}{dy}dydxdz. \end{aligned} \tag{11.11}$$

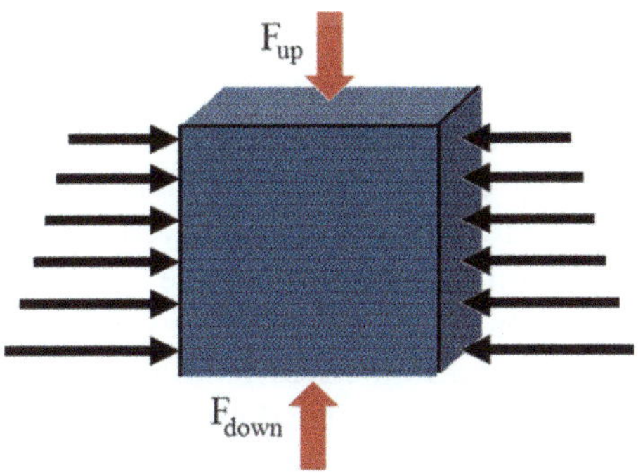

Figure 11.8. *P* is the pressure on the top face and $(p + dp)$ is the pressure on the bottom face of the submerged cube.

In equation (11.11) $(dy\ dx\ dz) = dV$ and $\frac{dp}{dy}$ is the rate of increase in pressure with depth and we know that $\frac{dp}{dy} = -\rho g$, since here y is the positive upwards. Then

$$dF_y = +\rho g dV$$

Integrating,

$$\int dF_y = \int_0^V \rho g dV$$

$$\Longrightarrow \boxed{F_y = \rho g V} \tag{11.12}$$

In equation (11.12), F_y is the buoyant force and this equation is the mathematical statement of Archimedes' principle.

11.6 Specific gravity

Specific gravity, also called relative density, is a dimensionless quantity. It is the ratio of the density of a substance to the density of a reference substance, usually taken to be water for a liquid or solid, and air for a gas. Sometimes specific gravity is also expressed as the ratio of the mass of a substance to the mass of a reference for the same given volume. The reference substance for liquids is nearly always water at 4 °C or 39.2 °F, whereas for gases it is air at room temperature (20 °C or 68 °F). Pressure is nearly always taken to be 1 atm (101.325 kPa).

Specific gravity varies with temperature and pressure, hence we have to take special care to make sure that the sample and reference are at the same temperature and pressure. Mathematically specific gravity is given as

$$\text{SG} = \frac{\text{mass}_{\text{sample}}}{\text{mass}_{\text{water}}} = \frac{\rho_{\text{sample}}}{\rho_{\text{water}}}. \tag{11.13}$$

Specific gravity is used to obtain information about the concentration of solutions of various materials such as syrups, juices, honeys, etc.

11.6.1 Hydrometer

A hydrometer, such as the one shown in figure 11.9, is an instrument used for measuring the specific gravity of liquids. Hydrometers use Archimedes' principle, which states that a solid suspended in a fluid is buoyed by a force equal to the weight of the fluid displaced by the submerged part of the suspended solid. A hydrometer with a given weight will sink more in fluids with lower densities and the stem of the hydrometer is calibrated to give a numerical reading.

Substances with a specific gravity greater than one are denser than water and will sink in water, those with specific gravity equal to one will be neutrally buoyant in water. Substances with specific gravity values that are less than one are less dense than water and will float in water.

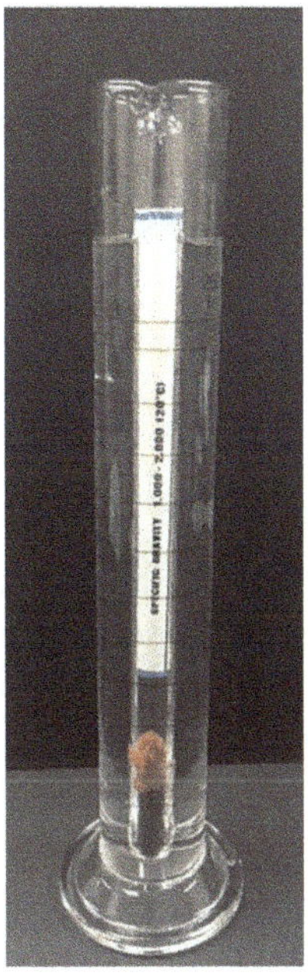

Figure 11.9. Hydrometer.

11.6.2 Specific weight

Specific weight is defined as force (weight) per unit volume. The SI unit for specific weight is [N m^{-3}] and can be expressed as

$$\text{specific weight} = \rho g, \tag{11.14}$$

where g is the acceleration due to gravity. For example, the specific weight for aluminum is 27 kN m^{-3}, and for regular water it is about 9.81 kN m^{-3}, whereas for sea-water it is 10.03 kN m^{-3}.

11.7 Pascal's principle

Pascal's principle was first stated in 1652 CE by Blaise Pascal. Squeeze a tube of tooth paste at any point and the tooth paste will flow out of the open top of the tube. It does not matter where you squeeze the tooth paste tube, the effect will be the same, the tooth paste will flow out of the top. This is also true in gases, for example when air flows out of a balloon. This is the action of Pascal's principle, formally stated as

> 'Pressure applied to an enclosed fluid is transmitted undiminished to every portion of the fluid and to the walls of the container equally'.

It may also be stated as

> 'when there is an increase in pressure at any point in a confined fluid, there is an equal increase at every other point in the container'.

Pascal's principle allows us to:

- Amplify forces, such as those used to lift cars in a mechanics workshop, or hairdresser or dentist chairs.
- Transmit forces over long distances, for example, in control mechanisms for the wing flaps in airplanes.
- Hydraulic force-transmitting mechanics such as the breaking system of cars, or earth-moving machinery.

11.7.1 Expression of Pascal's principle

To mathematically express Pascal's principle, let us consider a cylinder fitted with a piston filled with an incompressible liquid with density (ρ). Application of an external force, say P_{ext}, to the piston results in P_{ext} being applied uniformly throughout the fluid by the piston.

Then from equation (11.7) pressure at an arbitrary point (P) a distance (h) below the surface can upon rearranging be rewritten as

$$P = P_{ext} + \rho gh. \tag{11.15}$$

If we increase the applied external force by ΔP_{ext}, the pressure on the liquid will increase by a factor ΔP_{ext} as well, given by

$$\Delta p = \Delta P_{ext} + \Delta \rho g(\Delta h), \tag{11.16}$$

but density ρ remains constant since it is an incompressible liquid, $\Delta\rho = 0$. Similarly, there will not be any change in height, $\Delta h = 0$. So equation (11.16) reduces to

$$\Delta p = \Delta P_{ext}. \tag{11.17}$$

Equation (11.17) states that for an incompressible liquid the change in pressure at any point in the fluid is simply equal to the change in the externally applied pressure.

Pascal's principle holds for all real (incompressible and compressible) fluids, which include both liquids and gases. In gases the change in P_{ext} causes a change in density that quickly spreads throughout the fluid, but once equilibrium is re-established, we find that Pascal's principle still holds true.

11.7.2 Hydraulic lift

Hydraulic lift is one of the most common applications of Pascal's principle. A hydraulic lift works by using an incompressible liquid to multiply the effect of the force applied to lift something very large or very heavy by using a much smaller force in comparison.

A hydraulic lift system generally contains two pistons connected by a tube. The tubes are filled completely with an incompressible fluid such as oil or sometimes water. Since the fluid is incompressible, we know two things, that the pressure is constant and that very little energy is lost to friction. This means that the force is directly proportional to the surface area of the piston (figure 11.10).

Figure 11.10. Hydraulic lever.

If the pistons have the same diameter, then the force applied on one piston to push it down by a certain distance is the same as the force that lifts the second piston up by the same distance. If one of the pistons has a larger surface area than the other piston, the effect of the force can be multiplied by the same factor as the increase in area of the piston.

For example, in a particular hydraulic system one of the pistons is ten times larger than the other. Then moving the smaller piston over ten meters with a certain amount of force will allow the bigger piston to move by one meter. Hence, the same input force can be used to lift bigger objects.

11.7.2.1 Applied force

Pressure on the fluid at the smaller piston will be given as

$$P_i = \frac{F_i}{A_i}.$$

Pascal's principle says that:

$$P_{\text{in}} = P_{\text{out}}$$
$$\frac{F_{\text{in}}}{A_{\text{in}}} = \frac{F_{\text{out}}}{A_{\text{out}}}$$
$$F_{\text{in}} = F_{\text{out}}\frac{A_{\text{in}}}{A_{\text{out}}}$$

$$\boxed{F_{\text{in}} = (Mg)\frac{A_{\text{in}}}{A_{\text{out}}}.} \tag{11.18}$$

Usually, $(\frac{A_{\text{in}}}{A_{\text{out}}})$ is less than 1 and this implies that F_{applied} can be much smaller than the weight (Mg) to be lifted.

11.7.2.2 Distance moved

Downward movement of a smaller piston through a distance (d_{in}) displaces a volume of fluid equal to $V = A_{in}d_{in}$. When the fluid is incompressible this volume must be equal to the volume of fluid displaced by the upward motion of the larger piston, i.e. $V_{up} = V_{down}$. Let us suppose we need to lift a car up; then in this case we can assume $V_{down} = A_{in}d_{in}$ and $V_{up} = A_{out}d_{out}$. From these we can write

$$V_{down} = V_{up}$$
$$A_{in}d_{in} = A_{out}d_{out}$$

$$\boxed{d_{out} = (d_{in})\frac{A_{in}}{A_{out}}.} \tag{11.19}$$

If $(\frac{A_{in}}{A_{out}})$ is a small number then the distance moved by the larger piston is much smaller than the distance the applied force moves the smaller piston. The price we pay for gaining the ability to lift a large load is losing the ability to move it very far.

11.7.2.3 Net gain or loss in energy in a hydraulic system

Combining equations (11.18) and (11.19) we can write

$$\frac{F_{in}}{F_{out}} = \frac{d_{out}}{d_{in}}$$
$$F_{in}d_{in} = F_{out}d_{out}.$$

We know that [work ≡ force × displacement], then

$$\boxed{W_{in} = W_{out}.} \tag{11.20}$$

Work done by external force on the smaller piston is numerically equal to the work done by the fluid on the larger piston. This means that if we ignore friction and other dissipative forces there is no net gain or loss in energy in this hydraulic system.

Example 11.5

A hydraulic jack uses oil with density ($\rho = 800\ \text{kg m}^{-3}$). When you push down on a handle, of length 4 m, with force (F_h), it results in a force F_i applied to the smaller piston with diameter = 2 m. The distance x from pivot to piston is 1 m. The mass of the car plus the platform is 2000 kg and the large piston has a diameter of 20 m.

(a) What is the applied force needed to lift the car?

(b) For each downward stroke of the pump, in which the hand moves a vertical distance of 0.5 m, how far is the car raised?

Solution

(a) Use equation (11.18) given $r_1 = 1\,\text{m}$ and $r_2 = 10\,\text{m}$:

$$\begin{aligned}F_{\text{in}} &= (Mg)\frac{A_{\text{in}}}{A_{\text{out}}}\\ &= (2000\ \text{kg})(9.8)\frac{\pi(1)^2}{\pi(10)^2}\\ &= 196\ \text{N}.\end{aligned}$$

Then to find the force needed we can use equation (11.20),

$$\begin{aligned}F_h L &= F_i x\\ F_h &= F_i\frac{x}{L}\\ &= (196\ \text{N})\frac{1}{4}\\ &= 49\ \text{N},\end{aligned}$$

which is about 11 lbs and easily doable by a person.

(b) To calculate the distance a car is raised,
the smaller piston is moved by distance

$$\begin{aligned}d_i &= h\frac{x}{L}\\ &= (0.5\ \text{m})\frac{1}{4}\\ &= 2\ \text{m},\end{aligned}$$

correspondingly the larger piston will move by distance

$$\begin{aligned}d_o &= d_i\frac{A_i}{A_o}\\ &= (2\ \text{m})\frac{\pi 1^2}{\pi 10^2}\\ &= 0.02\ \text{m},\end{aligned}$$

which means the platform and hence the car moves by about 2 cm with each pump of the handle, which is a small distance. So you may need to pump the handle about 50 times to move the car up by 1 m.

11.8 Center of buoyancy

Gravity can be thought of as a pull on an object that acts on the center of mass, or center of gravity of the object. Center of mass is the hypothetical point where the entire mass of an object may be assumed to be concentrated for the purpose of gravitational forces, and about which all parts of the body balance in a uniform gravitational field.

In the same way, when objects are placed in water buoyant force acts on the object. The buoyant force can be thought of as acting on an object at its center of buoyancy. In other words, the center of buoyancy is the point through which the resultant of all buoyant forces on an immersed object appear to act, also called the center of mass of the displaced water.

An object whose center of mass is lower than its center of buoyancy will float stably in the fluid, while an object whose center of mass is higher than its center of buoyancy will tend to be unstable and will have a tendency to flip over.

11.8.1 Oscillating objects in water

Any object (say a boat or a ship) that is floating in a particular liquid will start oscillating about a point when it is given a small angular displacement. Displacement could be a result of high winds or waves. The point about which the body starts oscillating is called the *metacenter* of the object.

11.8.1.1 Metacenter (M)

The metacenter is the point at which an imaginary vertical line passes through the center of buoyancy and center of gravity when the ship is completely vertical and stable, as seen in figure 11.11(a).

When a ship or a boat leans over to one side, say under the influence of waves or winds, the center of buoyancy (*B*) also moves over, as seen in figure 11.11(b). However, for small oscillations (small heel angles) the metacenter (*M*) and center of gravity (*G*) remain fixed on the vertical line. For large oscillations (large heel angles) the metacenter (*M*) may not be considered fixed, and its actual location will affect the stability of the ship.

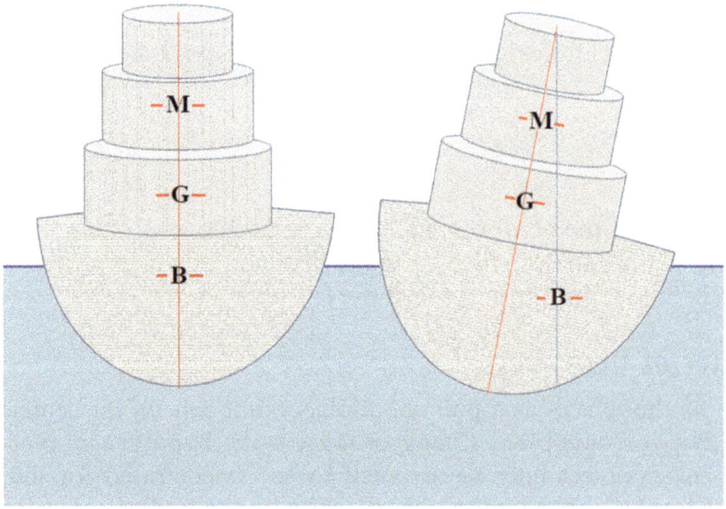

Figure 11.11. Ship under (a) upright condition and (b) heeled over to one side.

*11.8.1.2 Metacentric height (*H*)*

The metacentric height (H) is a measure of the stability of a floating object. It is the distance between the center of gravity (G) and metacenter (M) of the floating object. A larger metacentric height implies greater initial stability against overturning.

In practice points M and H depend on the angle of displacement (θ), but for relatively small displacements point M can be considered to be fixed. See figure 11.11.

11.8.2 Period of roll of the ship

The metacenter has a direct relationship with the natural period of the rolling ship's hull. To find the period of roll of the ship, assume the metacentric height (H) of a particular ship is positive and consider the motion about the center of mass of the ship. The torque about the center of gravity (G) will be given as

$$\begin{aligned} \Sigma\tau &= \vec{r} \times \vec{F} \\ I\alpha &= -\, H \times mg \text{ (negative sign for clockwise torque)} \\ &= -\, H(\rho V)g \sin\theta. \end{aligned}$$

For small angles ($\sin\theta \approx \theta$), the angular acceleration is $\alpha = \ddot{\theta}$, introduced in chapter 3. Then we can write the above equation as

$$\ddot{\theta} + \frac{H\rho Vg}{I}\theta = 0, \tag{11.21}$$

which is the equation of motion of simple harmonic motion, such as the ones we saw in chapter 7, with period (T) equal to

$$T = 2\pi\sqrt{\frac{I}{H\rho Vg}}. \tag{11.22}$$

It is clear from this equation that the period of the ship is inversely proportional to the square-root of the metacentric height (H). A very stable ship has a large metacentric height and a small period, which causes a ship to be too *stiff*, and it is a somewhat uncomfortable ride. A ship with a small metacentric height will have a long roll period and is considered less safe when flooded. An excessively low or negative value of (H) increases the risk of a ship capsizing in rough weather. Hence, a sufficiently high but not excessively high metacentric height is considered ideal for passenger ships. A passenger ship will typically have a long rolling period for comfort, about 12 s, while a tanker or cargo carrier might have a rolling period of 6–8 s.

The positions of G and M depend largely on the load carried by the ship (figure 11.11). When the payload is zero, the ships will generally carry ballast for safety. Also, the angle through which most ships are built to be able to safely roll is about 40°, which means that most ships will survive heavy wind or wave conditions that tend to roll the ship up to 40°, as shown in figure 11.12.

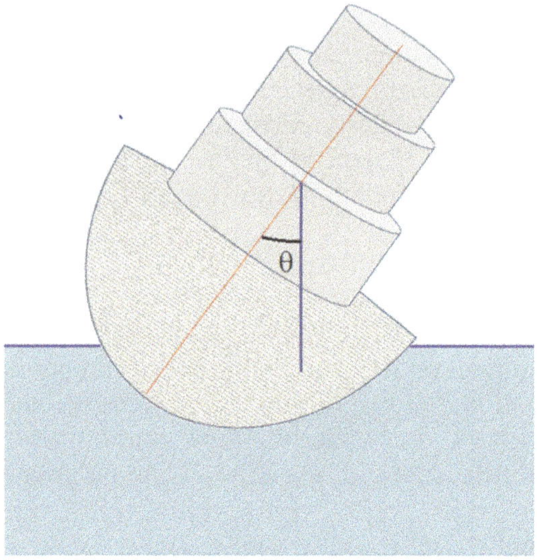

Figure 11.12. Ship roll.

Concept questions 11.2

1. A mass is attached to an inflated balloon by means of a string. Both are then thrown into a swimming pool. Explain what happens to the pressure inside the balloon and hence the size of the balloon as the mass sinks to the bottom of the pool?
2. You get on an elevator with a cup of water with one ice-cube in your hand. You do not wish to talk with other people in the elevator so you look at your cup. Keeping an eye on the floating ice-cube, explain what you observe:
   (a) When the elevator goes up (does the ice-cube float higher or lower than before?).
   (b) When the elevator comes down.
3. If soap bubbles reduce the surface tension of water, why are we able to blow soap bubbles and not water bubbles?
4. What fraction of the total volume of an iceberg is visible from the surface?
5. Why do ships made out of iron float but nails made out of iron sink?

Exercises 11.2

1. What is the diameter of the small piston of a hydraulic press when a force of 20 pounds on it produces a force of 40 tons on the large piston, whose diameter is 20 inches? Assume we can ignore friction for this case.
2. The mass of a rectangular bar of iron is 320 g. It is $2 \times 2 \times 10$ cm^3. What is its specific gravity?
3. Three Billy Goats Gruff decided to build a raft made up of logs, each of length 6 feet, instead of using the bridge, to avoid the ugly troll altogether. If each goat

Gruff weighs 80 lbs and each log is 1 foot in diameter, how many logs will be needed to keep them afloat and get them across the river at the same time? Assume the density of the wood is 47 lb ft^{-3}.

4. Katie blows a soap bubble of diameter 3 cm, using a soap solution with a surface tension of 0.025 N m^{-1}. How much energy is used in stretching the soap surface? *Hint*: remember, in a soap bubble there are two surface layers.

Further reading

Arya A 1990 *Introduction to Classical Mechanics* 2nd edn (Upper Saddle River, NJ: Prentice-Hall)

Chababy R and Sherwood B 2011 *Matter and Interactions* 3rd edn (New York: Wiley)

Fowles G 1986 *Analytical Mechanics* 4th edn (New York: CBS College Publishing)

Fowles G and Cassiday G 2005 *Analytical Mechanics* 7th edn (Belmont, CA: Thomson Learning)

Goldstein H 1980 *Classical Mechanics* 2nd edn (Reading, MA: Addison-Wesley)

Landau L D and Lifshitz E M 1960 *Mechanics* (Reading, MA: Addison-Wesley)

Ohanian H 1985 *Physics* (New York: WW Norton)

Resnick R, Halliday D and Krane K 2002 *Physics* vol 1, 5th edn (New York: Wiley)

Rutherford D E 1951 *Classical Mechanics* (New York: Interscience)

Stephenson R 1961 *Mechanics and Properties of Matter* 2nd edn (New York: Wiley)

Symon K 1971 *Mechanics* 3rd edn (Reading, MA: Addison-Wesley)

Chapter 12

Fluid resistance

12.1 Fluid resistance

So far we have talked about motion and the action of force without considering the medium that the object is moving through. But we know that on Earth all motion is inevitably a motion through a medium or fluid, whether it is through air or water. The effect of the presence of a medium is to reduce the speed of the moving object. This means that there is an extra force acting on the object due to its motion through the medium. We call that force *drag force* (F_B) or simply *drag* (D). Hence drag is a force that acts opposite to the relative motion of an object moving through a surrounding fluid. When the object moves through air, drag is called *air resistance*.

When considering the motion of cars or airplanes, drag is called *aerodynamic drag*, and for motion over water, say for ships or boats, drag is called *hydrodynamic drag*. In the case of fluid flow in a pipe, drag is called *viscous drag* and it decreases fluid velocity relative to the pipe. Drag also exists between two fluids or between layers of the same fluid as they move with respect to each other, or between a fluid and a solid surface.

Drag depends on the properties of the fluid and on the size, shape and speed of the object. You may think of drag as another type of friction.

12.2 Forces as a function of velocity: fluid resistance

Unlike other resistive forces, such as dry friction between two solids surfaces, which are nearly independent of velocity, drag forces are directly proportional to the respective velocities of the moving objects. For normal fluid resistance drag is not a simple function that can be evaluated by initial conditions, rather it must be found through experimental observations and measurements. However, for many situations it can be approximated as

$$D = D(v) = -b_1 v - b_2 v|v|, \tag{12.1}$$

doi:10.1088/2053-2563/aae1b9ch12

where constants b_1 and b_2 depend on the properties of the fluid and on the size, shape and speed of the object. The negative sign indicates that drag is opposite to the direction of motion. The absolute-value sign is necessary on the last term because drag is always opposite to the direction of v.

When the speed of the object is small the linear term in equation (12.1) will dominate and when the speed of object is large the quadratic term in $D(v)$ will dominate.

We will talk more about these in chapter 13, but I will introduce laminar and turbulent fluid flow here to explain the effect of drag forces. A *laminar flow* is very orderly flow of a fluid where all particles move in smooth curves, with no disruption between the layers (top of figure 12.1). Drag force is directly proportional to the velocity for a laminar flow,

$$D = D(v) = b_1 \, v, \tag{12.2}$$

where b is a constant of proportionality that depends on the shape and size of the object.

For *turbulent flows* drag is directly proportional to the square of the velocity. Turbulent flow, or turbulence, is characterized by chaotic pressure changes and vortices that randomly appear and disappear that can interact with each other and interrupt flow velocity, as seen in figure 12.1. Turbulence is commonly observed in everyday phenomena such as clouds and smoke:

$$D = D(v^2) = b_2 v^2. \tag{12.3}$$

12.2.1 Spheres moving through air

To get an idea of the values of constants let us look at an everyday example. Suppose we have a sphere, say a softball, with diameter (d) traversing through air. The approximate values of constants come out to be $b_1 = 0.00\,155\, d$ and $b_2 = 0.22\, d^2$. For a softball with an approximate diameter of 9.6 cm, we find that the linear term

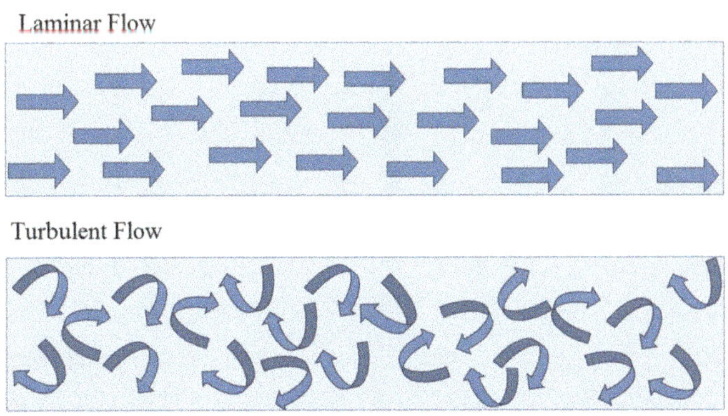

Figure 12.1. Laminar versus turbulent flow.

dominates for velocities less than 0.01 m s^{-1}, whereas the quadratic term dominates for speeds greater than 0.01 m s^{-1}.

12.2.2 Equation of motion under drag

To include drag force, the differential equation of motion has to be revised as

$$\Sigma F = F_{\text{ext}} - D(v) \tag{12.4}$$

$$\Rightarrow m\ddot{r} = mv\frac{dv}{dx} = F_{\text{ext}} - D(v), \tag{12.5}$$

where F_{ext} is the external or applied force acting on the object. F_{ext} may not be a function of velocity, however, drag will still be a function of velocity. From these equations we can find the trajectory of the particle by finding t and/or $x = x(v)$ by separating variables and integrating.

12.3 A falling object under linear drag

An object with mass (m) that drops from rest from a height (h) will have velocity given by equation (12.6) when it hits the ground, when we ignore drag:

$$v = \sqrt{2gh}, \tag{12.6}$$

assuming downwards y as positive.

In real life, however, we can never eliminate drag, in this case air resistance. Experience tells us that the final velocity is always smaller than that we calculate from equation (12.6). In fact, if the height (h) is large enough, the object does not continue speeding up as time goes by, but rather it will reach a *terminal velocity*, which is the velocity after which the speed of the object no longer increases, but rather remains constant.

Let us calculate the acceleration, velocity and position for the situation where we cannot ignore drag. This situation is shown in figure 12.2.

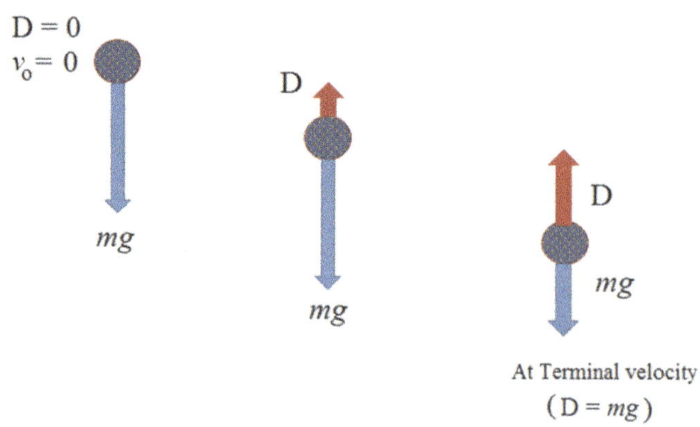

Figure 12.2. Linear drag.

12.3.1 Acceleration

The acceleration of the object when we cannot ignore drag will no longer be a constant but rather it will be a function of time. Starting from Newton's law we can write

$$\Sigma F_y = mg - D.$$

After a small time interval, at slow velocity the drag force will be given by $D = b_1 v$ and thus will give us the acceleration as

$$\boxed{a(t) = g - \frac{b_1 v}{m}.} \tag{12.7}$$

12.3.2 Velocity

As the object starts to fall, the velocity will increase as a function of time. Starting from the acceleration equation (12.7) and knowing $a = dv/dt$, we can integrate and solve to find velocity as a function of time as

$$a(t) = g - \frac{b_1 v}{m}$$

$$\frac{1}{g - \frac{b_1 v}{m}} dv = dt.$$

Integrating, we obtain

$$\int_0^v \frac{1}{g - \frac{b_1 v}{m}} = \int_0^t dt$$

$$-\left(\frac{m}{b_1}\right) \ln \frac{mg - b_1 v}{mg} = t$$

$$1 - \frac{b_1 v}{mg} = e^{-\left(\frac{b_1}{m}t\right)}.$$

Rearranging, we will obtain velocity as a function of time as

$$\boxed{v(t) = \frac{mg}{b_1}(1 - e^{-\frac{b_1}{m}t}).} \tag{12.8}$$

There are two main cases to consider when talk about the velocity of an object under resistive forces.

Case 1. A very small time (t) interval after the initial drop.
Since the object is dropped from rest, its initial velocity (v_o) will be zero, and at a very small time later equation (12.8) can be approximated as (because $e^{-x} \approx (1 - x)$):

$$v(t) = \frac{mg}{b_1}\left[1 - \left(1 - \frac{b_1}{m}t\right)\right]. \tag{12.9}$$

Then equation (12.9) with a bit of algebra can be further shown to reduce down to

$$\boxed{v(t) = gt.} \tag{12.10}$$

Recall we chose downwards y as positive. This implies that early in the motion, velocity is small, which means the drag force is small as well and the object is in an approximate free fall with acceleration equal to the acceleration due to gravity (g).

Case 2. A very large time after the drop.
After a significant time has passed we can say as $t \to \infty$ that $e^{-\infty} \to 0$. Then equation (12.8) can be rewritten as

$$\boxed{v = \frac{mg}{b_1}.} \tag{12.11}$$

If you put this value in equation (12.7) we see that acceleration of the object at this time is zero. This implies that the object will now have constant velocity, which we call the *terminal velocity* of the object.

Terminal velocity
Terminal velocity, which is also called the *settling velocity*, is the highest velocity an object can achieve as it falls through a fluid. It occurs in air when the drag force is equal to the downward force of gravity acting on the object, and the object stops accelerating. In a fluid, say in water, an object moves at the terminal velocity when its speed is constant under the effect of the restraining force exerted by the fluid.

If we denote the terminal velocity mg/b_1 by v_T and τ instead of m/b_1 we can rewrite equation (12.8) as

$$v(t) = v_T(1 - e^{-\frac{t}{\tau}}), \tag{12.12}$$

where τ is called the *characteristic time*. The characteristic time is very important to know because from equation (12.12) we see that after one characteristic time ($= 1\tau$) the speed of the object will be $(1 - e^{-1})$ times the terminal speed, after two characteristic times (2τ) the speed is the factor $(1 - e^{-2})$ of v_T and so on. After an interval of 5τ the speed of the object becomes within 1% of the $v_T = 0.993\ v_T$.

Some common values of terminal velocities are given in table 12.1.

An object moving downward faster than the terminal velocity (for example because it was thrown downwards, it fell from a thinner part of the atmosphere, or it changed shape) will slow down until it reaches the terminal velocity.

Table 12.1. Values of some terminal velocities in air.

Object	Mass	Radius	Terminal speed (v_T)	Distance to 95% of v_T
Ping-pong ball	2.7 g	2 cm	9 m s^{-1}	10 m
Small rain drops[a]	0.5 mg	0.15 cm	7 m s^{-1}	6 m
Tennis ball	58.5 g	3.33 cm	31 m s^{-1}	115 m
Basketball	0.626 kg	11.4 cm	20 m s^{-1}	47 m
Sky-diver (before opening parachute)	100 kg	3 feet	60 m s^{-1}	430 m
Sky-diver (after opening parachute)	100 kg	6 m	5 m s^{-1}	3 m

[a]Humphreys W J 1964 *Physics of the Air* (New York: Dover) p 279.

12.3.3 Position

The position as a function of time for the falling object is found simply by integrating the velocity equation (12.8) and knowing $v = dx/dt$,

$$\begin{aligned} v(t) &= \frac{mg}{b_1}(1 - e^{-(\frac{b_1}{m}t)}) \\ \int_0^y dx &= \frac{mg}{b_1}\int_0^t (1 - e^{-(\frac{b_1}{m}t)})dt \\ y(t) &= \frac{mg}{b_1}\left[t + \frac{m}{b_1}(e^{-\frac{b_1}{m}t} - 1)\right]. \end{aligned} \tag{12.13}$$

If we write $\frac{mg}{b_1} = v_T$, then in terms of the terminal velocity v_T the above equation can be rewritten as

$$\boxed{y(t) = v_T\left[t + \frac{v_T}{g}(e^{-\frac{g}{v_T}t} - 1)\right].} \tag{12.14}$$

Summarizing: vertical fall through a fluid—linear case (figure 12.3)

- Acceleration as a function of time: $a(t) = g - \dfrac{b_1 v}{m}$.
- Velocity as a function of time: $v(t) = \dfrac{mg}{b_1}(1 - e^{-\frac{b_1}{m}t})$.
- Position as a function of time:

$$Vy(t) = \frac{mg}{b_1}\left[t + \frac{m}{b_1}(e^{-\frac{b_1}{m}t} - 1)\right].$$

- Characteristic time: $\tau = \frac{m}{b_1} = \frac{v_T}{g}$.

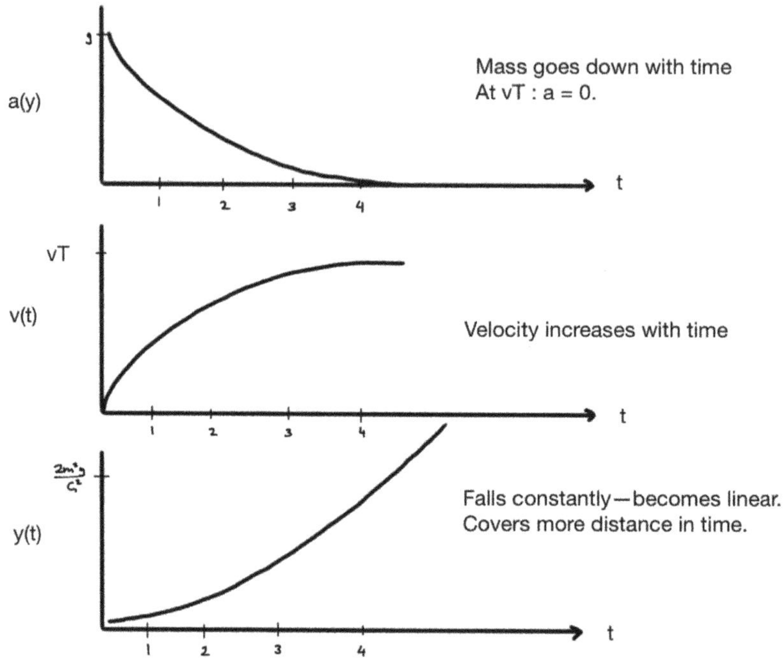

Figure 12.3. Summarizing linear drag.

12.4 Falling object: the quadratic case

In this case, the magnitude of $D(v)$ is proportional to v^2. We will solve this as always from Newton's equation and take the positive direction to be upwards:

$$\Sigma F_y = -mg \pm b_2 v^2. \tag{12.15}$$

To make sure that the drag force will always remain resistive, we must remember that the sign of the drag term depends on the direction of motion of the object. If we take upwards to be the positive direction in our reference frame then the sign of the drag force term will be positive for downwards motion and negative for upward motion. This technique holds for any resistive force proportional to an even power of velocity. A general solution usually involves treating the upward and downward motions separately, especially in cases where both motions happen, such as a ball thrown upwards which will inevitably come downwards as it falls back to the Earth.

Starting from equation (12.15) will give us acceleration as follows:

$$\begin{aligned} a(t) &= -g \pm \frac{b_2 v^2}{m} \\ \frac{dv}{dt} &= -g \pm \frac{b_2 v^2}{m} \\ \int_{v_o}^{v} \frac{m}{-mg \pm b_2 v^2} dv &= \int_{t_o}^{t} dt. \end{aligned} \tag{12.16}$$

There will be two separate situations to consider in this case. For the case where the object is rising, we will consider time interval $[t_o, t]$ and for the case where the object is falling the time interval $[t_o', t]$.

1. When the object is rising:

$$t - t_0 = \int_{v_o}^{v} \frac{m}{-mg - b_2 v^2} dv. \tag{12.17}$$

2. When the object is falling:

$$t - t_0' = \int_{v_o}^{v} \frac{m}{-mg + b_2 v^2} dv. \tag{12.18}$$

1. *Calculating velocity for the rising object.*
   Start from equation (12.17) and integrating we obtain

$$t - t_0 = \sqrt{\frac{m}{b_2 g}} \left[\tan^{-1} \frac{v_o}{\sqrt{\frac{mg}{b_2}}} - \tan^{-1} \frac{v}{\sqrt{\frac{mg}{b_2}}} \right].$$

   We call $\sqrt{\frac{m}{b_2 g}} = \tau$ the characteristic time and $\sqrt{\frac{mg}{b_2}} = v_T$ the terminal velocity and re-express the above equation as

$$t - t_0 = \tau \left[\tan^{-1} \frac{v_o}{v_T} - \tan^{-1} \frac{v}{v_T} \right]$$

$$\tan\left[-\frac{t - t_0}{\tau} + \tan^{-1} \frac{v_o}{v_T} \right] = \frac{v}{v_T},$$

   which is the equation of velocity when the object is going up

$$\boxed{v(t) = v_T \cdot \tan\left(-\frac{t - t_0}{\tau} + \tan^{-1} \frac{v_o}{v_T} \right).} \tag{12.19}$$

2. *Calculating velocity for the falling object.*
   Stating from equation (12.18) we can see that the velocity for when the object is falling down will be given as

$$\boxed{v(t) = -v_T \cdot \tanh\left(\frac{t - t_o'}{\tau} - \tanh^{-1}\left(\frac{v_o}{v_T}\right) \right).} \tag{12.20}$$

   The derivation is left as an exercise for the reader to attempt.

Example 12.1

(a) Find the terminal speed and characteristic time for a tiny drop, of diameter 1.5 μm and density (ρ) 800 kg $m^{-3,}$ of oil in the Millikan oil-drop experiment. Take the coefficient of drag to be approximately ≈0.000 15D, where D is the diameter of the spheres.

(b) Find the terminal speed and characteristic time for a small drop of water with diameter = 0.2 mm.

Solution

(a) To calculate the terminal speed from equation (12.11) we know we need the mass of the drop. However, we are only given the density and size. We can calculate the mass as

$$\rho = \frac{m}{V} \Rightarrow m = \rho\left(\frac{4}{3}\pi r^3\right),$$

From equation (12.11) terminal velocity will be

$$\begin{aligned} v_T &= \frac{mg}{b_1} \\ &= \frac{\rho\left(\frac{4}{3}\pi r^3\right)g}{b_1} \\ &= \frac{(800)\left(\frac{4}{3}\pi(0.75 \times 10^{-6})^3\right)9.8}{(0.000\ 15)(0.2 \times 10^{-3})} \\ &= 6.15 \times 10^{-5}\,\mathrm{m\,s^{-1}} \end{aligned}$$

and characteristic time

$$\tau = \frac{v_T}{g} = 6.3 \times 10^{-5}\,\mathrm{s}.$$

At 5τ the falling object attains 99% of terminal speed. This means that after falling for just 0.3 ms, the oil-drop attains 99% of its terminal speed.

(b) From equation (12.11) we can solve to obtain the terminal velocity for the water drop as

$$\begin{aligned} v_T &= \frac{\rho\left(\frac{4}{3}\pi r^3\right)g}{b_1} \\ &= \frac{(1000)\left(\frac{4}{3}\pi(0.1 \times 10^{-3})^3\right)9.8}{(0.000\ 15)(0.2 \times 10^{-6})} \\ &= 1.4\ \mathrm{m\,s^{-1}}. \end{aligned}$$

From the above solutions we can see that more massive drops will fall faster. The characteristic time is

$$\tau = \frac{v_T}{g} = 0.14 \text{ s}.$$

After about 0.7 s, the water drop attained 99% of its terminal speed.

Example 12.2

What are the dimensions of b when considering (a) linear drag or (b) quadratic drag?

Solution

(a) When drag force (D) varies linearly with speed,

$$b = \frac{D}{v}.$$

Then the dimensions of b are

$$\frac{\text{N}}{\text{m/s}} = \frac{[M][L][T]^{-2}}{[L][T]^{-1}} = [M][T]^{-1}.$$

In this case b will have units of kg s^{-1}.

(b) When drag force (D) varies quadratically with speed,

$$b = \frac{D}{v^2},$$

then the dimensions of b are

$$= \frac{[M][L][T]^{-2}}{[L]^2[T]^{-2}} = [M][L]^{-1}.$$

In this case b will have units of kg m^{-1}.

Concept questions 12.1

1. Under what conditions will your mass be zero? Your weight be zero?
2. A little boy is throwing a ball into the air while sitting in a train. Where will the ball land:
   (a) When the train is at rest?
   (b) When the train accelerates from rest?
   (c) When the train achieves constant velocity?
3. Referring to question 2, explain what happens in each situation with respect to:
   (a) His mother sitting next to him.
   (b) His friend looking in from the platform.

Exercises 12.1

1. Suppose a block of mass (m) slides on a floor with an initial speed (v_0). The block suffers a viscous resistance that varies as 2.5 power of speed given as

$$D(v) = -\beta v^{\frac{5}{2}},$$

where β is a known drag constant. What is the maximum distance the block will slide before coming to rest?
2. A professional thrower projects a football straight up in the air.
   (a) Assuming there is no air drag on the football, find the initial speed of the football as a function of height as the ball goes up.
   (b) Assuming the air drag on the football varies linearly with speed, find the speed of the football as a function of height as the ball goes up.
   (c) Assuming the air drag on the football varies quadratically with speed, find the speed of the football as a function of height as the ball goes up.
3. A ball of mass m is dropped from rest. Find the terminal speed assuming that the drag force is given by αv^2.

12.5 Projectile motion: air resistance

In this section we will consider the motion of a projectile when in addition to the force of gravity, the force of air resistance acts on it. In this case, the motion will not conserve total energy and the energy will decrease during the flight of the projectile. In this section we will discuss the effect it will have on the path of the projectile.

We know drag force acts in opposition to the direction of motion of the object. The magnitude of the resistive force is directly proportional to the instantaneous speed of the object under consideration.

12.5.1 Equation of motion: Newtonian mechanics

Similar to the no drag situation discussed in section 10.7, we will orient the coordinate system such that the x-axis is horizontal and lies along the projection of the initial velocity and the motion is confined to the xz-plane. Which means that $y_o = \dot{y}_o = 0$. Then the equation of motion in 3D can be rewritten as

$$\Sigma F = m\ddot{\vec{r}} = m\frac{d^2\vec{r}}{dt^2} = -b_1\vec{v} - mg\hat{k} \tag{12.21}$$

or as

$$\frac{d^2\vec{r}}{dt^2} = \ddot{\vec{r}} = -\gamma\vec{v} - g\hat{k}, \tag{12.22}$$

noting that here $\gamma = b_1/m$, where m is the mass of the projectile and b_1 is the linear drag coefficient.

12.5.2 Equation of motion: Lagrangian mechanics

The motion is defined in the xz-plane which implies that the generalized coordinates are (x, z). In chapter 4, section 4.4, we defined the Lagrangian for non-conservative forces. We will use them here and write the Lagrangian as

$$L = \frac{1}{2}m(\dot{x}^2 + \dot{z}^2) - mgz. \tag{12.23}$$

The equations of motion along the x- and z-axes will be given as

$$\frac{d}{dt}\left(\frac{\partial L}{\partial \dot{x}}\right) - \frac{\partial L}{\partial x} = - b_1\vec{\dot{x}} \tag{12.24a}$$

$$\frac{d}{dt}\left(\frac{\partial L}{\partial \dot{z}}\right) - \frac{\partial L}{\partial z} = - b_1\vec{\dot{z}}. \tag{12.24b}$$

Find the partials using equation (12.23) and put in equations (12.24a) and (12.24b) to obtain

$$\frac{d}{dt}(m\dot{x}) = - b_1\vec{\dot{x}} \tag{12.25a}$$

$$\frac{d}{dt}(m\dot{z}) + mg = - b_1\vec{\dot{z}}. \tag{12.25b}$$

The combination of equations (12.25a) and (12.25b) is identical to equation (12.22) given $\gamma = b_1/m$.

12.5.3 Acceleration as a function of time

We can rewrite equation (12.22) in component form and/or rewrite equations (12.25a) and (12.25b) to give the acceleration of the projectile as a function of time in

$$\begin{aligned} \ddot{x}(t) &= - \gamma\dot{x} \\ \ddot{z}(t) &= - \gamma\dot{z} - g. \end{aligned} \tag{12.26}$$

In chapter 10 section 10.6 we found that the horizontal motion of a projectile has a constant velocity and hence equation (10.47) gave $a_x = 0$. However, with air resistance we obtain $\ddot{x} \neq 0$, which means horizontal motion of a projectile will not have a constant velocity.

12.5.4 Velocity as a function of time

We note here that equations (12.26) are separated equations and do not depend on each other. So we can find the solutions individually. Since they are second order differential equations their solutions will be given as

$$\dot{x}(t) = \dot{x}_o e^{-\gamma t}$$
$$\dot{z}(t) = \dot{z}_o e^{-\gamma t} - \frac{g}{\gamma}(1 - e^{-\gamma t}). \tag{12.27}$$

These solutions give the velocity of the projectile as a function of time in component form.

12.5.5 Position as a function of time

We have defined that the motion is confined to the xz vertical plane and taken the initial position of the projectile at the origin, hence $x_0 = z_0 = 0$. Integrating once more, we will obtain the position coordinates

$$x(t) = \frac{\dot{x}_o}{\gamma}(1 - e^{-\gamma t})$$
$$z(t) = \left[\frac{\dot{z}_o}{\gamma} + \frac{g}{\gamma^2}\right](1 - e^{-\gamma t}) - \frac{g}{\gamma}t, \tag{12.28}$$

which we can represent in 3D vector form as

$$\vec{r} = \left[\left(\frac{v_0}{\gamma} + \hat{k}\frac{g}{r^2}\right)(1 - e^{-\gamma t})\right] - \frac{g}{\gamma}t\hat{k}. \tag{12.29}$$

It is clear from equation (12.29) that the path of the projectile is *not parabolic* in this case, which can be seen in figure 12.4.

12.5.6 Actual motion of a projectile

The actual motion of a projectile under the influence of a resistive force is definitely not just a linear function of velocity as we have supposed here. It is actually a really complicated function of the velocity. For accurate calculation of the trajectory we generally use numerical integration methods.

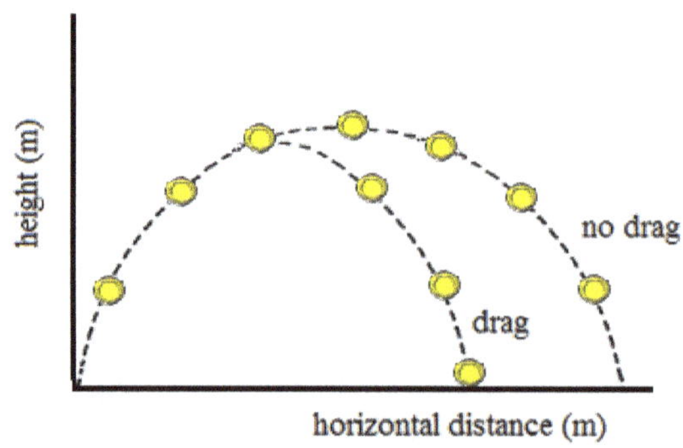

Figure 12.4. Projectile under drag conditions.

12.5.7 Horizontal range $R_{\max}$

The horizontal range of a projectile with linear air drag is found by setting $z = 0$ in the second equation in (12.28)

$$z(t) = 0 = \left[\frac{\dot{z}_o}{\gamma} + \frac{g}{\gamma^2}\right](1 - e^{-\gamma t}) - \frac{g}{\gamma}t \tag{12.30}$$

and from the first we obtain

$$x(t) = \frac{\dot{x}_o}{\gamma}(1 - e^{-\gamma t}), \tag{12.31}$$

and by knowing $e^{-x} = (1 - x)$ and $v = x/t$ and eliminating t between both equations, we obtain

$$t = -\frac{1}{\gamma}\ln\left(1 - \gamma\left(\frac{x}{\dot{x}_0}\right)\right). \tag{12.32}$$

Thus, the horizontal range is given by

$$\left(\frac{\dot{z}_o}{\gamma} + \frac{g}{\gamma^2}\right)\frac{\gamma R_{\max}}{\dot{x}_0} + \frac{g}{\gamma^2}\ln\left(1 - \frac{\gamma R_{\max}}{\dot{x}_0}\right). \tag{12.33}$$

This is a *transcendental equation* and must be solved by some approximation. We can expand the logarithmic term by use of the series $\ln(1 - y) = -y - \frac{y^2}{2} - \frac{y^3}{3} - \cdots$, which is valid for all values where $|y| < 1$. For our case $y = \gamma R_{\max}/\dot{x}_0$ and hence we obtain the following expression for $R_{\max}$:

$$R_{\max} = \frac{2\dot{x}_o\dot{z}_o}{g} + \frac{8\dot{x}_o\dot{z}_o^2}{3g^2}\gamma. \tag{12.34}$$

If the projectile is fired with initial speed v_0 at angle of elevation θ, then

$$\begin{cases} \dot{x}_o = v_0\cos(\theta), \\ \dot{z}_o = v_0\sin(\theta), \end{cases} \tag{12.35}$$

and multiplying equations (12.35) we obtain

$$\dot{x}_o\dot{z}_o = v_0^2\cos(\theta)\sin(\theta), \tag{12.36}$$

which can also be expressed as

$$2\dot{x}_o\dot{z}_o = 2v_0^2\cos(\theta)\sin(\theta) = v_0^2\sin(2\theta). \tag{12.37}$$

Then equation (12.34) can be rewritten as

$$R_{\max} = \frac{v_0^2\sin(2\theta)}{g} - \frac{4v_o^3\sin(2\theta)\sin(\theta)\gamma}{3g^2} + \cdots. \tag{12.38}$$

The first term on the right side is the range in the absence of air resistance. The remaining terms are due to air resistance and these are the reason why range with air resistance is less than range without air drag.

Example 12.3

Show that the terminal velocity of a falling spherical object is given by

$$v_T = \left[\left(\frac{mg}{b_2}\right) + \left(\frac{b_1}{2b_2}\right)^2\right]^{1/2} - \left(\frac{b_1}{2b_2}\right)$$

when both the linear and quadratic terms in the drag force are taken into account.

Solution

This question is very simple to solve. Balancing the gravitational force mg against the drag force $b_1v + b_2v^2$ gives the terminal velocity. Thus

$$b_1v + b_2v^2 = mg$$
$$b_1v + b_2v^2 - mg = 0,$$

which has the usual quadratic solution

$$v = \frac{-b_1 \pm [b_1^2 + 4b_2mg]^{1/2}}{2b_2}.$$

This can be rearranged to give

$$v_T = [(mg/b_2) + (b_1/2b_2)^2]^{1/2} - (b_1/2b_2).$$

Example 12.4

The drag experienced by baseball-sized objects with mass approximately equal to 140 g and diameter 7.5 cm traveling at normal speeds is nearly quadratic in v. For smaller objects, such as rain drops (with an approximate diameter 0.1 mm) falling through the atmosphere, the linear term of the drag equation dominates. Given that the above statements hold, calculate the following for the baseball and the raindrop:

(a) The terminal speed (v_T) in air.
(b) The characteristic time (τ).

Solution

Refer back to section 12.3 for the required formulas and then plug in the values to obtain the answers.

12.6 Damped harmonic oscillator in 1D

A simple harmonic oscillator, as we discussed in chapter 7, is an oscillator that oscillates with only the restoring force acting on the system. It is neither driven nor damped. This is an ideal condition where the motion will continue for ever.

Real oscillators are not as described above. Real oscillators are under the influence of restoring force ($F_R = -kx$) and frictional force (f_f). The additional frictional force results in damping of the system and slows the motion of the system. Under the influence of frictional forces, the velocity of the system decreases and this decrease is directly proportional to the frictional force as shown in equation (12.39)

$$F_f = -cv, \tag{12.39}$$

where c is called the *viscous damping coefficient*.

Newton's force equation for damped harmonic oscillators is then

$$\Sigma F = F_{\text{ext}} - kx - cv$$

$$m\ddot{x} = F_{\text{ext}} - kx - c\dot{x}. \tag{12.40}$$

When no external forces are present (i.e. when $F_{\text{ext}} = 0$), equation (12.40) can be rewritten as

$$\ddot{x} + 2\zeta\omega_0\dot{x} + \omega_0^2 x = 0, \tag{12.41}$$

where

$$\omega_0 = \sqrt{\frac{k}{m}} \tag{12.42}$$

is called the *undamped angular frequency* of the oscillator and

$$\zeta \equiv \frac{c}{2\sqrt{mk}} \tag{12.43}$$

is called the *damping ratio*. A damped harmonic oscillator can be over-damped, critically damped or under-damped.

12.6.1 Over-damped ($\zeta > 1$)

The system returns (exponentially decays) to steady state without oscillating. Larger values of the damping ratio ζ return to equilibrium more slowly.

12.6.2 Critically damped ($\zeta = 1$)

The system returns to steady state as quickly as possible without oscillating (although overshoot can occur). This is often desired for the damping of systems such as doors.

12.6.3 Under-damped ($\zeta < 1$)

The system oscillates (with a slightly different frequency than the undamped case) with the amplitude gradually decreasing to zero. The angular frequency of the under-damped harmonic oscillator is given by $\omega_1 = \omega_0\sqrt{1 - \zeta^2}$, and the exponential decay of the under-damped harmonic oscillator is given by $\lambda = \omega_0\zeta$.

12.6.4 The Q factor

The Q factor of a damped oscillator is defined as

$$Q = 2\pi \times \frac{\text{energy stored}}{\text{energy lost per cycle}} \tag{12.44}$$

Q is related to the damping ratio by the equation $Q = \frac{1}{2\zeta}$.

Concept questions 12.2

1. Which raindrops will fall faster, small ones or large ones under no air resistance and with air resistance cases?
2. How would you calculate the drag on a wood log floating downstream in a river?
3. Under what conditions would it be essential to consider the 3D motion of a projectile?

Exercises 12.2

1. For golf-ball sized objects (with mass about 0.046 kg and diameter (d) about 4 cm) traveling at normal speeds, the air drag is almost linear in v. Find the maximum range of a chip shot hit at an angle 60° with respect to the horizontal, with typical value of initial velocity, $v_o = 20$ m s^{-1}:
   (a) For the case where we ignore drag.
   (b) For the realistic case, where drag forces are included.
   How much is the difference in your answers to the above cases?
   Hint: take $b_1 = 0$ and $b_2 = 0.22\,d^2$ where d is the diameter of the object and drag constant $\gamma = \frac{b_1 + b_2 v_0}{m}$.

2. A stone of weight (W) is thrown vertically upward into the air with an initial speed v_o. Suppose that a (constant) air drag force f dissipates an amount fy as the stone travels a distance y.
   (a) Show that the maximum height reached by the stone is

   $$h = \frac{v_o^2}{2g\left(1 + \frac{f}{w}\right)}.$$

   (b) Show that the speed of the stone upon impact with the ground is

   $$v^2 = v_o^2\left[\frac{W - f}{W + f}\right].$$

Further reading

Arya A 1990 *Introduction to Classical Mechanics* 2nd edn (Upper Saddle River, NJ: Prentice-Hall)
Baierlein R 1983 *Newtonian Dynamics* (New York: McGraw-Hill)
Chababy R and Sherwood B 2011 *Matter and Interactions* 3rd edn (New York: Wiley)
Fowles G 1986 *Analytical Mechanics* 4th edn (New York: CBS College Publishing)
Fowles G and Cassiday G 2005 *Analytical Mechanics* 7th edn (Belmont, CA: Thomson Learning)
Goldstein H 1980 *Classical Mechanics* 2nd edn (Reading, MA: Addison-Wesley)
Landau L D and Lifshitz E M 1960 *Mechanics* (Reading, MA: Addison-Wesley)
Ohanian H 1985 *Physics* (New York: WW Norton)
Resnick R, Halliday D and Krane K 2002 *Physics* vol 1, 5th edn (New York: Wiley)
Stephenson R 1961 *Mechanics and Properties of Matter* 2nd edn (New York: Wiley)
Symon K 1971 *Mechanics* 3rd edn (Reading, MA: Addison-Wesley)

Chapter 13

Fluid dynamics

13.1 Fluid dynamics

Fluid dynamics is the branch of physics that is mainly concerned with the movement of fluids. It includes both the motion of liquids and of gases and focuses on the forces that impact these changes.

Liquids include many substances, however, the most commonly studied liquid is water. The study of liquid flow is called *hydrodynamics*. Most applications of hydrodynamics include managing water flows for delivery and extraction of drinking water to and from cities, controlling and maintaining flood areas, and building dams for energy production.

Basically, hydrodynamics deals primarily with the flow of water in pipes or open channels. Open-channel flows include stream or river flows, whereas closed channels may include water pipes or air ducts used for climate control in houses and other structures. In closed channels the boundaries are rigid and the flow can be considered to be always in contact with the boundary, whereas open channels are those whose boundaries are not entirely a solid and rigid material, such as the banks of rivers.

Boundaries play a very important role in fluid flow, since they affect the different types of forces involved in fluid flow. Fluid flows in open channels are driven by gravity alone, whereas in a closed pipe fluid flow is driven by gravity or pressure, which is determined primarily by the height of the fluid above the point of measurement.

13.1.1 History and development

The field of fluid dynamics has been around for a long time. We have historical evidence of the development of aqueducts that carried clean water from miles away to cities. Two main techniques of calculation with fluids were developed around the same time and are explained here:

doi:10.1088/2053-2563/aae1b9ch13

- The first technique was developed by Joseph Lagrange (1736–1813 CE). His idea was to divide the fluid into infinitesimal volume elements, call them *fluid particles*, and then follow the motion of each particle. From this he proposed to calculate the velocities, accelerations and finally calculate forces, etc. However, this was a very tedious proposition since there are too many particles to follow and it is too much work.
- Another technique was developed by Leonhard Euler (1707–1783 CE). In this method, Euler proposed that we do not look at each individual *fluid particle*, rather we should specify the density (ρ) and velocity (v) of the fluid at a particular point in space at each instant of time. Basically, we define the motion of the fluid by defining ρ and v at a particular physical point (x,y,z) as a function of time (t):

$$\begin{aligned} \rho &= \rho(x, y, z, t) \\ \vec{v} &= \vec{v}(x, y, z, t). \end{aligned} \tag{13.1}$$

  In other words, instead of looking at the fluid particles individually and writing equations of motion for each we can stand at a particular point and note the ρ and v of particles and hence find their trajectories as the particles pass by. Note that even if this method is definitely easier to use, we still have to follow the fluid particles for short intervals (dt) because the laws of mechanics apply on particles and not on points in space.

13.2 Fluid flow

Fluids have no fixed shape and are easily deformed, and when they are acted upon by a force they will start to flow. Generally flow is defined in terms of pressure, density and flow velocity at every point of the fluid. The most common force acting on fluids on Earth is the force of gravity but other forces also play a role in fluid flow.

Fluid flow due to the force of gravity
Fluids under the influence of gravity flow and hence can be poured from one container to another, which is easily observed in liquids, like water.

Influence of other forces
Suppose you have a container filled with a fluid and suppose you move it. The forces caused by acceleration, deceleration or change in direction will cause the fluid to flow or change its shape inside the container. Another example is the force of wind that acts on a body of water which causes surface waves on the water surface.

13.2.1 General characteristic fluid flow

Fluid flow is complicated but we can generally simplify and characterize it in terms of the following quantities to make it easier to handle various situations when dealing with fluid flow.

The various types of fluid flows:

- *Steady flow*. When variables, such as pressure, density, flow velocity are constant in time the flow is called *steady flow*, which happens at low speeds.
- *Non-steady flow*. Flow is called non-steady flow when velocities are a function of time.
- *Laminar flow*. When fluid flows in parallel layers it is called laminar flow. Laminar flow is useful and very desirable in various situations for liquids and gases, such as in drainage or in airflow over cars and airplane wings, because it is more efficient and has less energy loss. Laminar flow can be either steady or unsteady, but a steady flow is always laminar.
- *Turbulent flow*. In turbulent flow velocities vary erratically from point to point and from time to time, which makes these kinds of flows difficult to study and predict. Rapids and waterfalls are examples of turbulent flows. Turbulent flow can be useful for causing different fluids to mix together or for equalizing temperatures.

Fluid flow can be compressible or incompressible:

- *Compressible*. A fluid is called compressible if its density changes with the applied pressure. Most gases under normal conditions are considered compressible.
- *Incompressible*. When the density of the fluid is constant, i.e. the fluid density is independent of x, y, z, t, it is called incompressible. The flow of an incompressible fluid is also called incompressible flow. Usually liquids and highly compressed gases are considered incompressible.

Fluid flow can be viscous and non-viscous:

- *Viscous flow*. Viscosity in a fluid is like friction between solid objects. Friction hinders the motion of solids on other solids and viscosity hinders the flow of a fluid. A large viscosity means a large frictional component and hence a slow fluid flow. Friction could be due to the fluid itself, such as between the layers flowing at different rates, or it could be between the surface of the fluid and the container.
- *Non-viscous flow*. Non-viscous flow is fluid flow with no friction and hence with no energy loss. A non-viscous fluid would flow without slowing down because of a lack of a resistive force.

Fluid flow can be rotational or irrotational:

- *Irrotational flow*. The volume elements in irrotational flow do not rotate about their center of mass as they move along their path without rotating. Mathematically we can say that

$$\text{curl of } \vec{v} = (\nabla \times \vec{v}) = 0.$$

- *Rotational flow*. In rotational flow the volume elements rotate about their center of mass as they move. Here,

$$\text{curl of } \vec{v} = (\nabla \times \vec{v}) \neq 0.$$

Think of a small toothpick dropped in a vortex formed at a bathtub drain. If it moves with the stream in a circular path but does not rotate about an axis through its center of mass, the flow is called irrotational, but if as the toothpick moves down it also rotates about its center of mass the flow is called rotational.

13.2.2 Visualizing fluid flow

Physicists visualize flow using figures called streamlines and/or pathlines.

A streamline can be defined as

'A continuous line within a fluid such that the tangent at each point is the direction of the velocity vector at that point '.

A streamline shows the direction of the flow at any particular point in the flow. No two streamlines can ever cross each other. If they did the incoming particle will have a choice on where to go, which in classical physics never happens!

A pathline is the trajectory of an individual element of fluid. If the flow does not change over time, the pathline will be the same as the streamline. However, in the case of turbulent or unsteady flow, these lines can be quite different.

In a steady flow, the pattern of streamlines does not change with time. In principle, we can draw a streamline through every point in the fluid. For a steady flow we can select a finite number of streamlines to form a bundle, and the resulting tube will be called a *tube of flow*. The boundary of such a tube will be made of streamlines, which means that no fluid will cross it, which implies that the tube of flow behaves like a virtual pipe with no holes. Fluid will enter the tube at one end and will leave at the other end without any leakage on the way, as shown in figure 13.1.

13.2.3 Equation of continuity

An equation that describes the movement of quantities that flow from one point to another is called the *continuity equation*, also the *equation of continuity*. In fluid

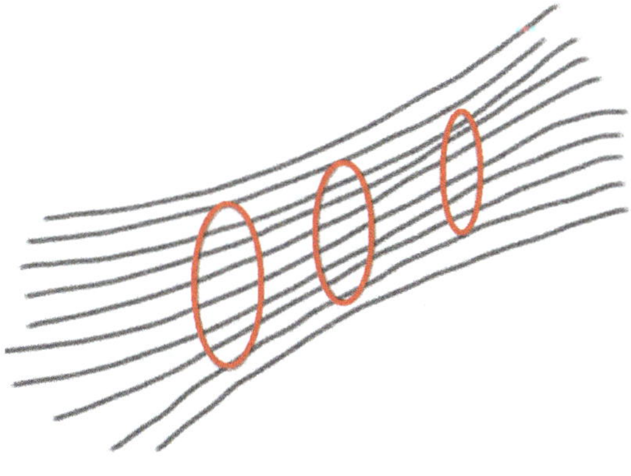

Figure 13.1. Tubular flow.

dynamics, the quantity that flows is the mass and hence the equation of continuity states that

> 'If in a volume element of space (not a volume of fluid) there are no sources, where additional matter can be injected, or sinks, where mass loss can occur, then the total mass within that volume element will remain constant'.

In other words, when a fluid is in motion, it must move in such a way that its mass is conserved. Suppose the fluid enters at one end of a tube defined by area (A_1) at speed (v_1) and it leaves at the other end defined by area (A_2) at speed (v_2). Fluid that crosses A_1 in time (dt) has volume

$$dV_1 = A_1 v_1 dt,$$

so the mass of fluid crossing area A_1 will be $dm_1 = \rho_1 A_1 v_1 dt$ and the mass flux will be defined as

$$\text{mass flux} = \frac{\text{mass of fluid}}{\text{unit time}} = \frac{dm_1}{dt} = \rho_1 A_1 v_1.$$

Similarly, mass flux at A_2 will be

$$\frac{dm_2}{dt} = \rho_2 A_2 v_2.$$

For a steady flow, conservation of mass says that mass flux at points 1 and 2 must be the same, hence

$$\frac{dm_2}{dt} = \rho_2 A_2 v_2 = \frac{dm_1}{dt} = \rho_1 A_1 v_1$$

$$\rho_2 A_2 v_2 = \rho_1 A_1 v_1, \tag{13.2}$$

which can also be written as

$$\rho A v = \text{constant}. \tag{13.3}$$

When the fluid is incompressible, we can write ($\rho_1 = \rho_2$) and equation of continuity may be rewritten as

$$A_2 v_2 = A_1 v_1 = \text{constant}. \tag{13.4}$$

Hence, we can define the *volume flow rate* or *volume flux* as

$$R \equiv Av = \text{constant}, \tag{13.5}$$

with the SI units = $m^3\ s^{-1}$ = (volume per time).

Equations (13.2), (13.3) and (13.4) are all various forms of the equation of continuity. Additionally, from the equation of continuity we can see that in the narrow part of the tube, the distance between the streamlines will decrease and they

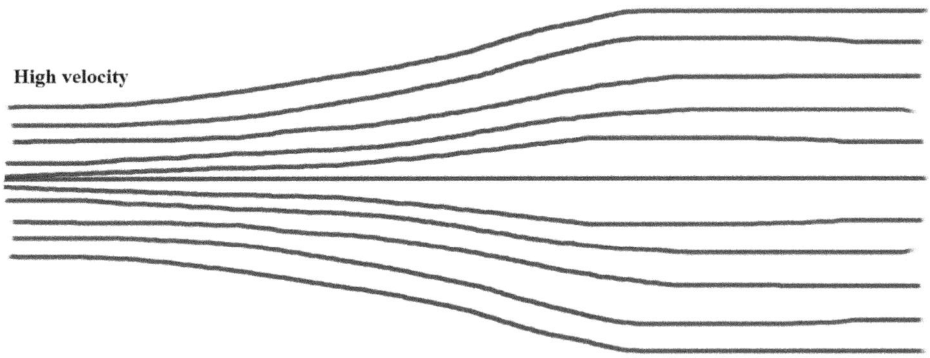

Figure 13.2. Wide streamlines versus narrow streamlines.

will crowd together, which can be seen in figure 13.2. This means that the speed of the streamlines must increase to keep the product (Av) constant:

- Wide streamlines correspond to regions of relatively low speeds and have higher internal fluid pressure.
- Narrow streamlines correspond to regions of relatively high speeds and have decreased internal fluid pressure.

For cases where there are sources and sinks present, the equation of continuity gives the mathematical representation of a reasonable assertion that the rate of outflow or inflow of matter is equal to the rate at which the mass contained within the volume element is changing.

Example 13.1

A stream of water from a faucet necks down as it falls, as seen in figure 13.3. The cross-sectional areas are $A_1 = 1\ \text{cm}^2$ and $A_2 = 0.5\ \text{cm}^2$. At what rate does the water flow from the tap, if the distance between the two levels is 20 cm?

Solution

The rate the water flows from the tap is given by equation (13.5). In order to use this equation we need to find the velocity v_1. To find v_1 use the equation of motion and equation of continuity as

$$v_f^2 = v_i^2 + 2gH,$$

and from the equation of continuity we obtain

$$A_1 v_1 = A_2 v_2$$

$$v_1 = \frac{A_2 v_2}{A_1}$$

$$v_1 = \frac{A_2 \sqrt{v_1^2 + 2gH}}{A_1}$$

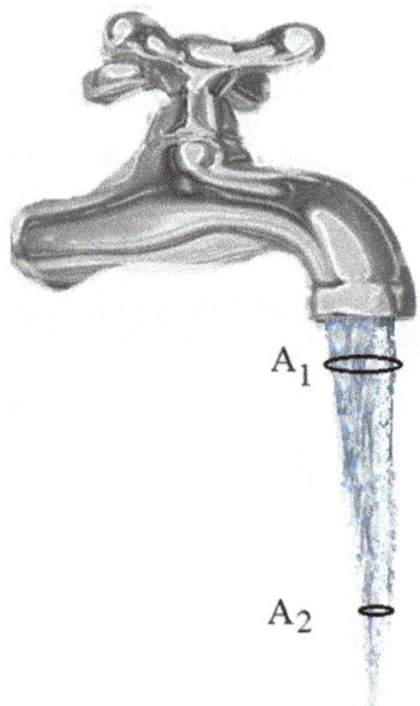

Figure 13.3. Stream of water from a faucet.

and we can rearrange to obtain

$$\left(\left(\frac{A_1}{A_2}\right)^2 - 1\right)v_1^2 = 2gH$$

from which, by putting in values, we obtain $v_1 = 1.14 \text{ m s}^{-1}$. Hence we obtain the rate of flow $= 1.14 \times 10^{-4} \text{ m}^3 \text{ s}^{-1}$.

13.2.4 Computational fluid dynamics

Most problems in fluid dynamics are too complex to be solved by direct calculations. Fluid dynamics problems are solved by numeric methods using computer simulations. This area of study is called numerical or computational fluid dynamics (CFD).

However, turbulent flow tends to be nonlinear and chaotic and so we need to take particular care in setting up the rules and initial conditions for these simulations. Small initial condition changes can result in large differences in the results.

13.3 Viscosity

Viscosity in fluid flow is similar to friction in solid body motion. Informally, viscosity is the quantity that describes a fluid's resistance to flow.

In order to move an object over another, we have to apply an external force ($\vec{F}_{\text{ext}}$) to get it started against frictional forces. For fluid flow, in the same way, we have to apply external forces to get the fluid to start to flow against viscosity. However, unlike in solids, there is no minimum force needed to get the fluid moving, at least for Newtonian fluids. Even the slightest force will cause fluid flow, no matter how high the viscosity is.

To handle fluid flow, we divide the fluid into parallel fluid plates, separated by a distance (D), resting one on top of another. When we apply F_{ext} to the top of the liquid it is like we are applying F_{ext} to the upper liquid plate. In response the upper plate will move at a constant $\vec{v}$ relative to the plate below it. This will continue for each respective plate until the lowest plate, which will remain almost at rest, as shown in figure 13.3.

Since we have divided the fluid into layers, viscosity acts between each layer of fluid and between the outermost layer and the contact surface. Suppose during the motion the speed of each layer differs by an amount (dv) from the one below it and the velocity of the upper plate is sufficiently small, then the liquid moves with horizontal streamlines, or with laminar flow. The magnitude of viscosity under these conditions depends on the temperature and pressure, but it does not depend on the distance between the plates and the magnitude of external force (F_{ext}).

Mathematical formula
The mathematical formula for viscosity is derived from experimental observations, and we find after many experimental observations that to set up laminar flow:

1. F_{ext} must be proportional to the area of the plate (A).
2. F_{ext} must be proportional to the change in velocity (dv) that occurs across each layer of thickness (dy).

We can write this as

$$F_{ext} \propto A\frac{dv}{dy} \tag{13.6}$$

and hence

$$F_{ext} = \eta A\frac{dv}{dy}, \tag{13.7}$$

where η is the coefficient of proportionality, which we call the coefficient of viscosity or simply *viscosity*.

The SI unit of viscosity is the pascal second [Pa s], which has no special name. However, Pa s is rarely used and the most common unit of viscosity is the dyne second per square centimeter [dyne s cm^{-2}], or the poise [P] after the French physiologist Jean Poiseuille (1799–1869 CE).

13.3.1 Factors affecting viscosity

- Viscosity is basically a function of the material under consideration. Most liquids have viscosities on the order of [1–1000] mPa s, while gases have viscosities on the order of [1–10] μPa s. The viscosity of water at 20° is 1.0020 mPa s and motor oil is about 0.11 mPa s at 0° and 0.03 mPa s at 20°.
- The viscosity of materials varies with temperature. In general, the viscosity of liquids decreases with increasing temperature because as temperature increases the kinetic energy of the molecules increases, which weakens the intermolecular forces. Honey flows better when heated and engine oils and hydraulic fluids thicken on cold days affecting the performance of cars.

  On the other hand, in gases the increase in temperature increases the kinetic energy of gas particles which increases the frequency of intermolecular collisions, hence increasing the viscosity of the gas. For gases the viscosity is found to be approximately proportional to the square root of temperature.
- Viscosity is normally independent of pressure. Since liquids are normally incompressible, an increase in pressure does not really bring the molecules closer. However, liquids under extreme pressure often experience an increase in viscosity.

13.3.2 Poiseuille's law

The flow of fluids through an IV catheter can be described by Poiseuille's law. In the nineteenth century during his study of flow in blood vessels, French physician Jean-Leonard Poiseullie discovered that the flow rate is proportional to the fourth power of the radius

$$\frac{\Delta V}{\Delta t} \propto R^4, \tag{13.8}$$

where $\Delta V/\Delta t$ is called the *volume flow rate.*

Poiseullie's law states that the flow of fluid is related to a number of factors, such as the density (ρ) and the viscosity (η) of the fluid, the length (L) and inner radius (R) of the tubing and ΔP, which is the pressure gradient between the two ends of the tubing. Poiseullie's law can then be mathematically given as

$$\frac{\Delta V}{\Delta t} = \frac{\pi \Delta P}{8\eta L} R^4. \tag{13.9}$$

Knowing the definition of density ($\rho = \frac{\text{mass}}{\text{volume}}$), this equation can also be written as

$$\frac{dm}{dt} = \frac{\rho \pi \Delta P}{8\eta L} R^4, \tag{13.10}$$

where the term ($\frac{dm}{dt}$) is called the *mass flux.*

In a pipe the slow moving molecules will be near the walls and fast moving molecules will be near the central axis, as seen in figure 13.4. The velocity at the center of the pipe is given as

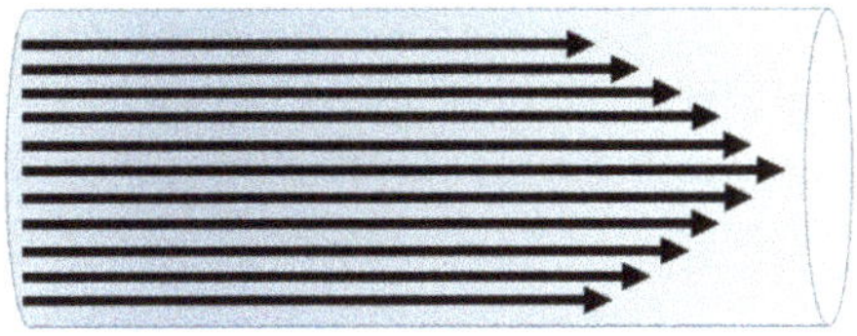

Figure 13.4. Poiseullie law.

$$v_{\text{center}} = \frac{R^2 \Delta P}{4\eta L}. \tag{13.11}$$

Please note that knowing η we can calculate the pressure difference (ΔP) that must be provided by an external force (say a pump) to maintain a mass flux through the cylinder. Vice versa, if we know the mass flux, we can find viscosity (η).

Example 13.2

Suppose a particular oil with a density of 7000 kg m^{-3} at room temperature is forced through a circular pipe with a diameter of 4 cm and length 1 m by a pump that maintains a gauge pressure of 1000 Pa. The oil coming out of the other end of the pipe at atmospheric pressure is collected. After 5 min, a total of 40 kg is collected. What is the η of this oil at room temperature?

Solution

The max flux = $(\frac{dm}{dt}) = \frac{40\ \text{kg}}{300\ s} = 0.13\ \text{kg s}^{-1}$. From equation (13.10) we know

$$\begin{aligned} \frac{dm}{dt} &= \frac{\rho\pi\Delta P}{8\eta L}R^4 \\ 0.13\ \text{kgs}^{-1} &= \frac{(7000)\pi(999\ \text{Pa})}{8\eta(1\ \text{m})}(0.02)^4 \end{aligned} \tag{13.12}$$

$$\eta = 3.4\ \frac{\text{Ns}}{\text{m}^2}. \tag{13.13}$$

13.3.3 Application of viscous flow

The dependence of the flow rate on radius of the pipe is very important when we talk about blood flow in the human body. Plaque deposits narrows the arteries in a person that suffers from cardiovascular disease, making it harder for blood to flow through. This means in order to maintain normal blood flow the heart has to pump harder and hence the blood pressure increases. For example, if a blood vessel contracted to half of its original diameter, we see from equation (13.9) that the flow rate will reduce to one-sixteenth of its original value. To compensate for that, the heart will pump harder, which places increased demand on the heart muscle and will result in many health problems.

Example 13.3

A man finds out that the radius of one of his arteries has reduced by 12%. To maintain normal blood pressure, what percent increase in blood pressure across this artery is required for this person? Assume the viscosity of the blood does not change.

Solution

Suppose we take r_1 as the original radius and r_2 as the reduced radius, which means $r_2 = 88\%\ r_1$. To maintain normal blood flow, the volume flow rate must stay the same. Then we can write equation (13.9) as

$$\frac{\pi \Delta P_1}{8\eta_1 L_1} r_1^4 = \frac{\pi \Delta P_2}{8\eta_2 L_2} r_2^4. \tag{13.14}$$

Solving the above equation, given $\eta_1 = \eta_2$,

$$\Delta P_1 r_1^4 = \Delta P_2 r_2^4. \tag{13.15}$$

Then we can solve for the pressure drop

$$\frac{\Delta P_1}{\Delta P_2} = \frac{r_1^4}{r_2^4} = 1.67. \tag{13.16}$$

This means that there is a 67% increase in blood pressure difference in that artery, which must be provided by the heart. The additional strain on the heart will not bode well for the man in the long term.

Suppose the normal pressure drop across this artery is 10 mm Hg, it will now be 16.7 mm Hg. The person's blood pressure either must increase by 6.7 mm Hg or there will be a reduction in the blood flow.

13.4 Bernoulli's principle

Daniel Bernoulli (1700–1782 CE), was a Swiss mathematician, physicist and physician. In 1738 CE, he applied the conservation of mechanical energy to the motion of incompressible liquid having a steady flow. As an ideal fluid flows through a pipe or a tube of flow, its condition changes and it can change in several ways:

- The cross-sectional area of pipe may change.
- The inlet and outlet of a pipe may be at different elevations.
- The inlet and outlet pressure may be different.

We know the equation of continuity relates changes in area to changes in velocity. Pressure difference and elevation change can both accelerate (or decelerate) fluids, which leads to a change in velocity. Changes of the types mentioned above are not independent. Bernoulli's principle relates the speed of a fluid to the pressure of the fluid. It states,

> 'Within a horizontal flow of fluid, points of higher fluid speed will have less pressure than points of slower fluid speed'.

In other words, Bernoulli's principle says that in a horizontal pipe when diameter decreases, the fluid will move faster and that will result in lower pressure. This is called the *Bernoulli effect*, and it seems counter-intuitive since we generally associate higher speeds with higher pressures. But think about it this way, as the velocity increases through the constriction the kinetic energy increases. This increase in kinetic energy occurs at the expense of pressure energy, hence pressure goes down.

In fact, Bernoulli's principle can be considered a statement of the conservation of energy appropriate for flowing fluids.

13.4.1 Derivation of Bernoulli's equation

Bernoulli's equation relates the pressure, speed and height of any two points (1 and 2) in a steady streamline flowing fluid of density ρ. The variables A_1, P_1, v_1, h_1 refer to the area of the tube, pressure, speed and height of the fluid at point 1, whereas the variables A_2, P_2, v_2, h_2 refer to the area of the tube, pressure, speed and height of the fluid at point 2 (figure 13.5).

Let $h = h_1 - h_2$ and if A_2 is smaller than A_1, it implies that v_1 is less than v_2. If A_1 is higher in altitude than A_2, then since the pressure is reduced from A_1 to A_2, the fluid will be accelerated going from A_1 to A_2. This means the liquid will gain kinetic energy at the expense of both pressure energy due to pressure and potential energy due to height.

If we suppose that in time (Δt), a mass ($\Delta m = \rho V = \rho A_1 \mathrm{x} = \rho A_1(v_1 \mathrm{t}) = \rho A_1 \mathrm{v}_1 \Delta\ t$) of liquid passes through A_1, then conservation of mass says that the same mass should pass through A_2. The work done by gravity on this mass as it moves from A_1 to A_2 is

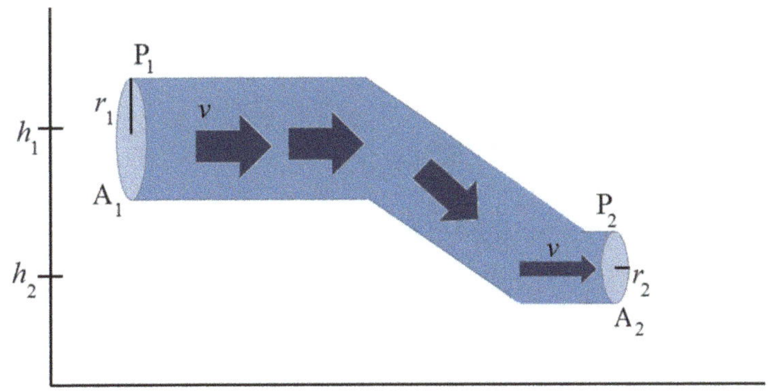

Figure 13.5. Bernoulli's equation.

$$W = (\Delta m)gh. \tag{13.17}$$

The work done by pressure forces on mass (Δm) having volume = $A_1 \mathrm{v}_1 \Delta t = A_2 \mathrm{v}_2 \Delta t$ will be

$$\begin{aligned} W &= \int_1^2 \rho dV \\ &= P_1(A_1 v_1 \Delta t) - P_2(A_2 v_2 \Delta t). \end{aligned} \tag{13.18}$$

Then from conservation of energy

$$W = \Delta T + \Delta V$$

$$P_1(A_1 v_1 \Delta t) - P_2(A_2 v_2 \Delta t) = \frac{1}{2}\Delta m(v_2^2 - v_1^2) + g\Delta m(h_2 - h_1),$$

dividing thoughout by Δm and rearranging, we obtain

$$P_1 + \frac{1}{2}\rho v_1^2 + \rho g h_1 = P_2 + \frac{1}{2}\rho v_2^2 + \rho g h_2 = \text{constant}, \tag{13.19}$$

where P_1 is the pressure energy term, $(\frac{1}{2}\rho v_1^2)$ represents kinetic energy per unit volume and $(\rho g h_1)$ is potential energy per unit volume. We can rewrite the above equation as

$$\boxed{P + \frac{1}{2}\rho v + \rho g h = \text{constant}}. \tag{13.20}$$

Equation (13.20) gives Bernoulli's equation for steady, incompressible, non-viscous and irrotational flow along the same streamline. Another way to express equation (13.19) is

$$\frac{P}{\rho} + hg + \frac{v^2}{2} = \text{constant}, \tag{13.21}$$

where the first term $(\frac{P}{\rho})$ expresses the potential energy per mass of liquid due to pressure and is called *pressure head*. The second term (hg) is the potential energy per mass of liquid due to gravity and is called *elevation head*. The term $(\frac{v^2}{2})$ is the kinetic energy per mass of liquid.

13.4.2 Features of Bernoulli's equation

Static pressure

Fluid statics is a special case of fluid dynamics. In equation (13.19) when fluids are not moving we can put, $v_1 = v_2 = 0$,

$$\begin{aligned} p_1 + \rho g h_1 &= p_2 + \rho g h_2 \\ p_1 - p_2 &= \rho g(h_2 - h_1) \end{aligned}$$

$$\boxed{\Delta p = \rho g \Delta h}. \tag{13.22}$$

which is called *static pressure.*

Dynamic pressure
Suppose the fluid flows horizontally. We know for a horizontally moving fluid $h_1 = h_2$ and we can write equation (13.19) as

$$p_1 + \frac{1}{2}\rho v_1^2 = p_2 + \frac{1}{2}\rho v_2^2$$

$$\boxed{p_1 - p_2 = \frac{1}{2}\rho g(v_2^2 - v_1^2)}. \tag{13.23}$$

When speed is large, the pressure must be small and vice versa. It is worth noting here that the term $(\frac{1}{2}\rho v^2)$ has the dimension of pressure and hence we call it *dynamic pressure.*

Compressible, viscous flow
If the fluid is compressible, its internal potential energy will change as the molecules come closer to each other or move away from each other. If the flow is viscous, the internal kinetic energy of molecules can change the same way as frictional forces between solid objects can increase their kinetic energy. The complete analysis of fluids using conservation of energy should therefore include an internal energy term:

$$\Delta E_{\text{int}} = \Delta T_{\text{int}} + \Delta V_{\text{int}}.$$

This might account for both of these effects mentioned above,

$$W_{\text{ext}} = \Delta E_{\text{int}} + \Delta T, \tag{13.24}$$

which is the generalized statement of conservation of energy. Note that Bernoulli's equation could be modified to account for energy transformation. However, if the flow is approximately incompressible and non-viscous these corrections are negligibly small.

13.4.3 Torricelli's theorem

Torricelli's law, also known as *Torricelli's theorem,* is a theorem in fluid dynamics relating the speed of fluid flowing out of an opening to the height of fluid above the opening.

Suppose you have a water tank sitting on a horizontal surface and it has a hole from which water can leak. The velocity of discharge of the water from the hole is given by Torricelli's theorem. To derive an expression for it, consider the hole is a distance (h) below the surface of the liquid. Using Bernoulli's equation

$$P_1 + \frac{1}{2}\rho g v_1^2 + \rho g y_1 = P_2 + \frac{1}{2}\rho g v_2^2 + \rho g y_2.$$

Since both the surface and the hole are open, we have $P_1 = P_2 = 1$ atm and $y_2 - y_1 = h$ and $v_2 = 0$, so the above equation reduces to

$$\frac{1}{2}\rho g v_1^2 + \rho g(h) = 0,$$

which can be shown by a bit of algebra to be

$$v_1 = \sqrt{2gh}.$$

This is called Torricelli's theorem. Note carefully that this is identical to the equation for a falling object when we do not account for drag.

Example 13.4

A water tank standing on the floor has two small holes vertically one above the another punched in one side. The holes are 6 cm and 12 cm above the floor. How high does the water stand in the tank when the jets from the holes hit the floor at the same point?

Solution

For the upper hole $v_2 = \sqrt{2g(y - y_2)}$.

The initial velocities are $v_{ox} = v_o$; $v_{oy} = 0$.

After time t,

$$\Delta x_2 = v_{ox2}t_2 = v_{o2}t_2$$

$$\Delta y_2 = v_{oy2}t_2 + \frac{1}{2}gt_2^2.$$

For the lower hole $v_1 = \sqrt{2g(y - y_1)}$.

Similarly, for the lower hole

$$\Delta x_1 = v_{ox1}t_1 = v_{o1}t_1$$

$$\Delta y_1 = v_{oy1}t_1 + \frac{1}{2}gt_1^2.$$

Given that the distances that jets from both holes hit the floor at the same points,

$$\Longrightarrow v_{o1}t_1 = v_{o2}t_2. \tag{13.25}$$

Putting in values from the equations above and simplifying we obtain

$$y = y_1 + y_2 = 18 \text{ cm}. \tag{13.26}$$

13.4.4 Application of Bernoulli's equation and equation of continuity

- *Venturi meter*. Venturi meters measure fluid speed in a pipe. The flow measurement is done by adding a constriction (of cross-sectional area A_2) into a normal pipe, of cross-sectional area (A_1). Two vertical tubes are

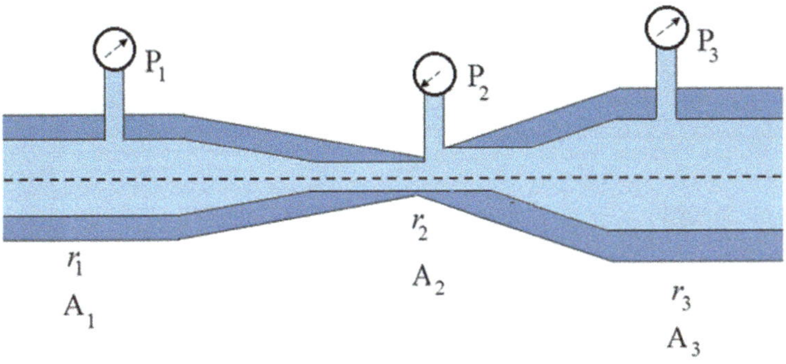

Figure 13.6. Venturi meter.

attached to each pipe. These tubes are open to the atmosphere and function as manometers[1] that measure pressures. When the fluid passes the constricted section it speeds up, which results in a corresponding pressure drop. Knowing the values of these quantities we can calculate the flow rate through the pipe. Venturi meters have been in use for many years, in the water supply (figure 13.6).

- *Pitot tube.* A pitot tube, based on the Bernoulli equation, can be used to measure fluid flow velocity by converting the kinetic energy in a fluid flow to potential energy. It is a simple instrument that measures the difference between static, total and dynamic pressure. The pressure difference (*dp*) can be measured and calculated with u-tube manometers, electronic pressure transducers, etc.

  A pitot tube is used in wind tunnel experiments and on airplanes to measure flow speed. It is a slender tube that has two holes on it. The front hole is placed in the airstream to measure the stagnation pressure[2]. The side hole measures the static pressure. By measuring the difference between these pressures, you obtain the dynamic pressure, which can be used to calculate airspeed (figure 13.7).
- *Dynamic lift.* A fluid flowing past the surface of a body exerts a force on it. Lift is the component of this force that is perpendicular to the oncoming flow direction. It contrasts with the drag force, which is the component of the surface force parallel to the flow direction. If the fluid is air, the force is called an *aerodynamic force.* In water, the same force is called a *hydrodynamic force.*

Example 12.5

Suppose we have a water pipe with cross-sectional area A_1 and we introduce a constriction into it, where the area of the constriction is half of the area of the pipe.

[1] A manometer is a device that measure pressures. A simple manometer consists of a U shaped tube of glass filled with some liquid, say mercury, because of its high density.

[2] The stagnation pressure, also called total pressure, p_0, is the pressure measured at the point where the fluid comes to rest. It is the highest pressure found anywhere in the flow field, and it occurs at the stagnation (rest) point.

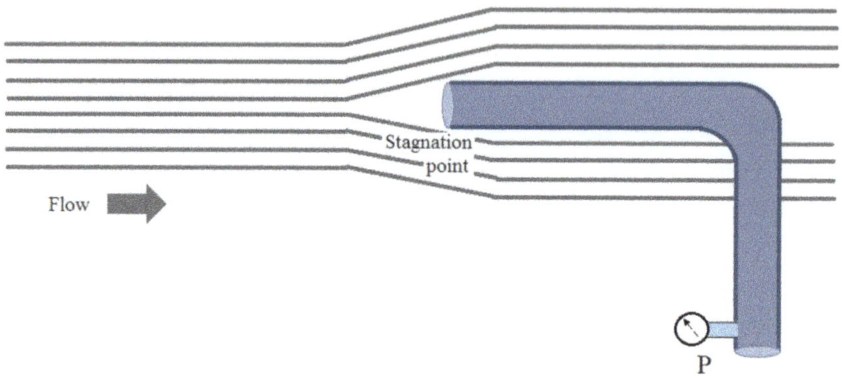

Figure 13.7. Pitot tube.

The corresponding fluid heights in the vertical tubes are $h_1 = 2$ m and $h_2 = 1$ m. Calculate the following:

(a) v_2 in terms of v_1.

Solution: From the equation of continuity:

$$\frac{v_2}{v_1} = \frac{A_2}{A_1} = 2.$$

This means that the water flows twice as fast in the constriction than the rest of the pipe.

(b) The gauge pressures for the vertical tubes. *Hint*: take the density of water to be approximately 1000 kg m^{-3}. *Solution*:

$$P_1 = \rho g h_1 = 19.6 \text{ kPa}$$
$$P_2 = \rho g h_2 = 9.8 \text{ kPa}.$$

Example 12.6

A water storage tower that supplies water to town is 6 m in diameter and 40 m in height supplies water to a three-story house and an eleven-story apartment building. A horizontal pipe at the base of the tower which has a diameter of 10 cm needs to supply water at the rate of 2.5×10^{-3} m^3 s^{-1} to both the house and the apartment building (figure 13.8).

(a) If the water were flowing at the maximum rate, what would be the pressure in the horizontal pipe at the base of both the house and the apartment building?

(b) A smaller pipe of diameter $d' = 5$ cm supplies the third floor of the house, a distance of 10 m above the ground level. What are the flow speed and water pressure in this pipe? Neglect the viscosity of water.

(c) Suppose the same diameter ($d'' = 5$ cm) pipe supplies water to the eleventh floor of the apartment building, a distance of 40 m above the ground level. What are the flow speed and water pressure in this pipe? Neglect the viscosity of water.

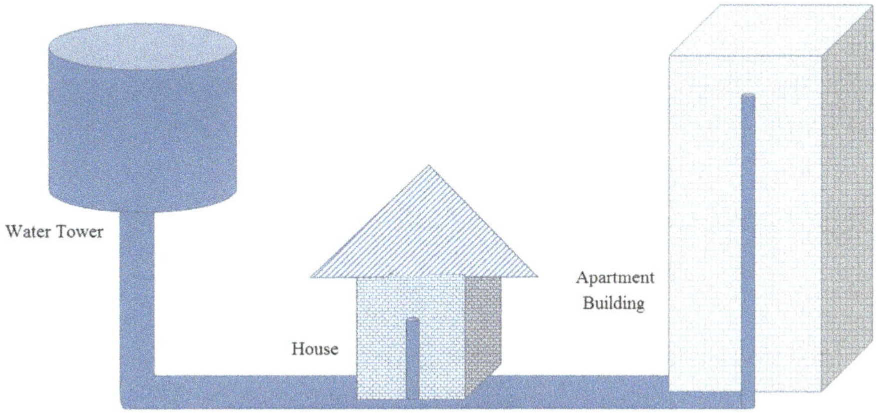

Figure 13.8. Water tower, house and apartment building.

Solution

(a) Bernoulli's equation along the streamline will be given by equation (13.19),

$$P_A + \frac{1}{2}\rho v_A^2 + \rho g h_A = P_B + \frac{1}{2}\rho v_B^2 + \rho g h_B. \tag{13.27}$$

At point A, pressure is the atmospheric pressure (P_o) and $y_A - y_B = 32$ m, velocities (v_A and v_B) are determined by the equation of continuity as

$$v_A = \frac{R}{A_A} = \frac{R}{\pi r_A^2} = 8.85 \times 10^{-5}\ \text{ms}^{-1}$$
$$v_B = \frac{R}{A_B} = \frac{R}{\pi r_B^2} = 0.318\ \text{ms}^{-1}.$$

Putting the values in equation (13.27), we obtain $P_B \cong 3.9$ atm.

It is interesting to note that under static conditions, i.e. when water is not flowing ($v_A = v_B = 0$), pressure at B will be higher:

$$P_B = P_o + \rho g h_{AB} = 4.9\ \text{Pa}.$$

(b) The smaller pipe needs to have the same flow rate (R), hence

$$v_c = \frac{R}{A_C} = \frac{R}{\pi r_C^2} = 1.27\ \text{ms}^{-1}.$$

Then Bernoulli's equation becomes

$$P_c = P_A + \frac{1}{2}\rho(v_A^2 - v_c^2) + \rho g(h_{AC}), \tag{13.28}$$

from which we obtain $p_c \approx 1.96$ atm.

Because of the larger flow velocity through the smaller pipe the dynamic contribution to the to the pressure is much larger than it is at point B. Both static and dynamic effects tend to reduce the pressure at this location relative to the pressure at point B.

(c) The smaller pipe needs to have the same flow rate (R), hence

$$v_D = \frac{R}{A_D} = \frac{R}{\pi r_D^2} = 1.27 \text{ ms}^{-1}.$$

Then Bernoulli's equation becomes

$$P_D = P_A + \frac{1}{2}\rho(v_A^2 - v_D^2) + \rho g(h_{AD}). \tag{13.29}$$

From which we obtain $p_D \approx 1$ atm.

Since the pressures at point A and point D are the same implies no water will get up to the eleventh floor, which will be a big issue for people living on that floor. It might be advisable to either not build the apartment building that high, or to increase the height of the water tower, or to install an external water pump to pump the water up to the eleventh floor.

Concept questions 13.1

1. Explain why rainwater falls as droplets whereas it flows in a continuous stream down a vertical pipe?
2. Why does a steam of water from a water tap becomes narrower as it falls?
3. If you were outside on the surface of Moon, would you be able to drink water from a regular glass with a straw? What about if you were in your pressurized habitat?
4. Think about why car companies recommend *multi-viscous* engine oil to be used in very cold weather?

Exercises 13.1

1. At two points on a horizontal tube of varying circular cross-section carrying water, the radii are 1 cm and 0.5 cm and the pressure difference between these points is 5 cm of water. How much liquid flows through the tube per second?
2. Models of torpedoes are sometimes tested in a horizontal pipe of flowing water, much as a wind tunnel is used to test model airplanes. Consider a circular pipe of internal diameter 25 cm and a torpedo model, aligned along the axis of the pipe, with a diameter of 5 cm. The torpedo is to be tested with water flowing past it at 3 m s^{-1}.
   (a) With what speed must the water flow in the unconstricted part of the pipe?
   (b) Find the pressure difference between the constricted and unconstricted parts of the pipe.
3. Using Poiseullie's law, as given in equation (13.9), show that the viscosity has SI units of Pascal seconds.

13.5 Velocity of the fall of a sphere through a viscous liquid

This problem was first solved by Sir George Stokes, around 1850. The methodology is beyond the scope of this book. However, we will discuss a partial solution by dimensional analysis. What do we need to do that? We need:

- Physical quantities that determine the force exerted on a sphere moving with constant velocity through a viscous fluid.
- Some intelligent guesswork and/or intuition.

Suppose a sphere moves through a viscous liquid with a constant but small velocity such that there is a streamline motion. Since we have assumed constant velocity, the applied force must be exactly equal and opposite to the frictional force exerted by the liquid on the sphere. The frictional force (F_f) depends on the following:

- The radius (r) of the sphere.
- Viscosity (η) of liquid.
- Velocity (v) with which the sphere moves.

This means that we can write

$$\boxed{F_f = k\ r^{\alpha}\eta^{\beta}v^{\gamma}}, \tag{13.30}$$

where k is the constant of proportionality. In terms of dimensional analysis we can write

$$[M]^1[L]^1[T]^{-2} = [L]^{\alpha}[ML^{-1}T^{-1}]^{\beta}[LT^{-1}]^{\gamma}$$
$$[M]^1[L]^1[T]^{-2} = [L]^{\alpha-\beta+\gamma}[M]^{\beta}[T]^{-\beta-\gamma}.$$

Equating powers

$$\text{mass } [M]\text{: } 1 = \beta$$
$$\text{length } [L]\text{: } 1 = \alpha - \beta + \gamma$$
$$\text{time } [T]\text{: } -2 = -\beta - \gamma.$$

Solving the above expressions and putting the values in equation (13.30) we obtain

$$F_f = k\ r\eta v. \tag{13.31}$$

The numerical value for the proportionality constant k is obtained by experimentation and it is found to be equal to 6π. Putting this value in equation (13.31) gives

$$\boxed{F = 6\pi\ r\eta v}. \tag{13.32}$$

This is called *Stokes' theorem*.

13.5.1 Velocity of a sphere through a viscous liquid

For a sphere falling at a constant velocity through a liquid, the effective weight of the sphere is balanced by the frictional force of the liquid. This effective weight is given by Archimedes' principle,

$$W_{\text{eff}} = \rho g V$$
$$W_{\text{eff}} = (\rho_s - \rho_l) g \left(\frac{4}{3}\pi r^3\right), \tag{13.33}$$

where ρ_s is the density and r is the radius of the sphere and ρ_l is the density of liquid. Then Stokes' theorem can be written as

$$F = 6\pi r \eta v = (\rho_s - \rho_l) g \left(\frac{4}{3}\pi r^3\right), \tag{13.34}$$

which can be solved to find the velocity of the sphere:

$$v = \left(\frac{2}{9\eta}\right)(\rho_s - \rho_l) g r^2. \tag{13.35}$$

We can use equation (13.35) to find η experimentally because all other quantities can be measured for the system.

However, we have to be careful when considering streamline verses laminar motion. For streamline motion, say a marble falling through glycerin, the motion will be in a straight line and will have constant velocity. In contrast, the same marble falling through water will have irregular (laminar) motion because turbulence may be created in the liquid, which makes it very difficult to analyze.

13.6 Turbulent motion and Reynolds number

At the end of the ninteenth century Sir Osborne Reynolds investigated the change from laminar flow to turbulent flow in liquids as he increased the velocity of the flow. He found that for all liquids there is a particular but different critical velocity at which there is a sudden change from laminar to turbulent flow.

In his experiment, Sir Reynolds allowed fluids to flow through a cylindrical tube. He inserted a narrow thread of colored liquid in it. He found that in all cases:

- Under laminar flow the narrow thread followed a straight line.
- As soon as the critical velocity is reached the thread breaks up and the colored liquid diffuses into the main body of the liquid.

13.6.1 Calculating critical velocity

To find critical velocity (v_c), where there is a sudden change from laminar to turbulent type of motion, we will use dimensional analysis. Our best guess about the quantities that critical velocity depends on include viscosity (η), density of fluid (ρ) and the diameter of the cylindrical tube (d),

$$v_c \propto \eta\rho d$$

$$v_c = R\eta^{\alpha}\rho^{\beta}d^{\gamma}. \tag{13.36}$$

Here, R is the constant of proportionality and is called the *Reynolds number*. It is dimensionless.

$$[L]^1[T]^{-1} = [ML^{-1}T^{-1}]^{\alpha}[ML^{-3}]^{\beta}[L]^{\gamma},$$

equating powers,

$$\begin{aligned}&\text{mass } [M]\text{: } 0 = \alpha + \beta\\ &\text{length } [L]\text{: } 1 = -\alpha - 3\beta + \gamma\\ &\text{time } [T]\text{: } -1 = -\alpha.\end{aligned}$$

Solving the above expressions and putting values in equation (13.36), we obtain the formula for critical velocity,

$$v_c = R\eta\rho^{-1}d^{-1} = \frac{R\eta}{\rho d}, \tag{13.37}$$

where v_c is the average velocity over the cross-section of the tube and R is the Reynolds number.

Note that critical flow speed increases with viscosity. The greater the viscosity (viscous friction) exerted by surrounding fluid the more likely that the flow will be steady.

13.6.2 Reynolds number

The Reynolds number is a dimensionless constant, numerically given by equation (13.38). In this interpretation R can be used to characterize *any* flow and its value can be determined experimentally and hence the Reynolds number is given as

$$R = \frac{\rho d v_c}{\eta}. \tag{13.38}$$

For cylindrical pipes, the critical R is about 2000 m s^{-1}. So if we want to calculate the critical speed for a typical garden hose of diameter of about 2 cm:

$$v_c = \frac{R\eta}{\rho d} = \frac{(2000)(0.001 \ \text{N sm}^{-2})}{(1000 \ \text{kg m}^{-2})(0.02 \ \text{m})}. \tag{13.39}$$

This comes out to be about 0.1 m s^{-1} or 10 cm s^{-1}. This is a very low critical velocity, which means the flow in the water pipes at home is nearly always turbulent.

13.6.3 Turbulence

When an object is pulled through a liquid in such a way that the resulting motion is laminar, the force will vary with velocity. When the motion becomes turbulent, the force will vary with the square of velocity.

With turbulent motion, eddies are set up behind the moving body such that the pressure behind the moving body is less than the front, as shown in figure 13.9, whereas with laminar or streamline motion there are no eddies, since there is not a pressure difference between the front and back of the body.

Hence in turbulent motion the resistive force is mainly due to the pressure difference and velocity and very little due to the viscosity of the liquid.

13.6.3.1 Force due to pressure difference in turbulent motion

The force due to pressure difference in turbulent motion depends on:

- The cross-sectional area (A) of the body perpendicular to the direction of motion
- The velocity (v) of the body relative to the fluid.
- The density (ρ) of the fluid.

Hence we write the resistance force due to pressure difference (F_P) as

$$F_P = kA^{\alpha}v^{\beta}\rho^{\gamma} \tag{13.40}$$

$$[M][L]^{1}[T]^{-2} = [L^{2}]^{\alpha}[LT^{-1}]^{\beta}[ML^{-3}]^{\gamma}.$$

Equating powers,

$$\begin{aligned}&\text{mass } [M]\text{: } 0 = \gamma\\ &\text{length } [L]\text{: } 1 = 2\alpha + \beta - 3\gamma\\ &\text{time } [T]\text{: } -2 = -\beta.\end{aligned}$$

Solving the above expressions and putting values in equation (13.40) we obtain the formula for resistance force due to pressure difference (F_P),

$$F_P = kA\rho v^{2}, \tag{13.41}$$

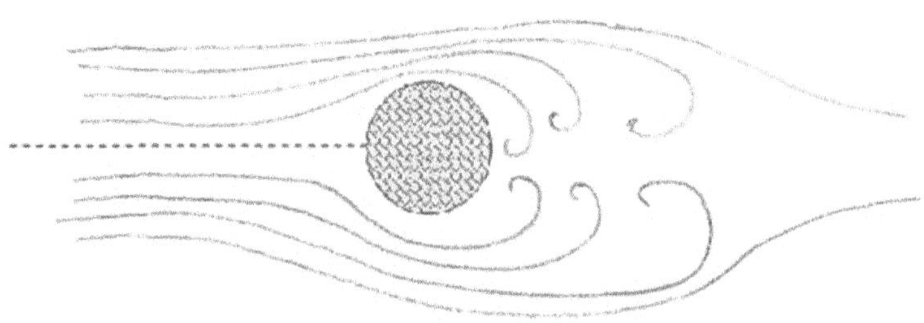

Figure 13.9. Turbulent motion around an obstacle (zoomed-in view).

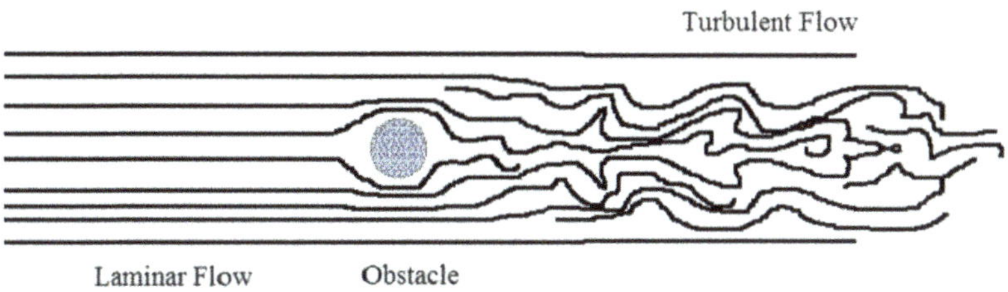

Figure 13.10. Laminar and turbulent motion around an obstacle.

and the value of proportionality constant $k = \frac{F}{A\rho v^2}$, which depends on the shape of the moving object relative to the fluid. For example, k for a flat disc is approximately 0.7, and for a pear-shaped object is approximately 0.3.

13.6.3.2 Examples of turbulence

- Cigarette smoke.

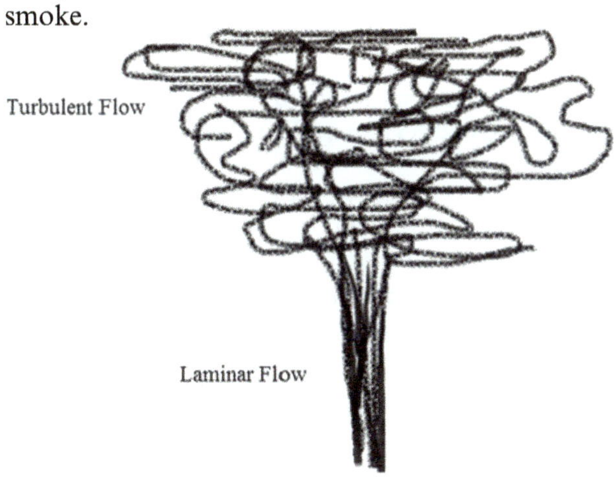

- A stream of fluid past an obstacle breaks into eddies and vortexes, which gives the flow irregular velocity components perpendicular to the flow direction (figure 13.10).
- Wakes left behind in water by moving ships or in air by moving cars and airplanes.
- Sounds produced by whistling and by wind instruments also result from turbulent flow of air.

Concept questions 13.2

1. Why are golf balls dimpled and tennis balls are not? Explain in your own words.
2. What is the difference between static pressure, total pressure and dynamic pressure?
3. What part of a tornado or a hurricane causes the most damage in their path?

Exercises 13.2

1. An engineer aboard a submarine 150 m below sea-level suddenly discovers a leak. Assuming air pressure in the sub remains the same as the pressure at sea-level, calculate the speed with which water enters the sub.
2. To find the electric charge of the electron, a fine mist of oil droplets, with density 860 kg m^{-3} and radius 2 μ m, is sprayed into air and observed as they fall. If the density of air is taken to be 1.2 kg m^{-3} and viscosity to be 1.8 x 10^{-5} Pa s^{-1}, find the terminal speed of the falling oil droplets.

Further reading

Arya A 1990 *Introduction to Classical Mechanics* 2nd edn (Upper Saddle River, NJ: Prentice-Hall)
Baierlein R 1983 *Newtonian Dynamics* (New York: McGraw-Hill)
Chababy R and Sherwood B 2011 *Matter and Interactions* 3rd edn (New York: Wiley)
Fowles G 1986 *Analytical Mechanics* 4th edn (New York: CBS College Publishing)
Fowles G and Cassiday G 2005 *Analytical Mechanics* 7th edn (Belmont, CA: Thomson Learning)
Goldstein H 1980 *Classical Mechanics* 2nd edn (Reading, MA: Addison-Wesley)
Landau L D and Lifshitz E M 1960 *Mechanics* (Reading, MA: Addison-Wesley)
Ohanian H 1985 *Physics* (New York: WW Norton)
Resnick R, Halliday D and Krane K 2002 *Physics* vol 1, 5th edn (New York: Wiley)
Rutherford D E 1951 *Classical Mechanics* (New York: Interscience)
Stephenson R 1961 *Mechanics and Properties of Matter* 2nd edn (New York: Wiley)
Symon K 1971 *Mechanics* 3rd edn (Reading, MA: Addison-Wesley)

IOP Publishing

Techniques of Classical Mechanics
From Lagrangian to Newtonian mechanics
Samya Zain

Chapter 14

Properties of solids

14.1 Solids

In this chapter we will discuss some properties of solids. In order to talk about the properties we will first define the different types of solids.

14.1.1 The main classes of solids

- *Isotropic solids.* Isotropic solids are made of materials having identical values of all properties in all directions. Glass and metals are examples of isotropic solids.
- *Anisotropic solids.* In anisotropic solids some or all properties, such as Young's modulus, change with direction along the object.

Sometimes it is difficult to differentiate between the words homogeneous and isotropic. Depending on the subject, properties and the classification, these terms can be distinguished as follows.

A *homogeneous* solid is a material that has uniform composition and uniform properties throughout it. Metals, ceramics and alloys, that are a mixture of two metals, are examples of homogeneous materials. Homogeneous fluid mixtures are called *solutions*. Salt solutions and air are examples of homogeneous fluid mixtures.

Isotropic means that all physical and mechanical properties of the material are the same in all directions and in all orientations. The isotropic nature of the material depends on its crystal structure. If the grains of the material are not oriented uniformly in all directions, it is not an isotropic material.

To help us understand the difference, note that a constant electric field is a homogeneous field because it is uniform throughout the field, but it is not isotropic because the electric field is directional.

The most simple situation is the case of a homogeneous and isotropic material, hence we will primarily be concentrating on such materials.

doi:10.1088/2053-2563/aae1b9ch14

But before we begin we need to understand a very important property of solids, which is called the *elasticity* of the solid.

14.1.2 Elasticity

Elasticity is the property of an object that allows the object to return to its original shape after it is subjected to a deforming force. Objects that can and do restore themselves to precisely their original configuration are called *perfectly elastic*. On the other hand, objects on which an imposed force causes a permanent deformation are called *plastic*.

A good example of an elastic object is a spring. When a spring is stretched, it exerts a restoring force that tends to bring it back to its original length. This restoring force is generally proportional to the amount of stretch, as described by Hooke's law, given in equation (7.1),

$$F = -kx. \tag{14.1}$$

All real *rigid bodies* are to some extent *elastic*, which means if we change their dimensions slightly by say pulling, twisting, pushing or compressing them, they will try to restore themselves once the force is removed.

The rigidity[1] of *real* rigid bodies is an illusion. Solids are composed of atoms in a lattice structure in which the atoms are not in rigid contact but are bound together by forces that behave like spring forces, as shown in figure 14.1. Each atom is in equilibrium under the influence of the springs that surround it. The effective spring constants are very large, so that it takes a large force to change the separation. This is the cause of our impression of rigidity.

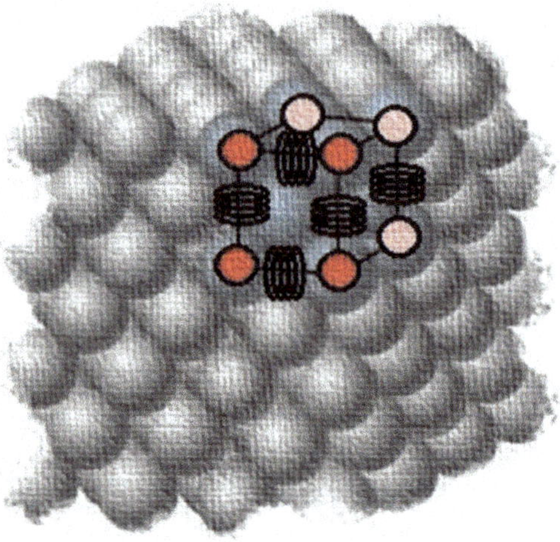

Figure 14.1. Atoms in a lattice structure bound by spring-like forces.

[1] The rigidity is the property of materials that the object made of the particular material will resist even infinitesimal motions.

All objects will stretch to some extent under the influence of forces. By how much the object stretches depends on the object. For example, rubber will stretch under minimal force, whereas a steel rod will require a lot of force to induce a change. In both cases, however, the materials will return to their original state when the force is removed.

If you hang a 1000 kg mass from a 1 m long steel rod, with a diameter of 1 cm, the rod will stretch about 0.5 mm or 0.05% and the rod will go back to its original length after you remove the weight. If you increase the mass to 2000 kg, the rod will be permanently stretched and will not go back to its original state after you remove the weight. Note here that at this point we have exceeded the elastic limit of the rod and it has deformed *plastically*. Next, if you increase the mass to 3000 kg the rod will break. Just before the break, the elongation in the rod will be less than 0.2%.

Tensile strength is the breaking stress that will cause permanent deformation or fracture of a material (figure 14.2).

14.2 Stress

Stress is a directional physical quantity that represents the internal forces that neighboring particles exert on each other in a material. Stress can vary with the

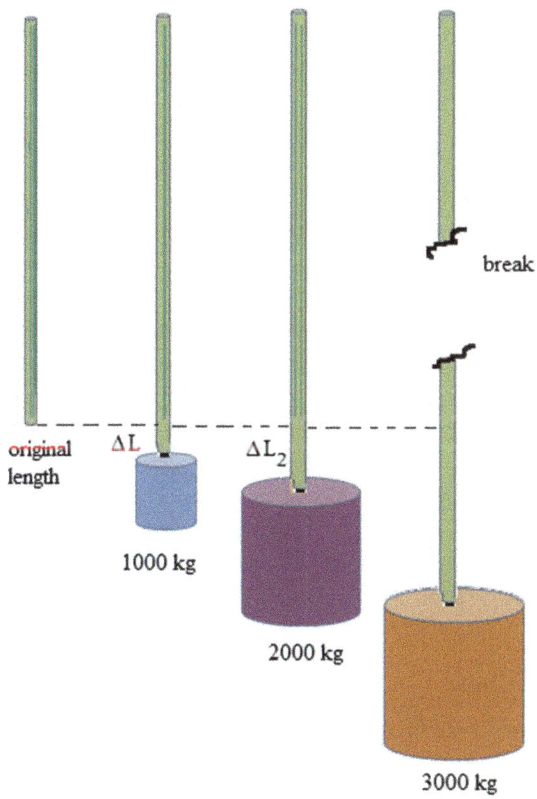

Figure 14.2. A cylindrical rod, under tensile stress, stretched by ΔL.

orientation of the area in question, and the force exerted on the area is itself a vector, which can have components perpendicular and parallel to the area.

Stress is defined as the force across a 'small' boundary per unit area of that boundary, for all orientations of the boundary. Mathematically, if force ΔF acts on an area ΔA at some point of a body, then stress will be defined as

$$\text{stress} \equiv \lim_{\Delta A \to \infty} \frac{\Delta F}{\Delta A}. \tag{14.2}$$

The dimensional formula for stress is $[M][L]^{-1}[T]^{-2}$ which is the same as the dimension of pressure, hence the SI unit of stress is the pascal (Pa $\equiv$ N m^{-1}), which is also the unit for pressure. However, since mechanical stresses easily exceed millions of pascals, the commonly used unit of stress is the megapascal (MPa).

Stress in a body may be due to multiple causes, including external forces such as gravity or internal forces or contact forces with other solids. The relation between stress and its effects or its causes can be quite complicated. However, in limited cases, such as the ones we discuss in this chapter, we can in principle find stresses for various solids.

Stresses can be of two kinds, normal or tangential. Normal stress on a body causes a change in length or a change in volume, whereas tangential stress produces changes in the shape of the object.

Stress that exceeds certain strength limits of the material will result in permanent deformation and can change the crystal structure and chemical composition of the material itself.

14.2.1 Stresses in a beam

If an imaginary surface cuts through any part of a solid structure (say a rod, string, cable, beam) then in general the material on one side of this surface will exert a force on the material on the other side and vice versa. These internal forces which act across any surfaces within the solid are called *stresses in a beam*.

There are mainly three types of stresses we are interested in studying when we talk about solids. These are given in figure 14.3 and are explained as follows:

- *Compression.* If the material on each side of any surface pushes on the material on the other side with a force perpendicular to the surface, the stress is called compression.
- *Tension.* If the stress is a pull perpendicular to the surface, it is called tension.
- *Shearing stress.* If the force exerted across the surface is parallel to the surface, is called shearing stress.

14.2.2 Stresses at an angle

A stress at an angle to a surface can be resolved into two components:

- A shear component.
- A tension or compression component.

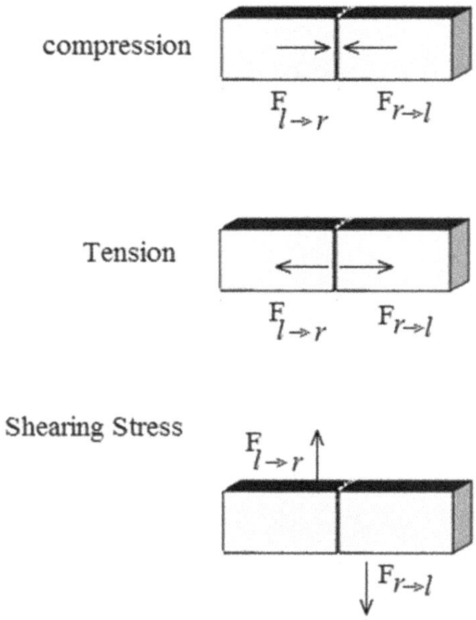

Figure 14.3. $F_{l\to r}$ is the force exerted by left half of the beam on the right side; $F_{r\to l}$ is the force exerted by the right half of the beam on the left side.

In general, stress may act in any direction relative to the surface. However, mostly we consider cases where the stress is either a pure compression or tension and is independent of the orientation of the surface, or cases where only one surface is of interest, so that only a single stress vector is needed to calculate stress as the force per unit area across the surface.

14.3 Strain

Any change from the normal configuration of a body in either size or shape is called a *deformation*. The corresponding fractional change is called a *strain*. In general such a change is the result of a force that acts on the body as a whole or just the surface of the body.

Mathematically the ratio of change produced in the dimensions of a body by a force to its original dimensions is called *strain*. Since strain is a ratio, it is dimensionless.

14.3.1 Types of strain

Strain can be categorized in three ways, depending upon the change produced in a body. The three types of strain are:

- *Longitudinal strain*. It is the ratio of the change in length of a body to the original length of the body. If L is the original length of a wire or a rod and the final length of the wire or the rod is $(L + \Delta L)$ under the action of a normal stress, the change in length is ΔL (figure 14.4):

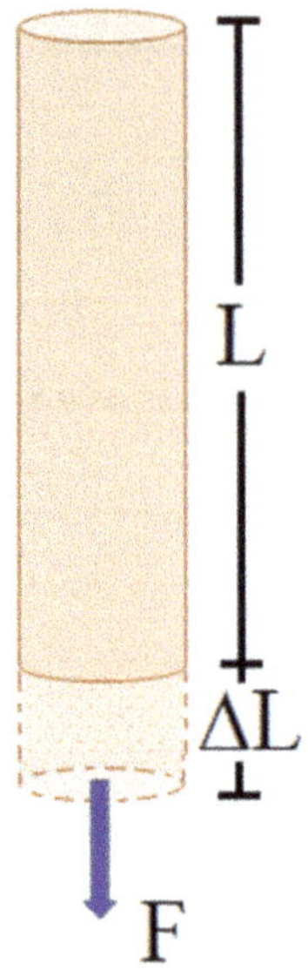

Figure 14.4. A cylindrical rod stretched by ΔL.

$$\text{strain} = \frac{\Delta L}{L}. \tag{14.3}$$

- *Volume strain.* Volume strain is the change in the volume of a body as a result of an external force. If the original volume of the object is V and the change in volume is $(V + \Delta V)$, then volume strain is given as:

$$\text{volume strain} = \frac{\Delta V}{V}. \tag{14.4}$$

- *Shearing strain.* If the angle of a face originally perpendicular to the fixed face is turned, the strain is called a shearing strain. It can also be defined as the ratio of the displacement of a layer to its distance from the fixed layer.

14.3.2 Stress–strain equation

Hooke's law gives a relationship between the stress and strain. According to Hooke's law, within the elastic limit, strain produced in a body is directly proportional to the stress produced (figure 14.5):

$$\frac{\text{stress}}{\text{strain}} = \text{constant} = \text{modulus of elasticity}. \tag{14.5}$$

14.3.3 Moduli of elasticity

Corresponding to the three types of strain, there are three moduli of elasticity:

- Young's modulus (Y or E).
- Bulk modulus (B or K).
- Rigidity modulus (η).

In the sections below we will discuss each one individually.

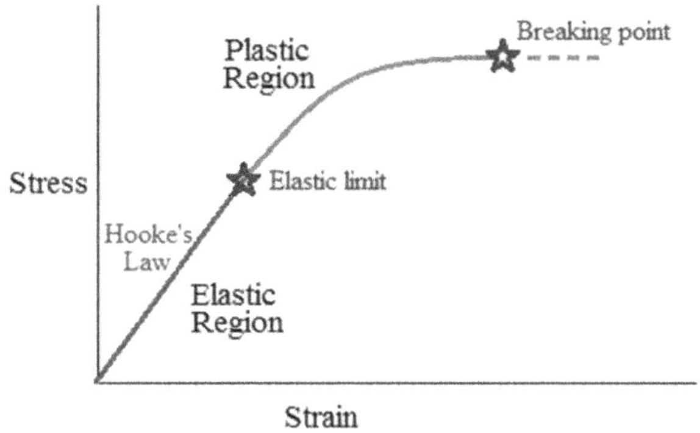

Figure 14.5. Stress versus strain for a material.

14.3.3.1 Longitudinal strain—Young's modulus (Y or E)

Young's modulus is a measure of the stiffness of a solid material. It defines the relationship between longitudinal stress (force per unit area) and strain in a material. The stress–strain curve is linear. Materials such as steel, carbon fiber and glass are usually considered linear, while materials such as rubber are considered non-linear.

Young's modulus is experimentally determined from the slope of a stress–strain curve created during tensile tests conducted on a sample of the material. Mathematically, Young's modulus is the ratio of stress $\sigma(\varepsilon)$ (which has units of pressure) to strain (ε) (which is dimensionless). Hence Young's modulus has units of pressure (pascals or N m^{-2}),

$$Y = \frac{\sigma(\varepsilon)}{\varepsilon} = \frac{F/A}{\Delta L/L_0}, \tag{14.6}$$

where F is the force exerted on an object under tension; A is the actual cross-sectional area through which the force is applied; L_0 is the original length of the object; and ΔL is the amount by which the length of the object changes.

14.3.3.2 Volume strain—bulk modulus (B or K)

The bulk modulus of a substance is a measure of the compressibility of the substance. Here the imposed stress is a pressure which is uniform in all directions. When we increase the pressure on a material by an amount ΔP, its density will go up accordingly. The fractional change in the volume of the material will be given as

$$\text{fractional change in the volume} = \frac{\Delta V}{V}. \tag{14.7}$$

If the fractional change is positive it represents an increase in volume and if it is negative it represents an decrease in the volume of the material. Bulk modulus is defined as the ratio of the infinitesimal pressure increase to the resulting relative decrease of the volume:

$$B = -\frac{\Delta P}{\Delta V/V}. \tag{14.8}$$

The negative sign is added in the definition to make B a positive quantity, since ΔP and $(\frac{\Delta V}{V})$ have opposite signs. The SI unit of bulk modulus (B) is the pascal.

When B is small

When B is small, it implies that small pressure changes produce large changes in the volume of a material. Such a material is called *compressible*. Gases are a good example with $B_{\text{gas}} \approx 10^5\,\text{Nm}^{-2}$. A small pressure change of 0.1 atm changes the value of a gas, $V_{\text{gas}} \approx 10\%$.

When B is large

When B is large, it implies that large pressure changes produce small changes in the volume of a material. This material is nearly *incompressible*. B_{solids} is usually higher than B_{liquids}, usually due to tighter coupling of atoms in solids; $B_{\text{water}} = 2.2 \times 10^9\,\text{Nm}^{-2}$.

Another way to define the bulk modulus formally is

$$K = -V\frac{dP}{dV}, \tag{14.9}$$

where P is pressure, V is volume and dP/dV denotes the derivative of pressure with respect to volume. Equivalently we can write

$$K = \rho\frac{dP}{d\rho}, \tag{14.10}$$

where ρ is density and $dP/d\rho$ denotes the derivative of pressure with respect to density (i.e. the pressure rate of change with volume). The inverse of the bulk modulus gives the compressibility of a substance.

14.3.3.3 Shearing strain (η)

The modulus of rigidity, which is also called the shear modulus (η) is the coefficient of elasticity for a shearing force. It is defined as the ratio of shear stress to the displacement per unit sample length (shear strain).

The modulus of rigidity can be experimentally determined from the slope of a stress–strain curve created during tensile tests conducted on a sample of the material.

The unit for the modulus of rigidity is the pascal and a typical value for aluminum is 24 GPa and for steel 79.3 GPa.

14.3.3.4 Difference between η and Y

The modulus of rigidity (η) measures the stiffness of a material. When an object is subjected to a strong force parallel to its surface the material either slides against the source of the stress, or it slides internally, causing small changes to its volume.

The Young's modulus measures the springiness of a material. When an object is pulled perpendicularly to its surface by strong forces, it stretches. If the forces are strong enough, it may exceed the elastic limit, but this has nothing to do with the difference between η and Y. Airplanes require materials with both high rigidity and elasticity to withstand high winds with little bending. Keep in mind that nothing is strictly unbendable and airplane wings do bend several feet at take-off.

Example 14.1.

A 10 m long steel bar with a rectangular cross-sectional area 0.02 m^2 is hung from a building at one end. On the other end of the bar a load of 1000 kg is attached. By how much is the bar stretched? *Hint*: take $Y_{\text{steel}} = 210$ GPa.

Solution

The Young's modulus (Y) of the metal bar is the ratio of longitudinal stress (F/A) to tensile strain ($\Delta L/L$),

$$Y = \frac{FL}{(\Delta L)A}. \tag{14.11}$$

Putting values in the above equation we obtain

$$210 \times 10^9\ (\text{Nm}^{-2}) = \frac{(1000\ \text{kg})(10\ \text{m})}{(\Delta L)(0.02\ \text{m}^2)}.$$

Hence, the bar is stretched by $\Delta L = (2.4 \times 10^{-6})$ m.

Example 14.2.

A cube has a volume of 3 l when the external pressure is 1 atm. The bulk modulus of the material is 4×10^{12} N m^{-2}.

(a) What is the change in volume when the body is subjected to a pressure of 6 atm?

(b) What additional energy per unit volume is now stored in the material?

Solution

(a) Bulk modulus is given by

$$B = -\frac{\Delta P}{\Delta V/V_0}. \tag{14.12}$$

Rearranging and putting in values, $\Delta P = 5\ \text{atm} \times 1.01 \times 10^5 = 5.05 \times 10^5$ Pa and $V_0 = 3 \times 10^{-3}\ \text{m}^3$,

$$\Delta V = -\frac{\Delta P V_0}{B} = -\frac{(5.05 \times 10^5)(3 \times 10^{-3})}{4 \times 10^{12}} = -3.8 \times 10^{-10}\ \text{m}^3.$$

(b) Additional energy per unit volume is stored in the material (E/V): for small change of pressure dP there will be a change in volume dV. We know $W = -\int PdV$. Then from equation (14.12) we can write

$$dV = -\frac{V_0}{B}dP. \tag{14.13}$$

Multiply both sides by P and integrating we obtain

$$\int PdV = \int \frac{V_0}{B} dP$$

$$W = \frac{V_0}{B} \int PdP$$

$$W = \frac{V_0}{2B}(P_2^2 - P_1^2)$$

$$\frac{W}{V_0} = 4.4 \times 10^{-7} \text{ Jm}^{-3}.$$

14.4 Waves in solids

Suppose a solid rod can be considered to be like a very, very stiff spring. This means that the rod can be compressed or stretched by force. The *elastic modulus* (k) depends on the material and the thickness of the rod and is proportional to the cross section of the rod. The force is mathematically given as

$$F = k\frac{\Delta L}{L}. \tag{14.14}$$

Young's modulus (Y)

We can write $k = YA$, where A is the cross section area (in m^2) and Y is the Young's modulus of the material (in N m^{-2}). Then Hooke's law can be rewritten as

$$\frac{F}{A} = Y\frac{\Delta L}{L}. \tag{14.15}$$

In this equation the left side is force per unit cross section, whereas the right side is Young's modulus multiplied by relative deformation. Equation (14.15) is the microscopic law of elasticity, which is independent of the specific shape of the rod.

Linear mass density (ρ)

Linear mass density is used when describing the characteristics of one-dimensional objects. Linear mass density is the mass per unit of length of the object under consideration and its SI unit is kg/m. For a long, thin uniform rod of mass M and length L, its linear mass density is given as

$$\lambda = \frac{M}{L}. \tag{14.16}$$

For a rod with varying mass, the linear mass density is given as the derivative of the mass function with respect to the position along its length, L:

$$\lambda = \frac{dM}{dL}. \tag{14.17}$$

14.4.1 Velocity of a longitudinal wave in an unlimited homogeneous isotropic medium

In an unlimited homogeneous isotropic medium, the velocity of longitudinal waves (v) is given by the equation

$$v = \sqrt{\frac{B + 4\eta/3}{\rho}}, \tag{14.18}$$

where B is the bulk modulus, η is the shear modulus and ρ is the density of the medium. The velocity of a longitudinal wave in a rod is given by the equation

$$v_l = \sqrt{\frac{Y}{\rho}}, \tag{14.19}$$

where Y is the Young's modulus.

Note that the difference between an unlimited homogeneous isotropic medium and a rod is the fact that when there is a longitudinal strain in the rod there is a lateral strain, whereas in the unlimited homogeneous isotropic medium no lateral strain is present. Also note that the ratio of the proportional decrease in a lateral measurement to the proportional increase in the length in a sample of material that is stretched elastically is called *Poisson's ratio*.

14.4.2 Velocity of a transverse wave in an unlimited homogeneous isotropic medium

In an unlimited homogeneous isotropic medium, the velocity of transverse waves (v_T) is given by the equation

$$v_T = \sqrt{\frac{\eta}{\rho}}. \tag{14.20}$$

We also see that in an unlimited solid

$$v_l > v_T. \tag{14.21}$$

This is important in seismology, where the waves created by earthquakes travel out in all directions. We have already discussed this in chapter 6 so I will only briefly discuss it here.

There are several different kinds of seismic waves. The two main types of waves are body waves and surface waves. Body waves can travel through the Earth's inner layers, but surface waves can only move along the surface of the planet like ripples on water.

The first kind of body wave is the P-wave or primary wave. This is the fastest kind of seismic wave. P-waves are also known as compressional or longitudinal waves. Particles move in the same direction that the P-wave moves in. The second type of body wave is the S-wave or secondary wave. An S-wave is slower than a P-wave and can only move through solid rock, not through any liquid medium. S-waves move rock particles up and down, or side-to-side, perpendicular to the direction that the wave propagates.

By measuring the time interval between the P- and S-waves, we can find the distance of the earthquake from the seismograph. Of course we need to know the difference between the velocities of the two kinds of waves.

Example 14.3.

The bulk modulus of water is 2.04×10^9 Pa. Find the wavelength of a wave with frequency 250 Hz.

Solution

The speed of a wave is given as

$$v = \lambda f. \tag{14.22}$$

We know the density of water is $\rho \approx 1000\ \text{kg m}^3$,

$$v = \sqrt{\frac{B}{\rho}} = \sqrt{\frac{2.04 \times 10^9}{1000}} = 1430\ \text{m/s}.$$

Putting values in equation (14.22) we can find the wavelength as

$$\lambda = \frac{v}{f} = \frac{1430}{250} = 5.7\ \text{m}.$$

14.4.3 Longitudinal wave in a solid rod

In longitudinal waves the displacement of the medium is parallel to the direction of propagation of the wave. A wave in a Slinky toy is a good visualization, which we have seen in chapter 6. Here we will talk about the velocity of propagation of a longitudinal wave in a uniform rectangular rod. Assume our rectangular rod has cross section A, density ρ and is oriented along the x-axis (figure 14.6).

Suppose the rod is subjected to a longitudinal force which varies with position and time. This varying force causes a cross section of the rod to undergo small longitudinal displacement about a mean position. These varying displacements make up the wave motion and result in a longitudinal wave that passes through the rod. For the sake of simplifying the situation we assume that:

- Displacements are within the elastic limit of the material.
- Any damping of the wave during motion can be neglected.

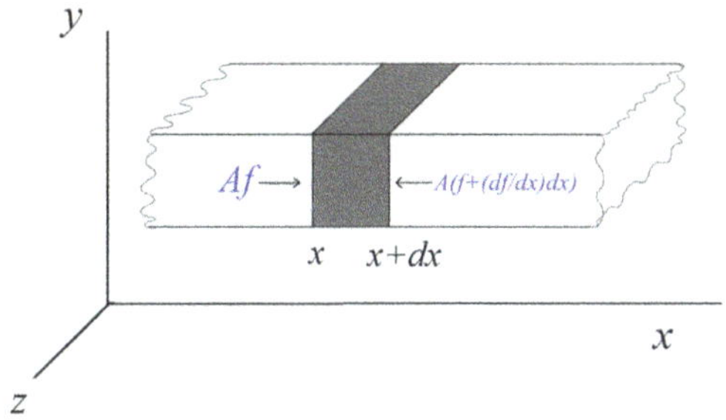

Figure 14.6. Longitudinal waves through a massive rod.

Consider a thin slab of this rod at position (x) and thickness (dx) at some time (t). Let the force per unit area transmitted to the slab by material of the rod at the left of the slab be given as f. Then the total force to the left is given as fA. The force exerted on the slab at ($x + dx$) by the material to the right to a sufficient degree of accuracy can be given by the first terms in Taylor's expansion as $-A(f + \frac{\partial f}{\partial x}dx)$. Then the total resulting force acting on the slab will be

$$F_{\text{ext}} = -A\left(\frac{\partial f}{\partial x}\right)dx. \tag{14.23}$$

Suppose ξ is the displacement from the equilibrium position of the cross section in the x-direction. This is the displacement of the slab with mass ($= \rho A dx$). Then, from Newton's second law,

$$\begin{aligned} \Sigma F &= F_{\text{ext}} \\ m\frac{\partial^2 \xi}{dt^2} &= -\ A\left(\frac{\partial f}{\partial x}\right)dx \\ \rho A dx\frac{\partial^2 \xi}{dt^2} &= -\ A\left(\frac{\partial f}{\partial x}\right)dx, \end{aligned}$$

which can be simplified to

$$\rho\frac{\partial^2 \xi}{dt^2} = -\frac{\partial f}{\partial x}. \tag{14.24}$$

Since the stress on the rod is different at different points and times, it follows that the strain must vary with position and time.

At position (x) the displacement is $= \xi$.

At position ($x + dx$) the displacement is $= \xi + \frac{\partial \xi}{\partial x}dx$.

The strain on the slab is given as

$$\text{strain} = \frac{\text{resultant displacement}}{\text{original length}} = \frac{\partial \xi}{\partial x}.$$

Then, from the definition of Young's modulus for a rod,

$$Y = \frac{\text{stress}}{\text{strain}} = \frac{-f}{\left(\frac{\partial \xi}{\partial x}\right)}.$$

When $(\frac{\partial \xi}{\partial x})$ is positive, the forces must be extension forces, whereas, when $(\frac{\partial \xi}{\partial x})$ is negative, then the forces are of compression. Differentiating w.r.t. x we can rewrite the above equation as

$$Y\frac{\partial^2 \xi}{\partial x^2} = -\frac{\partial f}{\partial x}.$$

Putting in values from equation (14.24) we obtain

$$Y\frac{\partial^2 \xi}{\partial x^2} = \rho\frac{\partial^2 \xi}{\partial t^2}. \tag{14.25}$$

This is the wave equation for a longitudinal wave traveling in the rod, with a solution of the form

$$\xi(x, t) = A\cos(\omega t), \tag{14.26}$$

with phase velocity

$$v = \sqrt{\frac{Y}{\rho}},$$

and the natural frequencies of the rod are given as

$$\boxed{f_n = \frac{2n-1}{4L}\sqrt{\frac{Y}{\rho}}},$$

where n is any positive integer. Please note that equation (14.25) for a longitudinal wave is a plane wave as long as the displacement ξ is the same at all points on any given cross section.

Concept questions 14.1

1. Can the concepts of Young's modulus and shear modulus apply to fluids?
2. Can Young's modulus have a negative value? What about the bulk modulus?
3. Can compressive stress be applied to a rubber band? Explain in your own words.
4. Why can a cat jump from a tree branch to the ground and run away unhurt but if a man tried to do that he would likely break a bone in the fall?

Exercises 14.1

1. If Young's modulus for steel is 210 GPa, how much force will be required to stretch a sample of wire 1 mm^2 in cross section by 0.2 % of its original length?
2. The density of a cubic meter of sea water at sea-level is 64 kg m^{-3}. Find the volume occupied by 100 kg of sea water at the sea bed where water pressure is 4000 Pa. Take the compressibility of sea water (k) = 36 $\times 10^{-7}$ in^2/lb. *Hint*: by definition: $k = \frac{\text{strain}}{\text{stress}} = \frac{\Delta V / V}{P}$.

Further reading

Arya A 1990 *Introduction to Classical Mechanics* 2nd edn (Upper Saddle River, NJ: Prentice-Hall)
Baierlein R 1983 *Newtonian Dynamics* (New York: McGraw-Hill)
Fowles G 1986 *Analytical Mechanics* 4th edn (New York: CBS College Publishing)
Fowles G and Cassiday G 2005 *Analytical Mechanics* 7th edn (Belmont, CA: Thomson Learning)
Goldstein H 1980 *Classical Mechanics* 2nd edn (Reading, MA: Addison-Wesley)
Landau L D and Lifshitz E M 1960 *Mechanics* (Reading, MA: Addison-Wesley)
Ohanian H 1985 *Physics* (New York: WW Norton)
Resnick R, Halliday D and Krane K 2002 *Physics* vol 1, 5th edn (New York: Wiley
Stephenson R 1961 *Mechanics and Properties of Matter* 2nd edn (New York: Wiley)
Symon K 1971 *Mechanics* 3rd edn (Reading, MA: Addison-Wesley)

Chapter 15

Rotation—motion of rigid bodies

For the mechanics of a many-particle system, we will have two situations, one in which the particles of the system all move independently of each another, called *free particles*, and the other where all particles are fixed in their positions with respect to each other, called *rigid bodies*. In this chapter we will concentrate on the dynamics of rigid bodies, but we will talk about free particles in chapters 16 and 17.

15.1 Rigid bodies

A body is called a point mass if it occupies one point in space and is called an extended body when it occupies many points in space. An extended body will be called a rigid body if the distance between any two points in the body always remains the same. The location of a rigid body can be described by the position of a particular point (P) inside it and angles that describes the rotation of the body. Note that one angle is needed to describe motion in 2D, three angles are required in 3D.

In reality no physical body is either a point mass or perfectly rigid, as no body can really be a single point with no extent, and no extended body can be exactly rigid. When forces, no matter how small or large are applied to an object, the object is always deformed slightly. However, objects tend to not want to deform and hence they provide the counter force to the external force. Calling a body rigid means that the changes in the dimensions of the body are small enough to be neglected.

15.2 Moment of inertia

The moment of inertia is for rotational motion what mass is for linear motion. The moment of inertia may also be called *angular mass*. The SI unit for the moment of inertia is the kilogram-square meter (kg m^2) and the usual symbol is (I).

The moment of inertia depends on the mass distribution of the object under consideration and the chosen axis. The moment of inertia for a point mass is mass times the square of perpendicular distance to the rotation axis, ($I = mr^2$). For extended bodies, the moment of inertia is given as ($I = \Sigma m_i r_i^2$) or

doi:10.1088/2053-2563/aae1b9ch15

$$\boxed{I = \int r^2 dm}, \tag{15.1}$$

where r is the perpendicular distance from the element of mass to the axis of rotation and the element of mass (dm) is given by a density factor multiplied by an appropriate differential for length, area or volume. For a length element, dm will be

$$dm = \lambda d\ell.$$

For an area element, dm will be

$$dm = \sigma da.$$

For a volume element, dm will be

$$dm = \rho dv.$$

In the case of a composite body, which is made up of many components, the moment of inertia will be given as

$$I = I_1 + I_2 + \ldots, \tag{15.2}$$

where I_1, I_2, … are the moments of inertia of the various parts about the chosen axis.

15.2.1 Moment of inertia for extended objects

We calculate the moments of inertia for some important special cases. In each of the cases consider m as the mass of the object under consideration.

15.2.2 Thin solid rod (1D)

For a thin, uniform rod of length L and mass m, we can find the moment of inertia starting from equation (15.1) as

$$\begin{aligned} I &= \int r^2 dm \\ &= \int r^2 \rho dV \\ &= \int r^2 \rho (A dx) \\ &= \int r^2 \frac{M}{AL} dA dx \\ \Longrightarrow I &= \frac{M}{L} \int x^2 dx. \end{aligned}$$

15.2.2.1 Thin solid rod (1D)—about its center

To find the moment of inertia for a thin, uniform rod of length L and mass m and an axis perpendicular to the rod about its center, we will take the limit of integration from $-L/2$ and $L/2$ if we place the origin at the center (figure 15.1). The moment of inertia will be calculated as

$$I = \frac{M}{L}\int_{-L/2}^{L/2} x^2 dx$$
$$= \frac{M}{L}\frac{x^3}{3}\Big|_{-L/2}^{L/2}.$$

After a bit of algebra we obtain

$$I = \frac{1}{12}ML^2. \tag{15.3}$$

15.2.2.2 Thin solid rod (1D)—about one edge

For a thin, uniform rod of length L and mass m and an axis perpendicular to the rod at one end, the moment of inertia will be calculated by taking the origin as the edge we need to calculate the moment of inertia, as (figure 15.2)

$$I = \frac{M}{L}\int_{0}^{L} x^2 dx$$
$$= \frac{M}{L}\frac{x^3}{3}\Big|_{0}^{L}.$$

After a bit of algebra we obtain

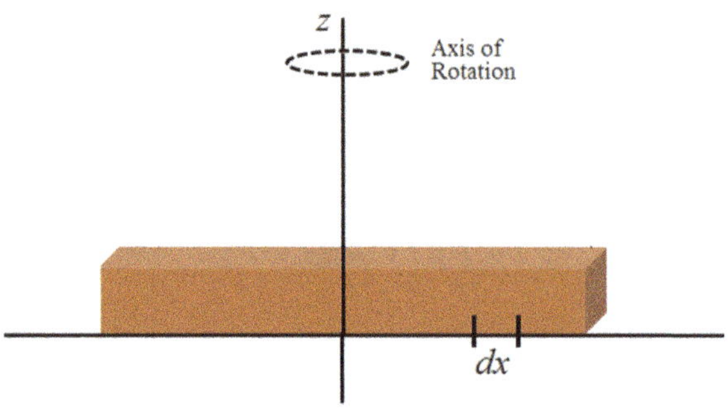

Figure 15.1. The moment of inertia for a thin solid rod, about its center.

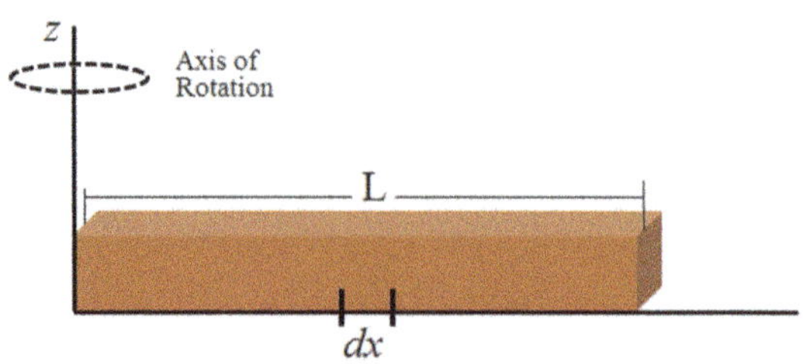

Figure 15.2. The moment of inertia for a thin solid rod, about one edge.

$$I = \frac{1}{3}ML^2. \tag{15.4}$$

15.2.3 Thin solid rod (1D)—at a distance H from one side

For a thin, uniform rod of length L and mass m and a perpendicular axis of rotation at a distance H from one side, the moment of inertia will be calculated as

$$\begin{aligned} I &= \frac{M}{L}\int_{-H}^{L-H} x^2 dx \\ &= \frac{M}{L}\frac{x^3}{3}\bigg|_{-H}^{L-H}, \end{aligned}$$

After a bit of algebra we obtain

$$I = \frac{1}{3}M[L^2 - 3H(L - H)]. \tag{15.5}$$

15.2.4 Uniform rectangular plate (2D)

We can use the idea we developed in section 15.2.2.1, equation (15.3), for the moment of inertia for a thin solid rod in 1D and expand it to find the moment of inertia for a uniform rectangular plate with sides of length a and b, about a perpendicular axis through the center, as follows,

$$\begin{aligned} I &= I_a + I_b \\ &= \frac{M}{12}a^2 + \frac{M}{12}b^2 \\ \Longrightarrow I &= \frac{M}{12}(a^2 + b^2), \end{aligned}$$

and use the idea we developed in section 15.2.2.2, equation (15.4), an axis perpendicular to one end:

$$\begin{aligned} I &= I_a + I_b \\ &= \frac{M}{3}a^2 + \frac{M}{3}b^2 \\ \Longrightarrow I &= \frac{M}{3}(a^2 + b^2). \end{aligned}$$

15.2.5 Uniform hoop (1D)

The derivation of the moment of inertia about the hoop axis for a uniform hoop starting from equation (15.1) is (figure 15.3)

$$
\begin{aligned}
I &= \int r^2 dm \\
&= R^2 \int dm \\
\Longrightarrow I &= MR^2.
\end{aligned}
$$

15.2.6 Uniform disk (2D)

For a 2D disk we can use polar coordinates $(r,\ \phi)$ and surface density σ and $dA = rdrd\phi$. We also know that the area of a disk is given as $A = \pi\ R^2$. The derivation of the moment of inertia for a uniform disk will be as follows:

$$
\begin{aligned}
I &= \int r^2 dm \\
&= \int r^2(\sigma dA) \\
&= \int\int r^2(\sigma rdrd\phi) \\
&= \sigma \int_0^R r^3 dr \int_0^{2\pi} d\phi \\
&= \sigma\ \frac{r^4}{4}\Big|_0^R (2\pi) \\
&= \frac{M}{(\pi R^2)} \frac{R^4}{4}(2\pi) \\
\Longrightarrow I &= \frac{1}{2} MR^2.
\end{aligned}
$$

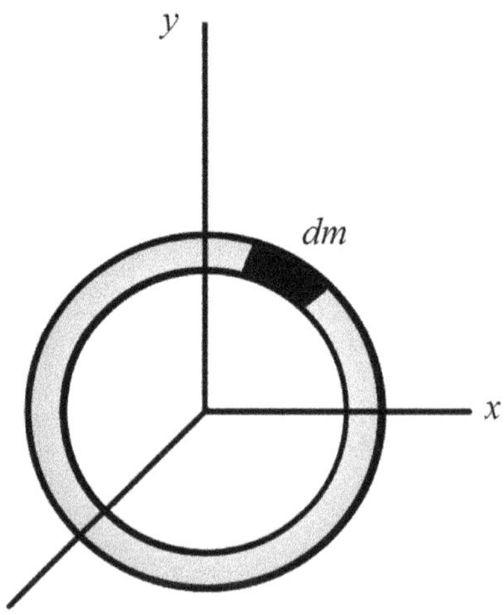

Figure 15.3. The moment of inertia for a uniform hoop.

The moment of inertia of a thin circular disk is the same as that for a solid cylinder, with respect to the axis of rotation.

15.2.7 Parallel-axis theorem

Parallel-axis theorem states that:

> 'The moment of inertia of a rigid body about any axis is equal to the moment of inertia about a parallel axis passing through the center of mass plus the product of the mass of the body and the square of the distance between the two axes.'

Uniform circular disk 1

We can use the parallel-axis theorem to calculate the moment of inertia of a uniform circular disk about an axis in the plane of the disk and tangent to an edge as shown in figure 15.4,

$$I = \frac{1}{4}MR^2 + MR^2 = \frac{5}{4}MR^2. \tag{15.6}$$

Uniform circular disk 2

We can also use the parallel-axis theorem to calculate the moment of inertia of a uniform circular disk about an axis perpendicular to the plane of the disk and passing through an edge, as shown in figure 15.5,

$$I = \frac{1}{2}MR^2 + MR^2 = \frac{3}{2}MR^2. \tag{15.7}$$

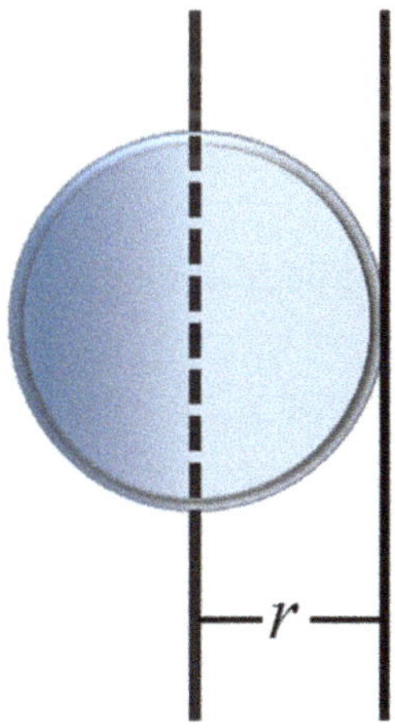

Figure 15.4. Moment of inertia of a uniform circular disk using the parallel-axis theorem.

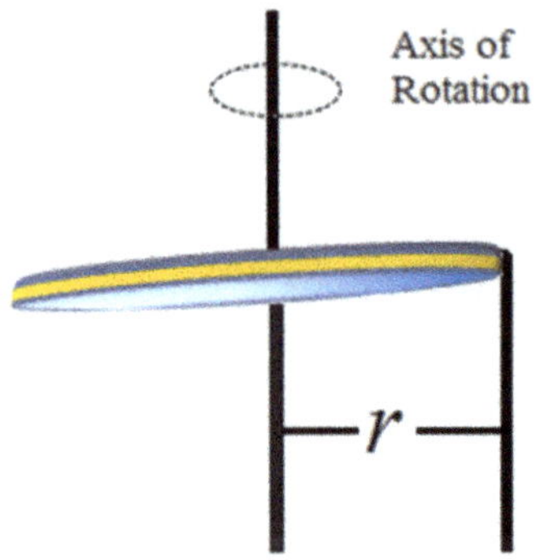

Figure 15.5. The moment of inertia of a uniform circular disk using the parallel-axis theorem.

15.2.8 Cylinder (3D)

For the derivation of the moment of inertia for a uniform cylinder we use cylindrical coordinates (r, ϕ, z), volume $\rho = \pi r^2 L$ and volume density $dV = rdrd\phi dz$. Then I about its central axis is

$$
\begin{aligned}
I &= \int r^2 dm \\
&= \int r^2 (\rho dV) \\
&= \rho \int\int\int r^2 (rdrd\phi dz) \\
&= \rho \int_0^R r^3 dr \int_0^{2\pi} d\phi \int_0^L dz \\
&= \rho \frac{r^4}{4}\Big|_0^R (2\pi)(L) \\
&= \frac{M}{(\pi R^2 L)} \frac{R^4}{4} (2\pi L) \\
\Longrightarrow I &= \frac{1}{2} MR^2.
\end{aligned}
$$

Note that the moment of inertia about one edge will be found by the parallel-axis theorem as

$$
\begin{aligned}
I &= I_{cm} + MR^2 \\
&= \frac{1}{2} MR^2 + MR^2 \\
\Longrightarrow I &= \frac{3}{2} MR^2.
\end{aligned}
$$

15.2.9 Sphere (3D)

For the derivation of the moment of inertia for a sphere we can use spherical coordinates (r, ϕ, θ), which makes it a bit easier to calculate and use the standard formulas for volume $V = \frac{4}{3}\pi r^3$ and volume element $dV = (r^2 dr)\sin\theta d\theta d\phi$. Taking the z-axis to be the axis of rotation, the radius from the z-axis is $R = r\sin\theta$ and the moment of inertia (I) about the z-axis of a uniform sphere will hence be given as (figure 15.6)

$$
\begin{aligned}
I &= \int (r\sin\theta)^2\, dm \\
&= \int (r\sin\theta)^2 (\rho dV) \\
&= \rho \int\int\int (r\sin\theta)^2 (r^2 \sin\theta dr d\theta d\phi) \\
&= \rho \int_0^R r^4 dr \int_0^\pi \sin^3\theta d\theta \int_0^{2\pi} d\phi \\
&= \rho\, \frac{r^5}{5}\Big|_0^R \left(\frac{4}{3}\right)(2\pi)
\end{aligned}
$$

$$
= \frac{M}{\left(\frac{4}{3}\pi R^3\right)} \frac{R^5}{5}\left(\frac{4}{3}\right)(2\pi)
$$

$$
\Longrightarrow I = \frac{2}{5} MR^2.
$$

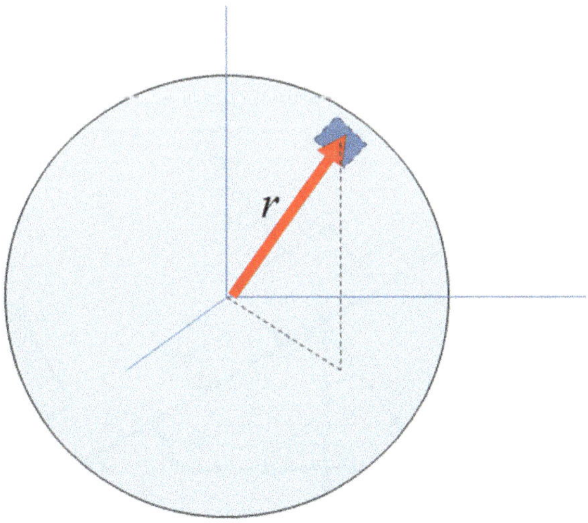

Figure 15.6. The moment of inertia for a sphere in 3D.

Note I have used ($\int_0^\pi \sin^3\theta d\theta = \frac{4}{3}$) in the above derivation, which can easily be shown if you write ($\sin^3\theta$) as ($\sin^2\theta \sin\theta$) instead and use trigonometric identity ($\sin^2\theta = 1 - \cos^2\theta$) before integration.

The moment of inertia about an axis tangent to the sphere will be found using parallel-axis theorem as

$$
\begin{aligned}
I &= I_{cm} + MR^2 \\
&= \frac{2}{5}MR^2 + MR^2 \\
\Longrightarrow I &= \frac{7}{5}MR^2.
\end{aligned}
$$

15.3 Mass on an incline

In order to talk about rotational dynamics, let us first consider the simple example of a mass sliding down an incline, as shown in section 15.3.1. Then we can change the scenario to an object rolling down the same incline and see how the approach changes to adjust for this seemingly minor change, as shown in section 15.4.

15.3.1 Block sliding down an incline

Consider a block that is free to slide down a smooth, frictionless plane that is inclined at an angle (θ) to the horizontal, as shown in figure 15.7. If h is the height of the plane and the block is released from rest at the top, what would its speed be when it reaches the bottom? Using

1. Newtonian mechanics for (a) a smooth surface and (b) a rough surface.
2. The energy method for (a) a smooth surface and (b) a rough surface.

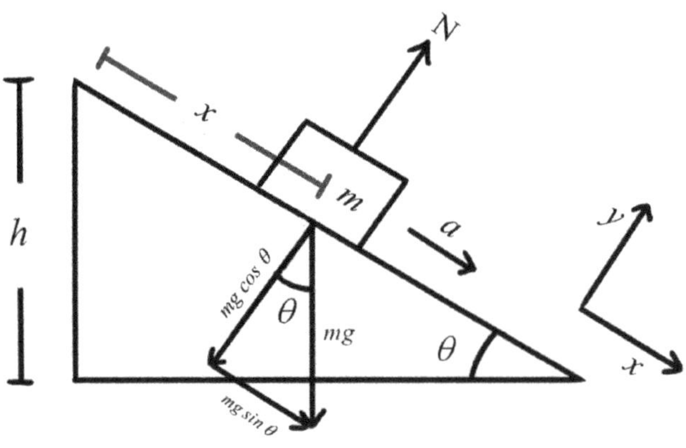

Figure 15.7. Block sliding down a plank.

1. *Newtonian mechanics.*
   Let us first tackle this question by using Newtonian mechanics.
   (a) For a smooth surface,
   writing Newton's force equations:

$$\begin{cases} \Sigma F_x = mg \sin\theta \\ \Sigma F_y = N - mg\cos\theta. \end{cases}$$

Taking the x-coordinate to be along the incline and the y-coordinate perpendicular to the incline, we can write the above equation as

$$\begin{cases} ma = mg \sin\theta \\ 0 \;\; = N - mg\cos\theta. \end{cases}$$

Solving, we obtain the value of acceleration as

$$\boxed{a = \ddot{x} = g\sin\theta}. \tag{15.8}$$

To find the velocity we can use trigonometric identities and the third equation of motion. Assuming that the object starts from rest ($v_o = 0$) at the top

$$v_f^2 - v_0^2 = 2a(x - x_o)$$
$$v_f^2 = 2(g\sin\theta)\left(\frac{h}{\sin\theta}\right)$$

$$\boxed{v_f = \sqrt{2gh}}. \tag{15.9}$$

(b) or a rough surface we can assume that the frictional force is given as (f_k). Then Newton's force equations can be written as

$$\begin{cases} \Sigma F_x = -f_k + mg\sin\theta \\ \Sigma F_y = N - mg\cos\theta. \end{cases}$$

This means $f_k = \mu_k N = \mu_k mg\cos\theta$,

$$ma = -\mu_k mg\cos\theta + mg\sin\theta$$
$$a = -\mu_k g\cos\theta + g\sin\theta.$$

Solving, we obtain the value of acceleration

$$\boxed{a = \ddot{x} = g(\sin\theta - \mu_k\cos\theta)}. \tag{15.10}$$

Using similar arguments as for the smooth case we can see that the speed of the object will be given as

$$\boxed{v_f = 2gh(1 - \mu_k \cot\theta)}. \tag{15.11}$$

Note that for a box moving up the plane the equation will become

$$\ddot{x} = g(\sin\theta + \mu_k \cos\theta). \tag{15.12}$$

2. *Energy method.*
   Next, we will solve this question using the energy method.

   (a) For a smooth surface,
   assume again that the object starts from rest from height (h_0) and at the bottom the height is 0. Then the energy conservation equation will give us

$$\begin{aligned} \Delta E &= 0 \\ \Delta T + \Delta V &= 0 \\ \frac{1}{2}mv_f^2 &= mgh_o \end{aligned}$$

$$\boxed{v_f = \sqrt{2gh}}. \tag{15.13}$$

   (b) For a rough surface we know the equation of motion from chapter 2. The energy is equal to the work done by non-conservative forces:

$$\begin{aligned} \Delta E = W_{\text{non-cons}} &= -f_k \cdot \Delta x \\ \Delta T + \Delta V &= -f_k \cdot \Delta x \\ \frac{1}{2}mv_f^2 - mgh_o &= -\mu_k(mg\cos\theta)\left(\frac{h}{\sin\theta}\right). \end{aligned}$$

   A few algebra steps lead us to

$$\boxed{v_f^2 = 2gh(1 - \mu_k \cot\theta)}. \tag{15.14}$$

15.4 Laminar motion of a rigid body

Laminar motion happens when the fixed plane is perpendicular to the axis of rotation of a rigid body. A rigid body will undergo both translational and rotational acceleration. In laminar motion the axis of rotation may change position but it does not change in direction. Special cases of laminar motion include:

- A physical pendulum (rotation of rigid body about a fixed axis).
- An object (say a sphere, disk or a cylinder) rolling down an inclined plane.

15.4.1 Object rolling down an incline

Let us consider the case of an object with radius (r) rolling down an incline in various methods that we have developed during the course of this book (figure 15.8).

15.4.1.1 Newtonian mechanics

The fundamental equation that governs the translation of a rigid body is

$$\Sigma F = m\frac{d^2 r_{cm}}{dt^2} = m\frac{dv_{cm}}{dt} = ma_{cm}, \tag{15.15}$$

where $\vec{F}$ is the vector sum of all the external forces acting on the body, m is its mass and a_{cm} is the acceleration of its center of mass. Forces acting on the body include:

- The downward force of gravity (mg).
- The normal reaction of the plane (N).
- The frictional force parallel to the plane (f_k).

For a rough surface we can assume that the frictional force is given as (f_k). Then Newton's force equations can be written as

$$\begin{cases} \Sigma F_x = -f_k + mg\sin\theta \\ \Sigma F_y = N - mg\cos\theta. \end{cases}$$

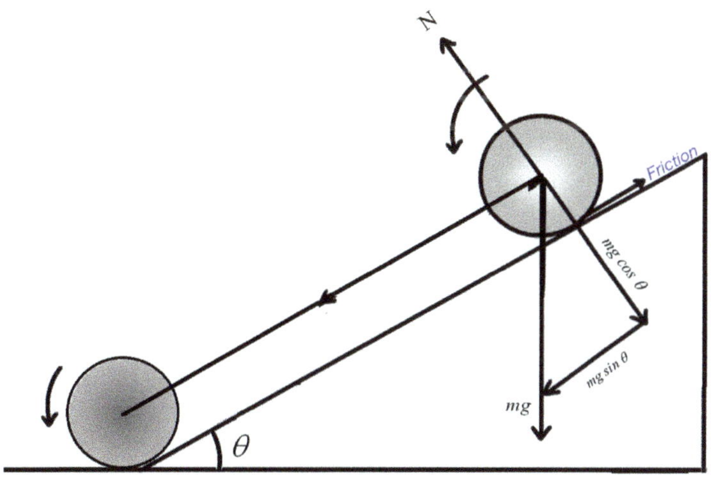

Figure 15.8. An object rolling down an inclined plane.

Taking the x-coordinate to be along the incline and the y-coordinate perpendicular to the incline, we obtain

$$\begin{cases} ma_{cm} = -f_k + mg\sin\theta \\ 0 \quad = N - mg\cos\theta. \end{cases}$$

The only force that exerts a moment about the center of mass is friction, which produces a torque, is given as $\tau = r \times f_k$ and also $\tau = I\alpha = I\ddot{\theta} = I\frac{\ddot{x}}{a}$, where α is the angular acceleration. Comparing these two, we can write

$$f_k = \frac{I\alpha}{r} = \frac{I\ddot{x}}{r^2}. \tag{15.16}$$

Putting this value in Newton's force equation, we can solve for the acceleration about the center of mass:

$$ma_{cm} = -f_k + mg\sin\theta$$

$$m\ddot{x}_{cm} = -\frac{I_{cm}\ddot{x}_{cm}}{r^2} + mg\sin\theta$$

$$m\ddot{x}_{cm} + \frac{I_{cm}\ddot{x}_{cm}}{r^2} = mg\sin\theta$$

$$\boxed{\ddot{x}_{cm} = \frac{mg\sin\theta}{m + \frac{I_{cm}}{r^2}}}. \tag{15.17}$$

Case 1. A uniform solid sphere

For a solid sphere we know that $I_{cm} = \frac{2}{5}MR^2$, so putting this value in equation (15.17) we see that the acceleration down the incline will be

$$\ddot{x}_{cm} = \frac{5}{7}g\sin\theta. \tag{15.18}$$

Case 2. A uniform solid cylinder

For a uniform solid cylinder we know that $I_{cm} = \frac{1}{2}MR^2$, so putting this value in equation (15.17) we see that the acceleration down the incline will be

$$\ddot{x}_{cm} = \frac{2}{3}g\sin\theta. \tag{15.19}$$

15.4.1.2 Energy method

The above results can also be obtained from considering the total energy of the system. The total energy will include kinetic energy due to translation (T_T), kinetic energy due to rotation (T_R) and gravitational potential energy (V_g):

$$\begin{aligned} E_{\text{total}} &= T_T + T_R + V_g \\ &= \frac{1}{2}mv_{cm}^2 + \frac{1}{2}I_{cm}\omega_{cm}^2 + mgh_{cm} \\ &= \frac{1}{2}mv_{cm}^2 + \frac{1}{2}I_{cm}\frac{v_{cm}^2}{r^2} + mg(-x_{cm}\sin\theta). \end{aligned}$$

Note that in the case of a pure rolling motion[1], the frictional force does not appear in the energy because mechanical energy is not converted to heat unless slipping occurs. In other words, for pure rolling motion the total energy is conserved. Differentiating the above equation w.r.t. time,

$$\begin{aligned} \frac{dE_{\text{total}}}{dt} &= \frac{1}{2}m(2v_{cm}\frac{dv_{cm}}{dt}) + \frac{1}{2}\frac{I_{cm}}{r^2}\left(2v_{cm}\frac{dv_{cm}}{dt}\right) - mg\left(\frac{dx_{cm}^2}{dt}\sin\theta\right) \\ 0 &= ma_{cm} + \frac{I_{cm}}{r^2}a_{cm} - mg\sin 0. \end{aligned}$$

Rearranging, we obtain

$$\boxed{\ddot{x}_{cm} = \frac{mg\sin\theta}{m + \frac{I_{cm}}{r^2}}}, \tag{15.20}$$

which is exactly what we saw before.

15.4.1.3 Lagrangian method

The above results can also be obtained from using Lagrangian equations. We can use the results above for kinetic and potential energies to write the Lagrangian ($L = T - V$) as

$$L = \frac{1}{2}m\dot{x}^2 + \frac{1}{2}I_{cm}\dot{\phi}^2 + mgx_{cm}\sin\theta, \tag{15.21}$$

where x represents the downward displacement of the center of mass of the cylinder parallel to the surface of the plane, and ϕ represents the angle of rotation of the cylinder about its symmetry axis.

We can write the Lagrangian equations as

$$\frac{d}{dt}\left(\frac{\partial L}{\partial \dot{x}}\right) - \frac{\partial L}{\partial x} = F_{\text{non-cons}} \tag{15.22}$$

and

$$\frac{d}{dt}\left(\frac{\partial L}{\partial \dot{\phi}}\right) - \frac{\partial L}{\partial \phi} = F_{\text{non-cons}}. \tag{15.23}$$

[1] A pure motion is a motion in which there is only a rolling component and no slipping.

Additionally, we know from the arc-length equation (10.7) that an object rolling without slipping implies

$$x = r\phi, \tag{15.24}$$

which means we can write $f(x, \phi) = x - r\phi = 0$ and its differentials as

$$\frac{\partial f}{\partial x} = 1 \text{ and } \frac{\partial f}{\partial \phi} = -r.$$

Then the $F_{\text{non-cons}} = A\frac{\partial f}{\partial x} = A$ and $F_{\text{non-cons}} = A\frac{\partial f}{\partial \phi} = -Ar$, where A is a function of time given as $A = A(t)$.

This will give us, from Lagrange's equations,

$$m\ddot{x} - mg\sin\theta = A \tag{15.25}$$

and

$$I\ddot{\phi} = -Ar. \tag{15.26}$$

Putting the value of A from equation (15.26) into equation (15.25) we see

$$m\ddot{x} - mg\sin\theta = -\frac{I\ddot{\phi}}{r}$$

$$m\ddot{x} - mg\sin\theta = -\frac{I\frac{\ddot{x}}{r}}{r}$$

$$m\ddot{x} - mg\sin\theta = -\frac{I\ddot{x}}{r^2}$$

$$m\ddot{x} + \frac{I\ddot{x}}{r^2} = mg\sin\theta.$$

From which we can see

$$\boxed{\ddot{x}_{cm} = \frac{mg\sin\theta}{m + \frac{I}{r^2}}}. \tag{15.27}$$

15.4.1.4 Torque method

The above results can also be obtained from using torques. For an object rolling down an incline, the friction at the contact point (P) between the surface and rolling object is static and the friction force vector points up the inclined plane. This friction produces a torque (τ_f). The torque produces an angular acceleration which increases the angular speed as the cylinder rolls down the incline. We can write the sum of torques at point P as shown in figure 15.9, as

$$\Sigma\tau = \tau_N + \tau_{mg} + \tau_f,$$

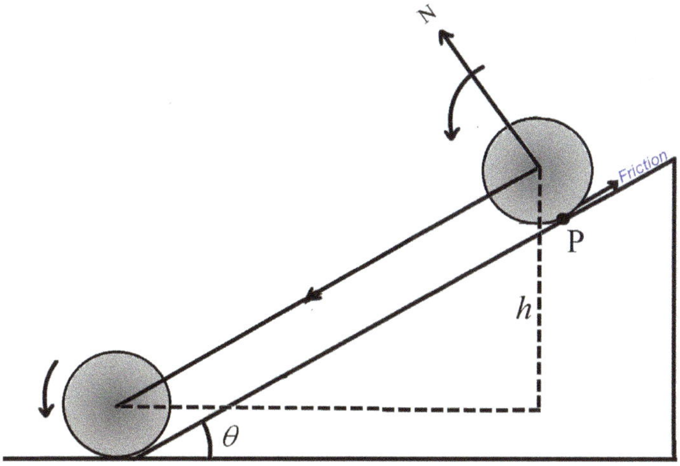

Figure 15.9. An object rolling down an inclined plane.

where τ_{mg} is the torque produced by weight and τ_N is the torque produced by the normal force, $\tau_N = 0$;

$$\begin{aligned} I_P\alpha &= 0 + \tau_{mg} + 0 \\ &= (r \times F_{mg}) \\ &= r \times (mg \sin\theta). \end{aligned}$$

Recall from equation (10.16) that $a_T = r\alpha$, hence we can write the above equation as

$$I_P\left(\frac{a_T}{r}\right) = mgr \sin\theta.$$

For a cylinder the moment of inertia will be given by parallel-axis theorem as

$$\begin{aligned} I_P &= I_{cm} + mr^2 \\ &= \frac{1}{2}mr^2 + mr^2 \\ I_P &= \frac{3}{2}mr^2. \end{aligned}$$

Putting this value in the above equation, we obtain

$$\frac{3}{2}mr^2\left(\frac{a_T}{r}\right) = mgr \sin\theta$$

$$\boxed{a_T = \frac{2}{3}g \sin\theta}, \tag{15.28}$$

which is the same expression as we saw in equation (15.19).

To find the velocity, we can use the above expression,

$$a = \frac{dv}{dt} = \frac{2}{3}g\sin\theta$$
$$\frac{dv}{dx}\frac{dx}{dt} = \frac{2}{3}g\sin\theta$$
$$vdv = \frac{2}{3}g\sin\theta dx$$
$$\frac{v^2}{2} = \frac{2}{3}g\sin\theta(\ell),$$

where ℓ is the distance that the object moves through;

$$\boxed{v^2 = \frac{4}{3}g\ell\sin\theta}. \tag{15.29}$$

Example 15.1

A rigid object that is free to swing about a point of suspension under its own weight is called a physical or a compound[2] pendulum. Suppose a particular physical pendulum is made up of a uniform rod of length ℓ and mass m. Write the equation of motion and find the period of the pendulum (figure 15.10).

Solution

Energy method

We can take the reference point to coincide with the location of the center of mass of the pendulum at rest. This means that gravitational potential energy will

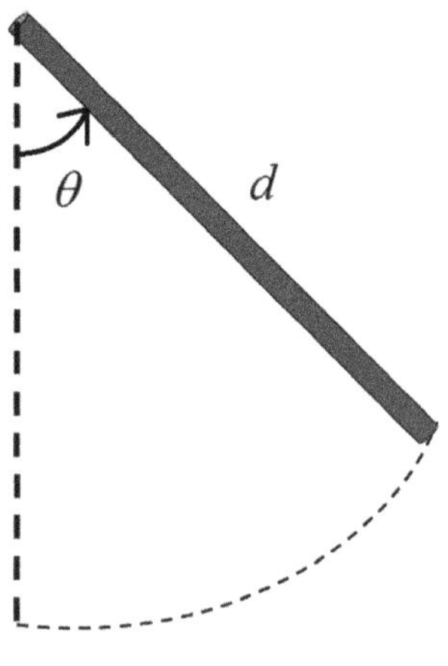

Figure 15.10. A physical pendulum.

[2] A compound physical pendulum consists of a disk of radius R and mass M fixed at the end of a rod of mass m and length l.

be zero here. At this point $\theta = 0$. When the pendulum is at an angle θ the potential energy is

$$V = mg\frac{d}{2}(1 - \cos\theta). \tag{15.30}$$

The kinetic energy of rotation about the pivot point is

$$T_{\text{rot}} = \frac{1}{2}I_P\omega^2. \tag{15.31}$$

Total energy will be given as

$$E = T + V$$
$$E = \frac{1}{2}I_P(\omega)^2 + mg\frac{\ell}{2}(1 - \cos\theta).$$

Using conservation of energy, we can say that total energy remains constant throughout this motion. This means the time derivative of energy is zero:

$$0 = I_P\omega\dot{\omega} + mg\frac{\ell}{2}\sin\theta\dot{\theta}. \tag{15.32}$$

Since $\dot{\theta} = \omega$ and $\omega \neq 0$, and for small angles $\sin\theta \approx \theta$, we obtain

$$0 = I_P\ddot{\theta} + mg\frac{\ell}{2}\theta\omega \tag{15.33}$$

$$\ddot{\theta} + \frac{mg\ell}{2I_P}\theta = 0. \tag{15.34}$$

Newtonian method

The angular acceleration of the pendulum is given by

$$\vec{\alpha} = \ddot{\theta}\hat{k}. \tag{15.35}$$

The torque about the pivot point (P) is given by

$$\vec{\tau}_P = \vec{r}_{cm} \times m\vec{g}$$
$$I_P\vec{\alpha} = \frac{\ell}{2}\hat{r} \times mg(\cos\theta\hat{r} - \sin\theta\hat{\theta})$$
$$I_P\ddot{\theta}\hat{k} = -\frac{\ell}{2}mg\sin\theta\hat{k}.$$

For small angles $\sin\theta \approx \theta$, which gives

$$\ddot{\theta} + \frac{mg\ell}{2I_P}\theta = 0. \tag{15.34}$$

This is a simple harmonic oscillator equation for the physical pendulum. The angular frequency of small oscillations for the pendulum is $\omega = \sqrt{\frac{mg\ell}{2I_P}}$, and the moment of inertia of the rod is $I_P = \frac{1}{3}m\ell^2$, so the period will be given as,

$$T = \frac{2\pi}{\omega} = 2\pi\sqrt{\frac{2\ell}{3g}}. \tag{15.37}$$

Keep in mind that this is the period of a simple pendulum of length 2/3 l.

Example 15.2

A cylinder of radius (a) and mass (m) rolls without slipping on a fixed cylinder of radius (b). Evaluate the Lagrangian and obtain the equation of motion for the short period of time before the cylinders separate.

Solution

The kinetic energy of the smaller cylinder is the kinetic energy of translation of its center of mass plus the kinetic energy of rotation about its center of mass,

$$T = \frac{1}{2}m(\dot{r}^2 + r^2\dot{\theta}^2) + \frac{1}{2}I\dot{\phi}^2,$$

where r is the distance between the centers of the cylinders ($r = a + b$) and moment of inertia $I = \frac{1}{2}ma^2$. Then we can write the Lagrangian as

$$\begin{aligned} L &= T - V \\ &= \frac{1}{2}m(\dot{r}^2 + r^2\dot{\theta}^2) + \frac{1}{2}I\dot{\phi}^2 - mgr\ \cos\theta. \end{aligned}$$

Note there are two constraints: ($r = a + b$) and ($a\phi = b\theta$). The derivatives of these constraints will be ($\dot{r} = 0$) and ($a\dot{\phi} = b\dot{\theta}$), so we can rewrite the Lagrangian as

$$L = \frac{1}{2}m(0 + (a + b)^2\dot{\theta}^2) + \frac{1}{2}\left(\frac{1}{2}ma^2\right)\left(\frac{b}{a}\dot{\theta}\right)^2 - mg(a + b)\cos\theta.$$

We can also rearrange the above equation as

$$L = \frac{1}{2}m\left(a^2 + \frac{3}{2}b^2 + 2ab\right)\dot{\theta}^2 - mg(a + b)\cos\theta. \tag{15.38}$$

Using this we can write the equation of motion:

$$\begin{gathered} \frac{d}{dt}\left(\frac{\partial L}{\partial \dot{\theta}}\right) - \frac{\partial L}{\partial \theta} = 0 \\ m\left(a^2 + \frac{3}{2}b^2 + 2ab\right)\ddot{\theta} - mg(a + b)\ \sin\theta = 0. \end{gathered} \tag{15.39}$$

Concept questions 15.1

1. Is there such a thing as a truly rigid object? Give examples.
2. Why do tightrope walkers carry balancing poles while walking on the tightrope?
3. How can you distinguish between a raw and a boiled egg by spinning each on a table?
4. A solid cylinder, a solid sphere and a solid box are let go from the top of an incline. What kind of motions (sliding, slipping or both) do each experience under the following conditions:
   (a) A frictionless surface?
   (b) The fictional force between the surface and object $= f_f$ has the coefficient μ_k?

Exercises 15.1

1. A uniform hollow cylinder has a density ρ, a length L, inner radius a and outer radius b. Show that its moment of inertia about the axis of the cylinder is $I = \frac{1}{2}\pi\rho L(b^4 - a^4) = \frac{1}{2}M(b^2 + a^2)$.
2. What will be the moment of inertia for a uniform-density (filled) cylinder about
   (a) The central axis of the cylinder.
   (b) About one edge of the cylinder.
   (c) Central diameter of the cylinder.
3. A string is wrapped around a yo-yo of mass M and radius R. The string is pulled vertically upwards to prevent the center of mass from falling as the yo-yo unwinds the string. Find:
   (a) The tension in the string.
   (b) The length of string unwound in the time it takes to reach angular speed (ω).

Further reading

Arya A 1990 *Introduction to Classical Mechanics* 2nd edn (Upper Saddle River, NJ: Prentice-Hall)

Baierlein R 1983 *Newtonian Dynamics* (New York: McGraw-Hill)

Fowles G 1986 *Analytical Mechanics* 4th edn (New York: CBS College Publishing)

Fowles G and Cassiday G 2005 *Analytical Mechanics* 7th edn (Belmont, CA: Thomson Learning)

Goldstein H 1989 *Classical Mechanics* 2nd edn (Reading, MA: Addison-Wesley)

Landau L D and Lifshitz E M 1960 *Mechanics* (Reading, MA: Addison-Wesley)

Ohanian H 1985 *Physics* (New York: WW Norton)

Resnick R, Halliday D and Krane K *Physics* vol 1, 5th edn (New York: Wiley)

Stephenson R 1961 *Mechanics and Properties of Matter* 2nd edn (New York: Wiley)

Symon K 1971 *Mechanics* 3rd edn (Reading, MA: Addison-Wesley)

IOP Publishing

Techniques of Classical Mechanics
From Lagrangian to Newtonian mechanics
Samya Zain

Chapter 16

System of particles

16.1 System of particles

In this chapter we will concentrate on the free particle case. A system is said to be composed of *free particles* when the particles of the system all move independently of each another.

For a general system of n particles, with masses m_1, m_2, m_3, … , m_n and with position vectors $\vec{r}_1$, $\vec{r}_2$, $\vec{r}_3$, … , $\vec{r}_n$, respectively, we can define the following important terms:

- *Center of mass.* The center of mass of a distribution is the unique point where the weighted relative position of the distributed mass sums to zero. The center of mass is also the point where an applied force causes the system to move in the direction of force without rotation. Equivalently, we can say that the center of mass is defined as the point where position vector in 3D is given as

$$\vec{r}_{cm} = \frac{m_1\vec{r}_1 + m_2\vec{r}_2 + \cdots m_n\vec{r}_n}{m_1 + m_2 + m_3 \cdots + m_n} = \frac{\Sigma m_i\vec{r}_i}{\Sigma m_i} = \frac{\Sigma m_i\vec{r}_i}{M}, \tag{16.1}$$

where $M = \Sigma m_i$ is the total mass of the system. The definition in equation (16.1) is equivalent to the following three equations for the system consideration in 3D:

$$x_{cm} = \frac{\Sigma m_i x_i}{M}$$
$$y_{cm} = \frac{\Sigma m_i y_i}{M}$$
$$z_{cm} = \frac{\Sigma m_i z_i}{M}.$$

The center of mass is a useful reference point for calculations in mechanics that involve masses distributed in space, such as the linear and angular momentum of planetary bodies and rigid body dynamics.

- *Linear momentum.* Linear momentum is a vector quantity defined as the product of an object's mass, m, and its velocity, $\vec{v}$,

$$\vec{p} = m\vec{v}_{cm}. \tag{16.2}$$

- *Conservation of momentum.* The conservation of linear momentum is based on the principle of Newton's first law of motion. It expresses that the total momentum of a particle or a system of particles remains the same unless an external force is applied to it. In an isolated system momentum of an object always remains constant and is mathematically given as

$$\Delta\vec{p} = 0. \tag{16.3}$$

- *Newton's second law.* Newton's second law of motion can be formally stated as

'The acceleration of an object due to a net force is directly proportional to the magnitude of the force and is in the same direction as the applied force, and inversely proportional to the mass of the object',

$$\Sigma F_i = \frac{d\vec{p}}{dt} = m\vec{a}_{cm}. \tag{16.4}$$

Acceleration of the center of mass of a system of particles is the same as that of a single particle having a mass equal to the total mass of the system and acted on by the same sum of external forces. For example, a school of fishes moving together, as seen in figure 16.1.

Figure 16.1. A school of bluestripe snapper fish swims near a tropical coral reef off the islands of Palau in Micronesia. Credit: School of Bluestripe Snapper fish photo via Shutterstock.

For a system of n particles, we can write Newton's force equation as

$$\Sigma F_i = m\vec{g}$$
$$\Sigma m_i a_{cm} = m\vec{g}.$$

Here the sum is over the number of particles, $i = 1, 2, \ldots, n$. Hence,

$$a_{cm} = g. \tag{16.5}$$

- *Angular momentum.* Angular momentum is the rotational analogue of linear momentum. It is also sometimes called moment of momentum or rotational momentum:

$$L = \sum_{i=1}^{n} \vec{r}_i \times m_i\vec{v}_i. \tag{16.6}$$

- *Torque.* The time rate of change of the angular momentum of a system is equal to the total moment of all the external forces acting on the system

$$\frac{dL}{dt} = \sum \tau_i. \tag{16.7}$$

- *Conservation of angular momentum.* For an isolated system, we know the sum of torques on the system is zero, mathematically defined as $\sum\tau = 0$, and we know from equation (16.7) $\sum\tau = \frac{dL}{dt}$. Combining these equations means $\frac{dL}{dt} = 0$, hence angular momentum (L) is a constant vector.
- *Kinetic energy of a system.*

$$T = \frac{1}{2}mv_{cm}^2 + \sum_i \frac{1}{2}m_i\vec{v}_i^{\,2}, \tag{16.8}$$

where $(\frac{1}{2}mv_{cm}^2)$ is the kinetic energy of translation of the whole system and $(\sum_i \frac{1}{2}m_i\vec{v}_i^{\,2})$ is the kinetic energy of motion relative to the center of mass.

Example 16.1

There are three masses, $m_1 = 2$ kg located at $x = 1$ m, $m_2 = 1$ kg at $x = 1$ m and $m_3 = 0.5$ kg at $x = 2.5$ m. Find the center of mass of the system.

Solution

To find the center of mass, we use equation (16.1) and since all particles lie along the x-axis, we can reduce it to the 1D case:

$$x_{cm} = \frac{m_1x_1 + m_2x_2 + m_3x_3}{m_1 + m_2 + m_3}. \tag{16.9}$$

Putting in values and solving, we obtain $x_{cm} = 1.2$ m.

Example 16.2

An ingenious method to measure the weight of any object is by hanging it with known masses until all things balance. Suppose you went fishing with your friends and want to weigh the catch of the day to see who caught the biggest fish. You do not have a weighing scale, but you have a perfectly balanced meter stick and two weights, $m_1 = 2\,\text{kg}$ and $m_2 = 1\,\text{kg}$. How will you find the weight of your fish?

Solution
Center-of-mass method
Suppose you hang $m_1 = 2\,\text{kg}$ at the $x = 0\,\text{cm}$ mark and $m_2 = 1\,\text{kg}$ at $x = 100\,\text{cm}$ on the meter stick and find that the fish balances the system when placed at $x = 75\,\text{cm}$:

$$
\begin{aligned}
x_{cm} &= \frac{m_1x_1 + m_2x_2 + m_{\text{fish}}x_3}{m_1 + m_2 + m_{\text{fish}}} \\
0.50\text{ m} &= \frac{(2\text{ kg})(0\text{ m}) + (1\text{ kg})(1\text{ m}) + m_{\text{fish}}(0.75\text{ m})}{2 + 1 + m_{\text{fish}}}. \\
\Longrightarrow m_{\text{fish}} &= 2\text{ kg}
\end{aligned}
\tag{16.10}
$$

Torque's method

$$\Sigma\tau = \vec{r}_1 \times \vec{F}_1 + \vec{r}_2 \times \vec{F}_2\vec{r}_{\text{fish}} \times \vec{F}_{\text{fish}} \tag{16.11}$$

When the system is balanced, recalling that clockwise torques are negative and anti-clockwise torques are positive, we obtain

$$
\begin{aligned}
0 &= -\, x_1(m_1g) + x_2(m_2g) + x_{\text{fish}}(m_{\text{fish}}g) \\
0 &= (0.5\ \text{m})(2\ \text{kg})g + (0.5\ \text{m})(1\ \text{kg})g + (0.25\ \text{m})m_{\text{fish}}g \\
\Longrightarrow m_{\text{fish}} &= 2\ \text{kg}.
\end{aligned}
$$

16.2 Two-particle system

As a simple application of many-body kinematics, let us start with a discussion of the motion of just two particles. For a N particle system we need $3N$ position variables to describe the system. For a two-particle system we will need six position variables to define the position vectors (figure 16.2).

We can simplify the problem by using the definitions we discussed in section 16.1.

The position vector of the center of mass:

$$\vec{r}_{cm} = \frac{m_1\vec{r}_1 + m_2\vec{r}_2}{m_1 + m_2}. \tag{16.12}$$

The velocity of the center of mass:

$$\vec{v}_{cm} = \frac{m_1\vec{v}_1 + m_2\vec{v}_2}{m_1 + m_2}. \tag{16.13}$$

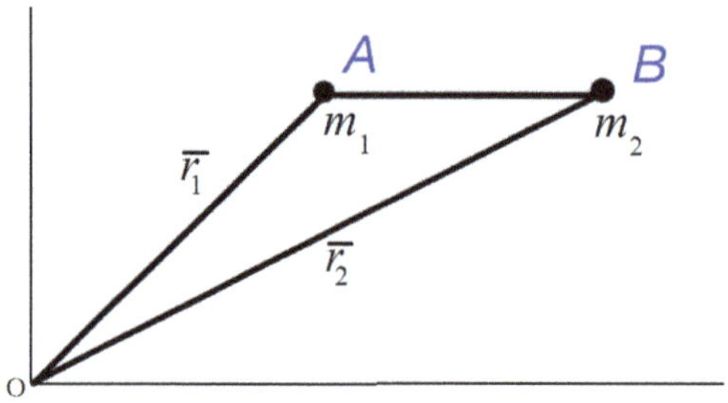

Figure 16.2. A two-body system.

The acceleration of the center of mass:

$$\vec{a}_{cm} = \frac{m_1\vec{a}_1 + m_2\vec{a}_2}{m_1 + m_2}. \tag{16.14}$$

Newton's second law:

$$\sum \vec{F}_{\text{ext}} = (m_1 + m_2)\vec{a}_{cm}. \tag{16.15}$$

16.2.1 The reduced mass

Assume an isolated system and consider the motion of a system consisting of two particles, with masses m_1 and m_2, that interact with each other via a central force. Suppose the distance between m_1 and m_2 is defined as $\vec{r}$.

Since we have an isolated system, the center of mass will move with constant velocity. For simplicity, we can assume the center of mass to be located at origin (O). This means $\vec{r}_{cm} = 0$, which means that we can rewrite equation (16.12) as

$$m_1\vec{r}_1 + m_2\vec{r}_2 = 0, \tag{16.16}$$

where $\vec{r}_1$ and $\vec{r}_2$ are position vectors of mass 1 and mass 2 with respect to the origin.

From figure 16.3 and equation (16.16) we obtain $\vec{r}_2 = -\frac{m_1\vec{r}_1}{m_2}$ and can write

$$\vec{r} = \vec{r}_1 - \vec{r}_2$$

$$\vec{r} = \vec{r}_1\left(1 + \frac{m_1}{m_2}\right)$$

$$\vec{r}\frac{1}{\left(1 + \frac{m_1}{m_2}\right)} = \vec{r}_1. \tag{16.17}$$

Newton's law for particle 1 w.r.t. to the origin can be written as

$$\sum F = F_{\text{central force}},$$

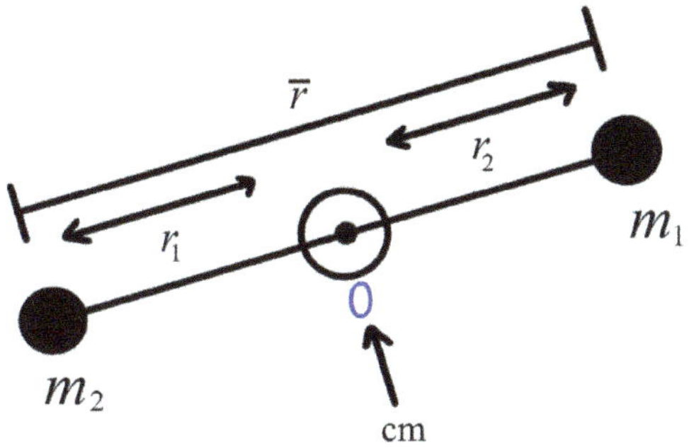

Figure 16.3. Reduced mass—a two-particle system.

where the central force is $F_{\text{central force}} = \frac{-Gm_1m_2}{r^2}$ when particles are interacting via gravitational force;

$$m_1\frac{d^2\vec{r}_1}{dt^2} = F_{\text{central force}}\ \hat{r}$$

$$m_1\frac{d^2\vec{r}_1}{dt^2} = F_{\text{central force}}\left(\frac{\vec{r}}{|\vec{r}|}\right).$$

Putting the value of $\vec{r}_1$ from equation (16.17) we obtain

$$m_1\frac{1}{\left(1+\frac{m_1}{m_2}\right)}\frac{d^2\vec{r}}{dt^2} = F_{\text{central force}}\ \frac{\vec{r}}{r} \tag{16.18}$$

and simplify to obtain

$$\left(\frac{m_1m_2}{m_1+m_2}\right)\frac{d^2\vec{r}}{dt^2} = F_{\text{central force}}\left(\frac{\vec{r}}{r}\right) \tag{16.19}$$

$$\boxed{\mu \equiv \frac{m_1m_2}{m_1+m_2}}, \tag{16.20}$$

where μ is called *reduced mass*. We can use the definition of reduced mass to rewrite equation (16.19) as

$$\mu\ddot{r} = F_{\text{central force}}\ \hat{r} \tag{16.21}$$

and for the gravitational field it will become

$$\mu\ddot{r} = \frac{-Gm_1m_2}{R^2}\hat{r}. \tag{16.22}$$

It is worth noting that equation (16.21) is exactly the same as the ordinary equation of motion of a single particle of mass μ moving in a central field of a force given by $F(r)$. The fact that both particles are moving relative to the center of mass is automatically accounted for by replacing m_1 by the reduced mass μ.

Special cases to consider when we talk about the reduced mass case include the following situations:

- When bodies have same mass, $m_1 = m_2 = m$, then equation (16.20) will give

$$\mu = \frac{m}{2}.$$

- When $m_2 \gg m_1$ (which implies that $\frac{m_1}{m_2}$ is negligible) this means

$$\mu \cong m_1.$$

For this case gravitational field equation (16.21) may also be written as

$$m_1 \ddot{r} = \frac{-G(m_1 + m_2)m_1}{r^2}\hat{r}. \tag{16.23}$$

16.2.2 Correction to Kepler's third law

We found the periodic time of orbital motion of a planet (m) moving in the Sun's gravitational field in chapter 8, equation (8.43) as $T_t^2 = \frac{4\pi^4}{GM}a^3$ where a is the semi-major axis of the elliptical orbit.

In order to derive the expression, we assumed that the Sun was stationary and was located at the origin of our system. But the Sun is not stationary; it actually moves about the common center of mass, so the correct equation to use will be equation (16.23), in which $m_1 = m$ (the mass of the planet) and $m_2 = M$ (the mass of the Sun) and the force constant $K = GMm$ should now become $G(M + m)m$. Hence the correct equation for the period is

$$T_t^2 = \frac{4\pi^2}{G^2(M + m)^2}a^3. \tag{16.24}$$

If m_1 and m_2 are expressed in units of the Sun's mass and a is expressed in astronomical units (mean distance from Earth to Sun = 1 AU). Then the orbital period in years is given as

$$T_t^2 = \frac{1}{(M + m)^2}a^3. \tag{16.25}$$

For most planets in our solar system the added mass term of the mass of the planet makes little difference. As we can see, the mass of Earth $\approx \frac{1}{330\,000}$ the mass of the Sun, it will have little effect on the calculation for period. The most massive planet in our solar system is Jupiter; where Jupiter's mass = $\frac{1}{1000}\ M_{\text{Sun}}$. Here the effect of reduced mass is to change the earlier calculation in the ratio $\frac{1}{1.001} = 0.995$ for the period of Jupiter around the Sun.

Example 16.3

Binary star systems. Binary star systems are made-up of two stars revolving around each other. A particular binary star system, say m_1 has orbital speed $v_1 = 1$ AU yr^{-1} and m_2 has orbital speed $v_2 = 3$ AU yr^{-1} and a period of 7 years. Assuming circular orbits, calculate the mass of each star in solar mass units M_{Sun}.

Solution
To find the masses, let us start from equation (16.25) to find $(m_1 + m_2)$ as

$$m_1 + m_2 = \frac{a^3}{T^2}. \tag{16.26}$$

We know $a = r_1 + r_2$, to find r_1 use, for circular orbits,

$$\text{speed} = \frac{\text{distance}}{\text{time}} = \frac{2\pi r_1}{T}$$
$$r_1 = \frac{v_1 T}{2\pi} = \frac{(1 \text{ AU/year})(7 \text{ years})}{2\pi} = 1.114 \text{ AU}$$

Similarly we can find $r_2 = 3.34$ AU and hence $a = 4.48$ AU. Putting values in equation (16.26) we obtain

$$m_1 + m_2 = 1.83\ M_{\text{Sun}}.$$

In order to find individual masses use (16.16) and recognize $\dot{r} = v$, as

$$\frac{m_1}{m_2} = \left|\frac{v_2}{v_1}\right| = 3.$$

Put this value in the above equation and solve to obtain

$$m_1 = 1.83\ M_{\text{Sun}}$$
$$m_2 = 0.457\ M_{\text{Sun}}.$$

16.3 Many-particle systems

The physical world consists of interacting many-particle systems. For a system consisting of N particles of masses $m_1, m_2, \ldots, m_N$, the total mass of the system is

$$M = \sum_{n=1}^{N} m_n = m_1 + m_2 + \cdots + m_N. \tag{16.27}$$

Each particle in the system is represented by:

- Its mass (m_n).
- Its location at the coordinate $(\vec{r}_n)$, whose coordinates are (x_n, y_n, z_n).
- Its velocity $(\vec{v}_n)$, whose components are (v_{xn}, v_{yn}, v_{zn}).
- Its acceleration $(\vec{a}_n)$, whose components are (a_{xn}, a_{yn}, a_{zn}).

The net force on each particle is $\Sigma\vec{F}_n$, which in general differs from one particle to another.

16.3.1 Center of mass of a many-particle system

The position of the center of mass of a many-particle system is given as

$$\vec{r}_{cm} = \frac{m_1\vec{r}_1 + m_2\vec{r}_2 + \cdots + m_N\vec{r}_N}{m_1 + m_2 + \cdots + m_N} = \frac{1}{M}\sum_{n=1}^{N} m_n\vec{r}_n \tag{16.28}$$

and the coordinates will be given as

$$\begin{aligned} x_{cm} &= \frac{m_1x_1 + m_2x_2 + \cdots + m_Nx_N}{m_1 + m_2 + \cdots + m_N} = \frac{1}{M}\sum_{n=1}^{N} m_nx_n \\ y_{cm} &= \frac{m_1y_1 + m_2y_2 + \cdots + m_Ny_N}{m_1 + m_2 + \cdots + m_N} = \frac{1}{M}\sum_{n=1}^{N} m_ny_n \\ z_{cm} &= \frac{m_1z_1 + m_2z_2 + \cdots + m_Nz_N}{m_1 + m_2 + \cdots + m_N} = \frac{1}{M}\sum_{n=1}^{N} m_nz_n. \end{aligned} \tag{16.29}$$

We can find the velocity of the center of mass by taking the derivative of equation (16.28) as

$$\vec{v}_{cm} = \frac{d\vec{r}_{cm}}{dt} = \frac{m_1\vec{v}_1 + m_2\vec{v}_2 + \cdots + m_N\vec{v}_N}{m_1 + m_2 + \cdots + m_N} = \frac{1}{M}\sum_{n=1}^{N} m_n\vec{v}_n. \tag{16.30}$$

We can find the acceleration of the center of mass by taking the derivative of equation (16.30) as

$$\vec{a}_{cm} = \frac{d\vec{v}_{cm}}{dt} = \frac{m_1\vec{a}_1 + m_2\vec{a}_2 + \cdots + m_N\vec{a}_N}{m_1 + m_2 + \cdots + m_N} = \frac{1}{M}\sum_{n=1}^{N} m_n\vec{a}_n. \tag{16.31}$$

We see from equation (16.31), knowing Newton's second law, that

$$\begin{aligned} M\vec{a}_{cm} &= m_1\vec{a}_1 + m_2\vec{a}_2 + \cdots + m_N\vec{a}_N \\ M\vec{a}_{cm} &= \Sigma\vec{F}_1 + \Sigma\vec{F}_2 + \cdots + \Sigma\vec{F}_N. \end{aligned} \tag{16.32}$$

Equation (16.32) is just the statement of Newton's second law for a system of N particles. This shows that the motion of a system of N particles can be treated as a single particle of mass M located at the center of mass, moving with velocity $\vec{v}_{cm}$ and experiencing $\vec{a}_{cm}$.

Summarizing this result we can state:

> 'The overall translational motion of a system of particles can be equivalently analyzed using Newton's laws as if all the mass were concentrated at the center of mass and that the net external force is applied at the center of mass'.

A special case

horizontal. What is the location of $\Sigma\vec{F}_{\text{ext}} = 0$, we note that in this case the center of mass of the system will not have any accelerating motion and hence will move with a constant velocity.

Concept questions 16.1

1. Is the center of mass of a solid object always located within the object? If not, give examples.
2. Where is the center of mass of the Earth located? Where is the center of mass of the Earth's atmosphere located?
3. Where is the center of mass of a 1000 page book, when:
   (a) The book is closed lying flat on the table?
   (b) The book is open to page 500 and is lying flat on the table?
   (c) The book is open to page 10 and is lying flat on the table?

Exercises 16.1

1. Assume we have a system of five particles with masses 1 kg, 2 kg, 3 kg, 4 kg and 5 kg located at (0, 0, 0), (2, 2, 1), (−3, 1, 2), (1, −2, 2) and (2, 1, −2), respectively. Find the location of the center of mass assuming that the particles are all at rest.
2. A canon ball of mass 20 kg is fired at 50° w.r.t. horizontal with initial velocity 24 m s^{-1}. After launch, the canon ball splits into two pieces, one with mass (m_1 = 12 kg) and the other with mass (m_2 = 8 kg). At 2 s after launch m_1 has moved diagonally in a straight line 26 m from the launch site and is located at 12 m from the horizontal surface. At this time what will be the location of m_2? *Hint*: you need of find x_1 and y_1 for mass 2.

16.4 Conservation of momentum in a system of particles

Consider a special case of an isolated system of particles, where $F_{\text{ext}} = 0$. This implies that the rate of change of the total momentum of a system does not change, i.e. total linear momentum is constant.

In other words, when there is no net external force acting on a system of particles the total momentum of the system will be conserved. It is that simple. It does not matter about the nature of the interactions that occur within a system, the total momentum of the system will remain the same.

16.4.1 Conservation of linear momentum in action

Consider a system that is comprised of a cannon and a cannonball. Initially, both cannon and ball are at rest, hence the total momentum of the system is zero. What happens when the cannon is fired? Well, the cannonball shoots out with a large velocity, and thus a large momentum. Since there are no net external forces acting on the system, this momentum has to be compensated by a momentum in the opposite direction so that total final momentum remains zero. Thus the cannon obtains a backward velocity called a *recoil.* This also accounts for the *kick* associated with firearms. Any time a gun or a cannon fires a projectile, it must move in the direction opposite the projectile to conserve momentum.

We can account for a lot of factors when considering the motion of a system of particles if we understand the center of mass of a system of particles and in turn develop equations for conservation of linear momentum.

16.5 Collisions

A collision is an interaction between two or more objects that lasts for a very short interval of time and results in a change in the motion of the interacting objects.

A collision can also be thought of as an event in which momentum (p) or kinetic energy (T) is transferred from one object to another. A large eighteen-wheeler truck with a mass of about 10 000 kg moving at $2\,\mathrm{m\ s^{-1}}$ has the same momentum as a 1000 kg compact car moving at $20\,\mathrm{m\ s^{-1}}$, they both have $p = 20\,000\,\mathrm{kg\ m\ s^{-1}}$.

Kinetic energy is linearly proportional to the mass of the object; doubling the mass results in doubling the kinetic energy of the object. In contrast, the relationship between kinetic energy and velocity is quadratic, which means as you increase the speed, kinetic energy increases quadratically.

16.5.1 Types of collisions

In general types we can divide collisions into two broad categories:

- Elastic.
- Inelastic/completely inelastic/explosive.

We know all collisions adhere to conservation laws:

- Law of conservation of momentum: $\Delta p = 0$.
- Law of conservation of mass: $\Delta m = 0$.
- Law of conservation of energy: $\Delta E = 0$; or $= Q$.

  There will be two types of collisions corresponding to the two cases as given below:

  (a) An *elastic collision*, $\Delta E = 0$, occurs when the two objects *bounce* apart when they collide, as shown in figure 16.4. Collisions between pool balls are a good example of elastic collisions.

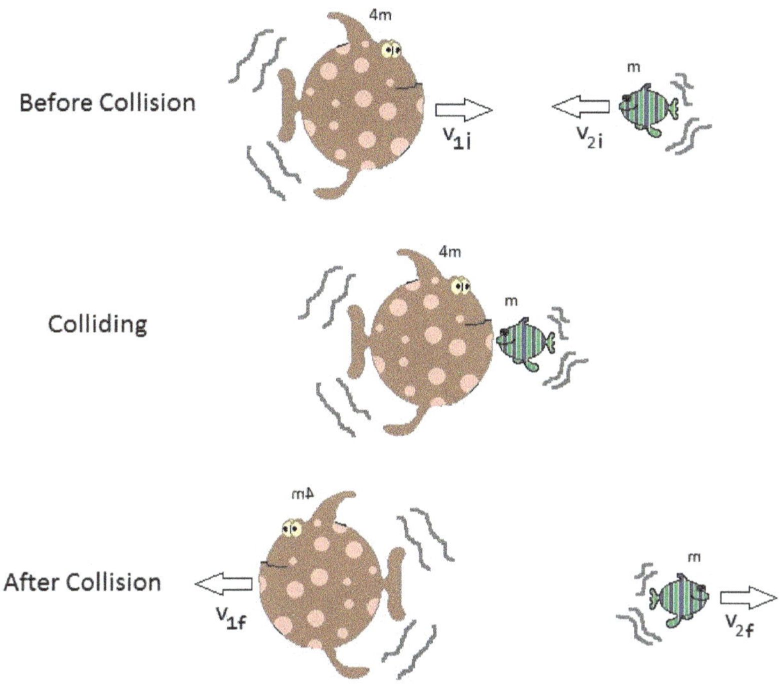

Figure 16.4. Elastic collision.

In an elastic collision, both momentum and kinetic energy are conserved. Almost no energy is lost to sound, heat or deformation. This could be because the colliding objects are either very rigid and they do not deform during the collision, or it could be because the colliding objects are elastic and they deform during the collision. In the latter case, ball 1 will deform upon collision but then quickly bounce back to its former shape and transfer almost all the kinetic energy to the second ball.

Conservation of momentum is

$$m_1 v_{1i} + m_2 v_{2i} = m_1 v_{1f} + m_2 v_{2f}.$$

Conservation of kinetic energy is

$$\frac{1}{2} m_1 v_{1i}^2 + \frac{1}{2} m_2 v_{2i}^2 = \frac{1}{2} m_1 v_{1f}^2 + \frac{1}{2} m_2 v_{2f}^2,$$

where subscripts 1 and 2 are for mass 1 and mass 2; and i and f represent the initial and final states, i.e. before and after the collision.

(b) An *inelastic collision* occurs when colliding objects do not bounce away from each other. An example of a completely inelastic collision is shown in figure 16.5. A high speed car collision is a good example of inelastic collision.

In inelastic collisions momentum is conserved, because the total momentum of both objects before and after the collision is the same.

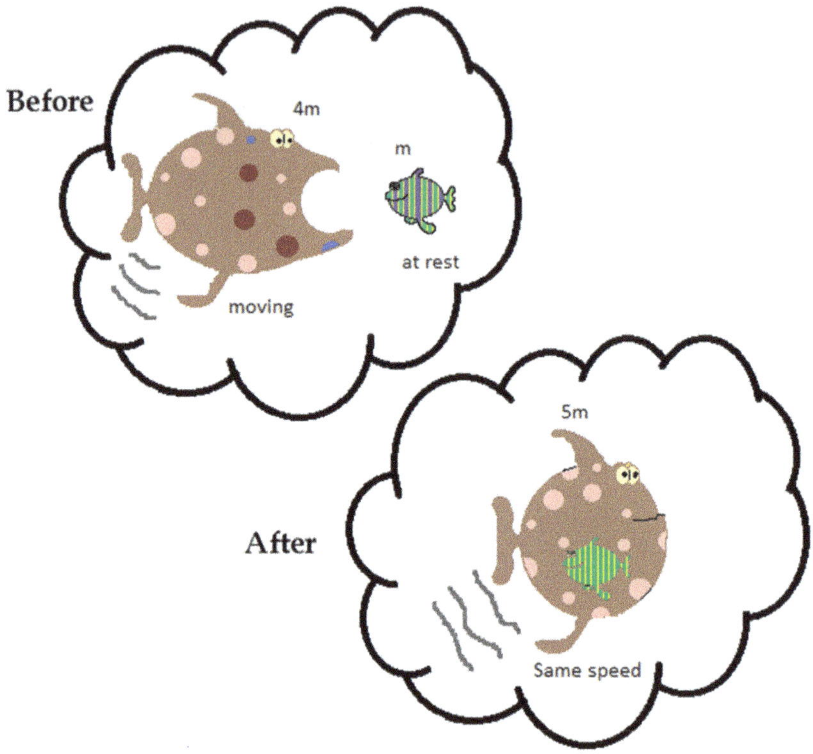

Figure 16.5. Inelastic collision.

However, kinetic energy is not conserved. Some of the kinetic energy is converted into sound or heat or it goes towards the deformation of the objects.

In inelastic collisions $\Delta E = Q$, where Q indicates net loss or gain of kinetic energy as a result of the collision. When energy is lost, Q will be negative and the collision is called an *exoergic collision*, and when kinetic energy is gained Q will be positive and the collision is called an *endoergic collision*. An example of an endoergic collision is a collision when an explosive device is placed on one or both of the colliding masses and it explodes upon impact.

16.5.2 Another way to distinguish collisions

Another way to distinguish collisions is to divide collisions depending on the direction of approach and the directions of retreat of the colliding particles.

- *Head-on collision.* A collision will be called head-on if the collision occurs along a straight line. In other words, a head-on collision is said to occur when the point of impact and the velocities of the incoming and outgoing particles all lie in a straight line that connects the center of mass of each of the objects (figure 16.6).

For head-on collisions the center of mass of the colliding particles must be traveling in the same straight line before and after impact. Note that the velocity vectors may be parallel or anti-parallel to each other but must lie in the same line.

- *Glancing collision.* Glancing collision is a collision that takes place under a small angle, with the incident body being nearly parallel to the surface, as shown in figure 16.7.

16.5.3 Collision factors

If we know the incoming velocities of colliding particles, we can easily find the velocities of outgoing particles provided we know the value of Q. Usually we

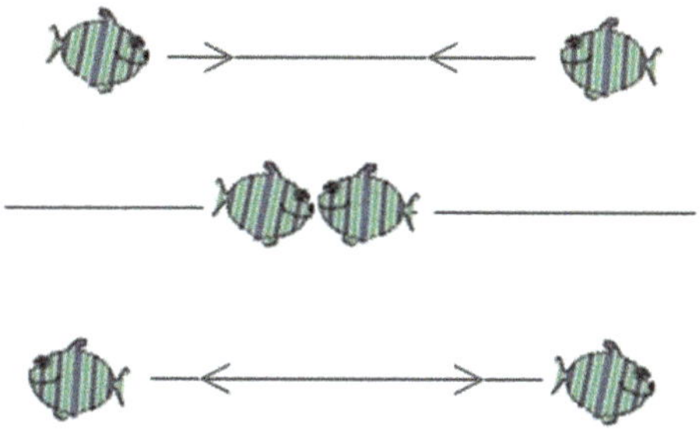

Figure 16.6. Head-on collision.

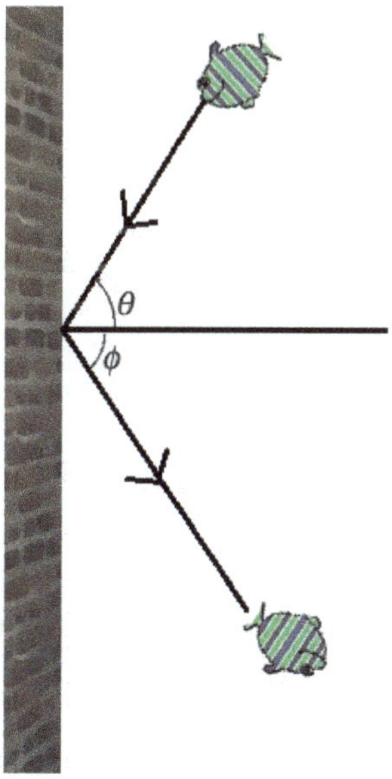

Figure 16.7. Glancing collision.

introduce another parameter (ϵ) called the *coefficient of restitution*, defined as the ratio of speed of separation to the speed of approach as

$$\epsilon = \frac{|v_{2f} - v_{1f}|}{|v_{2i} - v_{1i}|}, \tag{16.33}$$

where the subscript f represents final velocities of the particles and subscript i gives values for the initial velocities. Also keep in mind that we are using the absolute values for the initial and final relative velocities. The numerical value of ϵ depends mainly on:

- The composition of the colliding objects.
- The physical make-up of the objects, e.g. colliding spheres versus colliding cubes will have different properties.
- The speed of approach.

In the general case of a direct inelastic collision, the energy loss (Q) is related to ϵ as

$$Q = \frac{1}{2}\mu v_i^2(1 - \epsilon^2), \tag{16.34}$$

where μ is the reduced mass (for two colliding particles is given by $\mu = \frac{m_1 m_2}{m_1 + m_2}$) and v_i is the relative speed before impact.

Range of ϵ

The values of ϵ range from 0 to 1. We have $\epsilon = 1$ for elastic collisions and $\epsilon = 0$ for perfectly inelastic collisions. In reality, for most objects ϵ lies between the two extremes, where some kinetic energy will always be lost during a collision.

It is interesting to note that for exploding collisions $\epsilon > 1$. Some recent studies have described super-elastic collisions, where it is argued that ϵ has a value greater than one in a special case of oblique collision. That is discussed in detail elsewhere.

16.5.4 Impulse in collisions

Forces that last for an extremely short interval of time, say during collisions, are called *impulsive forces*. For an object the equation of motion in differential form can be written as $d(mv) = Fdt$. Taking the time integral over the interval $t = t_1$ to t_2 during which the force is considered to act, we obtain

$$\vec{J} = \int \vec{F}dt = \int \frac{dp}{dt}dt = \Delta p. \tag{16.35}$$

The SI unit of impulse is the newton second (N s), and the dimensionally equivalent unit of momentum is the kilogram meter per second (kg m s^{-1}).

Ideal impulse

An ideal impulse is produced by a force that tends to ∞ but lasts for a time interval that approaches zero in such a way that integral $\int \vec{F}dt$ remains finite. An impulse produces instantaneous change in momentum and velocity of a body without producing displacement.

16.5.5 Deformation phase

In any collision where ϵ is in the range between 0 and 1, there is a very short period of time when the colliding objects have the same velocity along the line of impact. At this instant they are momentarily at rest with respect to each other. This is also the instant when kinetic energy is lost in maximum fraction as heat, sound and light (figure 16.8). For this short interval of time, this collision $\epsilon = 0$ and this moment may be referred to as the *inelastic phase*. To address this phase, we generally divide the impulse into two parts:

- *The impulse of compression* (P_C)—just before impact.
  For mass 1 $P_C = m_1v_o - m_1v_{1i}$ and
  for mass 2 $-P_C = m_2v_o - m_1v_{2i}$,
  where v_o is the common velocity of both particles at the time when their relative speed is zero.
- *The impulse of restitution* (P_R)—just after impact.
  For mass 1 $P_R = m_1v_{1f} - m_1v_o$ and
  for mass 2 $-P_R = m_2v_{2f} - m_1v_0$.

Relation between impulse and ϵ

From the above equations we can eliminate v_o to obtain

$$m_1m_2(v_{2i} - v_{1i}) = P_C(m_1 + m_2)$$
$$m_1m_2(v_{1f} - v_{2f}) = P_R(m_1 + m_2).$$

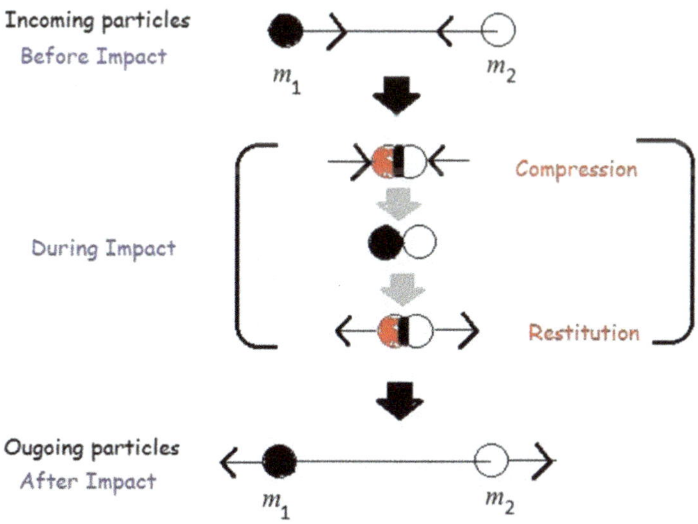

Figure 16.8. Compression and restitution during impact.

Dividing these equations we obtain

$$\left| \frac{v_{2f} - v_{1f}}{v_{1i} - v_{2i}} \right| = \frac{P_R}{P_C}.$$

The left side of this equation is just the definition of the coefficient of restitution (ϵ), hence we can write

$$\epsilon = \frac{P_R}{P_C}. \tag{16.36}$$

Thus the coefficient of restitution is the ratio of the impulse of restitution to the impulse of compression.

16.6 1D collision in the center-of-momentum reference frame

Center-of-momentum (c.o.m.) frame

The center-of-momentum (c.o.m.) frame of a system is the unique inertial frame in which the total momentum of the system equals zero.

A special case of the c.o.m. frame is the center-of-mass frame. The center-of-mass frame is an inertial frame in which the physical center of mass is considered to be always located at the origin (figure 16.9).

Laboratory frame

In physics, the laboratory frame of reference, or lab frame, is a frame of reference centered on the laboratory, which is at rest, in which the experiment is performed. Most measurements are made in the lab frame. For observing and analyzing collisions, we often use the lab frame which is fixed in the lab in which the collision is observed (figure 16.10).

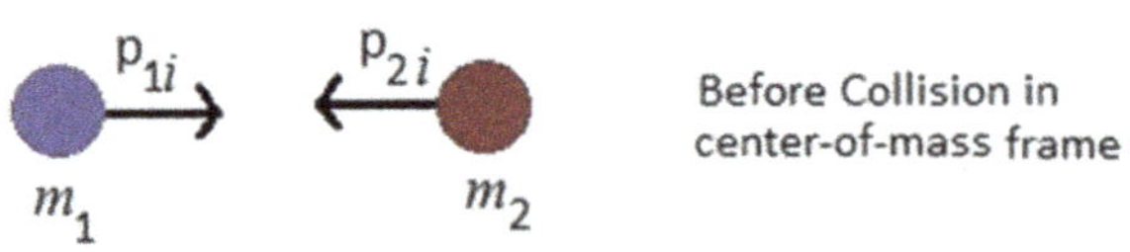

Figure 16.9. Collision in a center-of-momentum frame.

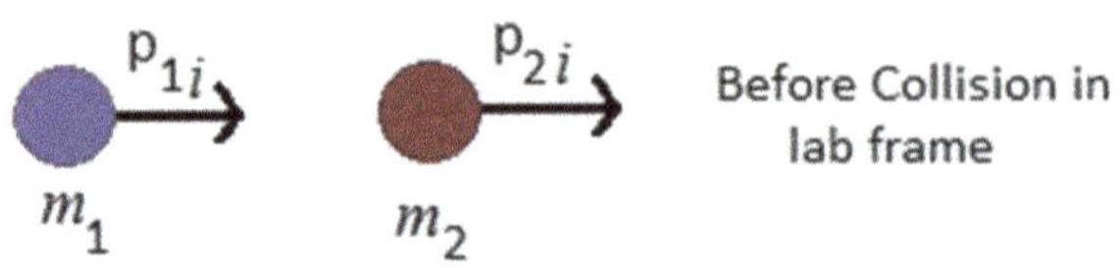

Figure 16.10. Collision in lab frame.

Analysis in the center-of-momentum frame
Assuming that all velocities are in the x-directions, we can write the relative motion as

$$\vec{v}'_{s'} = \vec{v}_s - \vec{v}_{ss'}, \tag{16.37}$$

where v_s represents the velocity of the particle in the lab frame and $v_{s'}$ represents the velocity of the particle in the c.o.m. frame and $v_{ss'} = v$ gives the velocity of the c.o.m. frame relative to the lab frame (figure 16.11).

In this case, to an observer in the center-of-mass frame, the relationship for initial velocities (v_i) in all frames of reference will be given as follows.

For mass 1 (m_1),

$$v'_{1i} = v_{1i} - v_{ss'}. \tag{16.38}$$

For mass 2 (m_2),

$$v'_{2i} = v_{2i} - v_{ss'}, \tag{16.39}$$

where (v'_i) are measured in the s' frame and (v_i) are measured in the lab frame. $v_{ss'} = v$ is the velocity of the c.o.m. to lab frame.

16.6.1 Evaluate v

For the sake of simplicity, I will call the call relative velocity of the c.o.m. frame to lab frame v. Then to find v, we start with the total initial momentum of the two colliding particles, which is the sum of the initial momentum of mass 1 and mass 2:

$$p'_i = m_1 v'_{1i} + m_2 v'_{2i}.$$

Put in values from the above equations,

$$p'_i = m_1(v_{1i} - v) + m_2(v_{2i} - v).$$

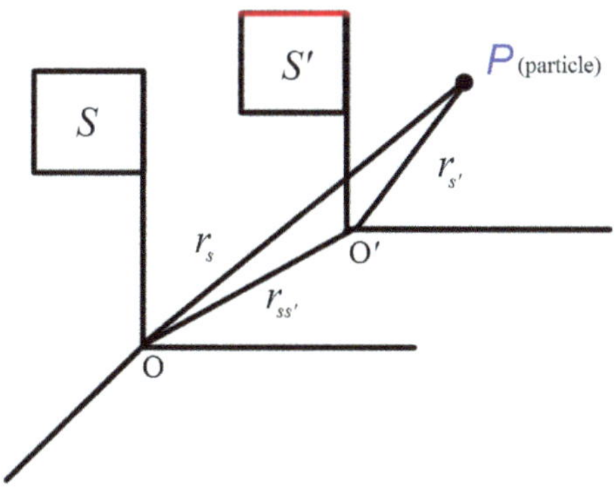

Figure 16.11. Center-of-momentum frame versus lab frame.

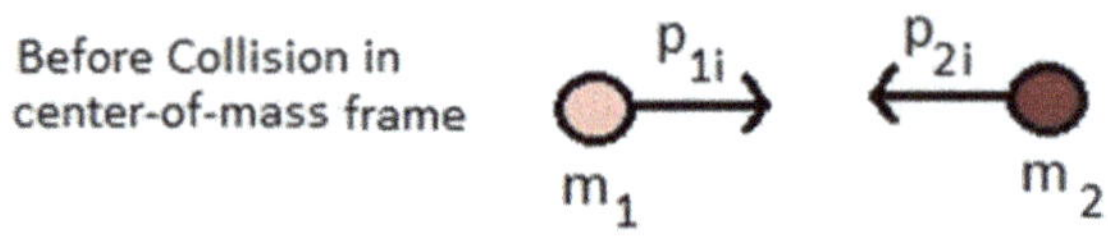

Figure 16.12. Collision in center-of-momentum frame.

In a c.o.m. frame, we know that the initial momentum of a two-body system is zero, i.e. $p_i' = 0$, hence we can write the above equation as

$$0 = m_1(v_{1i} - v) + m_2(v_{2i} - v)$$
$$m_1 v_{1i} - m_1 v = - m_2 v_{2i} + m_2 v$$
$$m_1 v_{1i} + m_2 v_{2i} = (m_1 + m_2) v$$

$$\boxed{v = v_{ss'} = \frac{m_1 v_{1i} + m_2 v_{2i}}{m_1 + m_2}}. \tag{16.40}$$

This implies that if we were traveling at this velocity and observed the collision, then the two bodies before the collision will appear as shown in figure 16.12.

When $p_i' = 0$, the conservation of momentum says that p_f' must also be zero, hence we can say

$$|p_{1f}'| = |p_{2f}'|$$
$$p_{1f}' = - p_{2f}'.$$

This equation shows that they are equal in magnitude but opposite in direction. Please note that these equations will hold true regardless of the type of collision in the center-of-mass frame. The outcome of the collision will depend on the particular properties of the colliding bodies and the nature of the forces they exert on each other.

16.6.2 Elastic collisions

For an elastic collision in the center-of-mass frame, the velocity of each object remains constant in magnitude but changes in direction. So in the c.o.m. frame,

for mass 1 (m_1)

$$v_{1f}' = -v_{1i}', \tag{16.41}$$

for mass 2 (m_2)

$$v_{2f}' = -v_{2i}'. \tag{16.42}$$

Hence the final velocities in the two frames are related as

$$v_{1f}' = v_{1f} - v, \tag{16.43}$$

where v_{1f} is the velocity in the lab frame and v is the relative velocity between the two frames derived in equation (16.40).

Calculating v_{1f} for elastic collisions

Starting from equation (16.43) and using equation (16.41) we can solve for v_{1f}:

$$\begin{aligned} v'_{if} + v &= v_{if} \\ -v'_{1i} + v &= v_{if} \\ -(v_{1i} - v) + v &= v_{if}. \end{aligned}$$

Using values from equation (16.40),

$$\begin{aligned} -v_{1i} + 2v &= v_{if} \\ -v_{1i} + 2\left(\frac{m_1 v_{1i} + m_2 v_{2i}}{m_1 + m_2}\right) &= v_{if}. \end{aligned}$$

Using a bit of algebra we can simplify to obtain

$$\boxed{v_{1f} = \frac{m_1 - m_2}{m_1 + m_2} v_{1i} + \frac{2m_2}{m_1 + m_2} v_{2i}}. \tag{16.44}$$

Calculating v_{2f} for elastic collisions

To find the final velocity of mass 2 (v_{2f}), repeat the above analysis but change the subscripts from 1 to 2, to obtain

$$\boxed{v_{2f} = \frac{2m_1}{m_1 + m_2} v_{1i} + \frac{m_2 - m_1}{m_1 + m_2} v_{2f}}. \tag{16.45}$$

Equations (16.44) and (16.45) are the general result for the 1D elastic collision.

16.6.2.1 Special cases

There are some special cases that are important for understanding collisions:

- *Equal masses.* When the target and projectile are of equal masses, $m_1 = m_2$, then equations (16.44) and (16.45) reduce to give

$$\begin{aligned} v_{1f} &= v_{2i} \\ v_{2f} &= v_{1i}. \end{aligned} \tag{16.46}$$

  The interacting particles will just exchange velocities.
- *Target mass at rest.* When the target particle is at rest, $v_{2i} = 0$, and equations (16.44) and (16.45) give us

$$\begin{aligned} v_{1f} &= \frac{m_1 - m_2}{m_1 + m_2} v_{1i} \\ v_{2f} &= \frac{2m_1}{m_1 + m_2} v_{1i}. \end{aligned} \tag{16.47}$$

- *Equal masses and target mass at rest.* When the target and projectile are of equal masses, $m_1 = m_2$, and the target particle is at rest, will see that the first particle will instantly come to rest and the second (target) particle will take off with the initial velocity (v_{1i}) of mass 1 (m_1). This can easily be seen in collisions of non-rotating billiard balls:

$$\begin{aligned} v_{1f} &= 0 \\ v_{2f} &= v_{1i}. \end{aligned} \tag{16.48}$$

- *Massive target.* When the target particle is much more massive than the incident particle (projectile), i.e. $m_2 \gg m_1$, then equations (16.44) and (16.45) reduce to

$$\begin{aligned} v_{if} &\approx -v_{1i} + 2v_{2i} \\ v_{2f} &\approx \ 2v_{2i}. \end{aligned} \tag{16.49}$$

- *Massive target at rest.* If the target (massive particle) is at rest then the equations reduce down to

$$\begin{aligned} v_{if} &\approx -v_{1i} \\ V_{2f} &\approx \ 0. \end{aligned} \tag{16.50}$$

  Here the velocity of the lighter particle is reversed and the velocity of the massive target remains at rest or moves very slowly. A good example of this is α-particle scattering incident on gold foil, called Rutherford scattering, that we will discuss in chapter 17.
- *Massive projectile.* When the incident particle (projectile) is much more massive than the target, i.e. $m_1 \gg m_2$, then equations (16.44) and (16.45) reduce to

$$\begin{aligned} v_{if} &\approx v_{1i} \\ V_{2f} &\approx 2v_{1i} - v_{2i}. \end{aligned} \tag{16.51}$$

  If the target, which is the lighter particle, is at rest or moving very slowly, then the target particle will move with two times the speed of the incident particle. The motion of the incident particle will virtually be unchanged (figure 16.13).

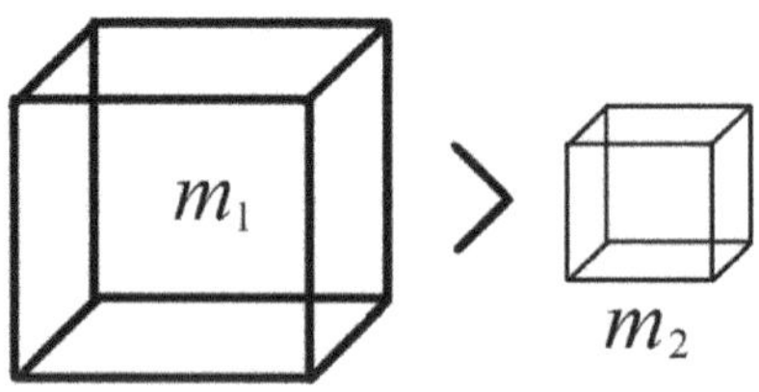

Figure 16.13. Massive projectile.

Example 16.4

Suppose mass m_1 (= 2 kg) is projected towards m_2 = 4 kg at rest with velocity v_{1i} (= 4 m s^{-1}). After the collision, m_1 recoils with velocity 2 m s^{-1}. What is the final velocity of m_2?

Solution

We can use momentum conservation in 1D:

$$m_1 v_{1i} + m_2 v_{2i} = m_1 v_{1f} + m_2 v_{2f}. \tag{16.52}$$

Since m_1 recoils after the collision, $v_{1f} = -2$ m s^{-1}, then we can solve for v_{2f},

$$\begin{aligned} v_{2f} &= \frac{m_1 v_{1i} + m_1 v_{1f}}{m_2} \\ &= \frac{(2)(4) + (4)(2)}{4} \\ \Longrightarrow v_{2f} &= 3 \text{ ms}^{-1}. \end{aligned}$$

Example 16.5

In example 16.4, suppose the two masses stick together upon collision. What is the final velocity of the $(m_1 + m_2)$ combination?

Solution

If the masses stick together, equation (16.52) can be rearranged to give

$$\begin{aligned} m_1 v_{1i} + m_2 v_{2i} &= (m_1 + m_2) v_{12} \\ v_{12} &= \frac{m_1 v_{1i} + m_1 v_{1f}}{m_1 + m_2} \\ &= \frac{(2)(4) + (4)(0)}{6} \\ \Longrightarrow v_{12} &= 1.3 \text{ ms}^{-1}. \end{aligned} \tag{16.53}$$

Concept questions 16.2

1. How does an airbag protect the driver and passengers in the case of a head-on collision?
2. When firing a shotgun, why is it recommended to press the stock tightly against your shoulder?
3. Assuming that neither penetrates the skin, and the mass, initial speed and time of contact are identical, why do rubber bullets hurt more than lead bullets?

Exercises 16.2

1. A bullet of mass (m) is fired horizontally at a block of wood of mass (M), which is initially at rest. The bullet becomes embedded in the block and the impact causes the system to slide a distance d before coming to rest. Given that the coefficient of sliding friction is μ_k, find the initial speed (muzzle velocity) of the bullet.
2. A hemispherical object of uniform density such as that shown in figure 16.14 has a radius R. Use integration to find the location of its center of mass. (*Hint*: symmetry arguments help resolve where the center of mass will be located in x and y, so you only need to integrate to find z.)

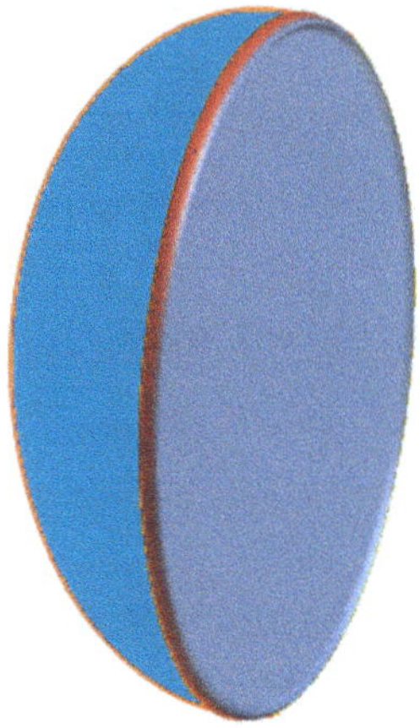

Figure 16.14. Hemisphere.

Further reading

Arya A 1990 *Introduction to Classical Mechanics* 2nd edn (Upper Saddle River, NJ: Prentice-Hall)

Baierlein R 1983 *Newtonian Dynamics* (New York: McGraw-Hill)

Chababy R and Sherwood B 2011 *Matter and Interactions* 3rd edn (New York: Wiley)

Fowles G 1986 *Analytical Mechanics* 4th edn (New York: CBS College Publishing)

Fowles G and Cassiday G 2005 *Analytical Mechanics* 7th edn (Belmont, CA: Thomson Learning)

Goldstein H 1980 *Classical Mechanics* 2nd edn (Reading, MA: Addison-Wesley)

Landau L D and Lifshitz E M 1960 *Mechanics* (Reading, MA: Addison-Wesley)

Ohanian H 1985 *Physics* (New York: WW Norton)

Resnick R, Halliday D and Krane K 2002 *Physics* vol 1, 5th edn (New York: Wiley)

Stephenson R 1961 *Mechanics and Properties of Matter* 2nd edn (New York: Wiley)

Symon K 1971 *Mechanics* 3rd edn (Reading, MA: Addison-Wesley)

IOP Publishing

Techniques of Classical Mechanics
From Lagrangian to Newtonian mechanics
Samya Zain

Chapter 17

Scattering theory

17.1 Cross-section

In scattering experiments, the most important information is to know the likelihood of an event to occur. Scattering experiments have beams of particles, that we call incident particles and target particles. The incident beam could comprise elementary particles, nuclei or even light, whereas the target could be made of atoms, nuclei or gas particles. The cross-section is an effective area that quantifies the essential likelihood of a scattering event when a beam is incident on a target object.

17.1.1 The idea behind cross-sections

Suppose a particle (X) finds itself in the path of a beam of incident particles. What happens? Well, most of the atoms in the incident beam pass particle X with no impact. However, some atoms in the incident beam will be seen to bounce off particle X, as seen in figure 17.1.

The next logical question to ask is, how many scatterings will occur? Then, the next question will be, how many scatterings will occur per second? To answer this question, we first need to find how many incoming particles have the initial trajectories that can even result in a collision with the target.

Let us suppose that particle X has a radius (R) and the atoms in the incident beam have radius R_H, as shown in figure 17.2. We know that the incident atom must be in a specific volume if it is to strike particle X. Let us try to find that volume.

In time Δt, the incident atom moves a distance given by $\Delta x = v\Delta t$ and in time Δt, the incident atom must be in the cylindrical volume given as $A = \pi(R + R_H)^2$ to the left of particle X in order to collide with particle X. This is called the *impact volume*, given mathematically as

$$\text{impact volume} = V = A\Delta x = Av\Delta t = \pi(R + R_H)^2 v\Delta t. \tag{17.1}$$

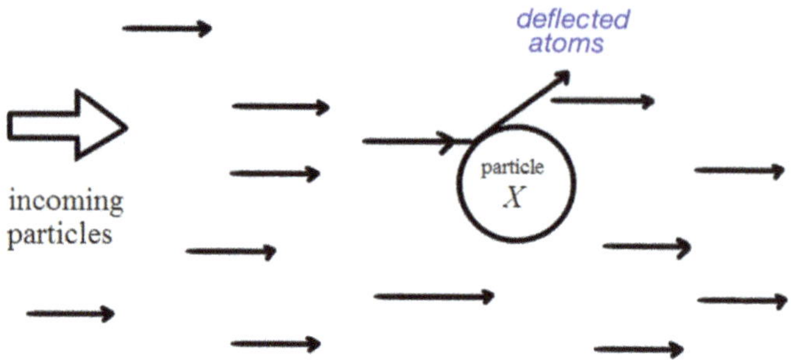

Figure 17.1. Particle X in the path of the incident beam.

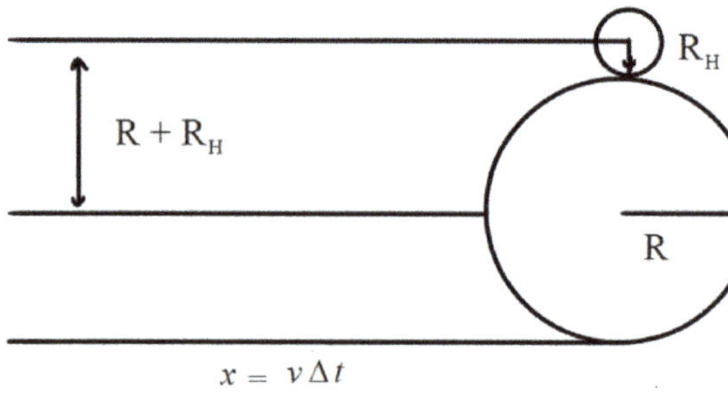

Figure 17.2. Impact area—the incoming particle must be in this surface area for a possible collision.

If the number of incident atoms per unit volume is given as n_H, then the number of scatterings per Δt will be the product of the number of incident atoms per unit volume and the impact volume, given as

$$\text{number of scatterings in time } \Delta t = (n_H)\pi(R + R_H)^2 v\Delta t \tag{17.2}$$

and

$$\text{number of scatterings per second} = n_H\pi(R + R_H)^2 v, \tag{17.3}$$

where the number of scatterings per second is usually called the *scattering rate*, given by equation (17.3).

17.1.2 Scattering effectiveness

Now, what happens if the incident beams becomes more intense, say by increasing the speed (v) of incoming particles or by increasing the number of incoming particles (n_H) or both. Then the *scattering rate* will go up, even though the size of the particles has not changed during this time. What changes is that the incident and target particles just have more chances to interact. In this case we have to talk about *scattering effectiveness*, which can be defined as

$$\text{scattering effectiveness} = \frac{\text{scattering rate}}{\text{scattering opportunities}}. \tag{17.4}$$

In defining the scattering effectiveness we have a term in the denominator called scattering opportunity. What is a scattering opportunity? A good choice when defining scattering opportunity is to talk about *incident flux*, which is the number of incoming particles that come within the impact area per second, mathematically given as $= vn_H$. Hence, scatting effectiveness can be defined as

$$\text{scatting effectiveness} = \frac{\text{number of scatterings per second}}{\text{incident flux}}$$
$$= \frac{\pi(R + R_H)^2 v\, n_H}{v\, n_H}$$

$$\text{scatting effectiveness} = \pi(R + R_H)^2. \tag{17.5}$$

In equation (17.5), note that $\pi(R + R_H)^2$ has units of area and is called the *cross-section for scattering*. Additionally, cross-section depends on both the radius of the incident particles (R_H) and the radius (R) of the target. It is not a property of either separately, and is usually represented in most textbooks as (σ_{scatt}).

Think of cross-section as an area that the scatterer presents to the projectile beam. Any particle whose incident trajectory is within the area will be scattered.

Why cross-section is a good thing to know:

- Measuring the cross-section (σ_{scatt}) tells us information about both the projectile and target properties, especially in cases where we may not know either R or R_H or both.
- To test theories about the behavior of the particular projectile and target particles under experimental scattering conditions. This strategy has been a major theme in physics since Rutherford in the early 1900s.
- To compare the expected event rate to the experimental rate for any specific incident flux.

17.2 Types of scattering

Particles in nature are of two basic types, neutral and charged particles. We can differentiate the various types of cross-sections on the basis of the particles involved in the interaction as:

- When both interacting particles are neutral: when both the incident and the scattered particles are charge neutral, the interaction is called *scattering* and the cross-section is called the *neutral* cross-section or just the cross-section, discussed in detail in section 17.3.

- When the interacting particles are oppositely charged or one particle in the interaction is neutral and the other is charged: in the case where one particle is positively charged and the other is negatively charged, we call that interaction *attraction scattering* and the cross-section is called the *capture cross-section*, discussed in detail in section 17.4.
- When the interacting particles are similarly charged particles: in the case where both particles are of the same charge we call the interaction *repulsive scattering*, and it corresponds to a *repulsive cross-section*, discussed in detail in section 17.5.

17.3 Neutral cross-section

Let us first consider the situation where both the target, radius (R), and the incident particles, radius (R), are charge neutral. In this case the incident particles will be scattered in all directions. Some will be deflected a little, others will go off at angles nearly at right angles and a few will even be deflected backwards. The angle of scatter (θ_s) will depend on the *impact parameter* (b).

Impact parameter

The impact parameter (b) is the perpendicular distance from the initial trajectory to a parallel line through the center of the target, as seen in figure 17.3. It is a very valuable tool in scattering analysis.

Angle of scatter (θ_s)

The angle between the incoming ($\vec{v}_0$) and final velocity ($\vec{v}_f$) of the projectile is called angle of scatter (θ_s).

Relationship between θ_s and b

The distance between the centers at the instant of collision equals ($R + R_H$). Assume that both interacting particles act like hard smooth spheres and that during impact the component of velocity v along the connecting line ($v_{//}$) is reduced to zero and is

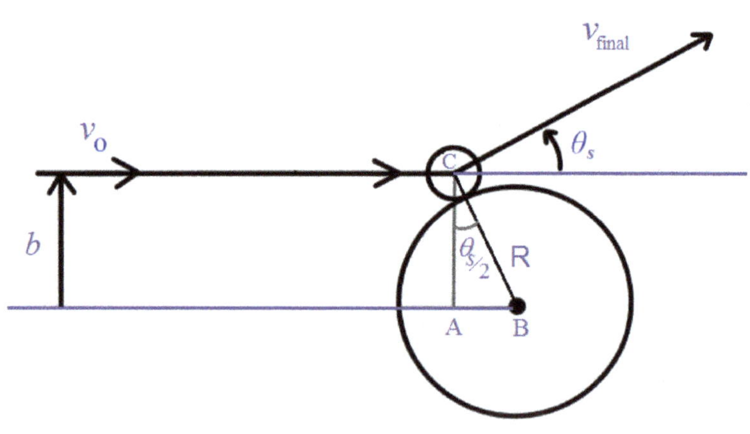

Figure 17.3. Relationship between angle of scatter (θ_s) and impact parameter (b).

then reversed, whereas the perpendicular component ($v_\perp$) remains unchanged during the interaction. The angle between ($v_\perp$) and incoming (v_0) is given by ($\theta_s/2$).

From figure 17.3 we also see that in $\triangle ABC$:

$$\cos\frac{\theta_s}{2} = \frac{b}{R + R_H}$$
$$(R + R_H)\cos\frac{\theta_s}{2} = b.$$

Differentiating the above equation we obtain

$$(R + R_H)\left(-\sin\frac{\theta_s}{2}\right)\frac{1}{2}d\theta = db$$

$$\boxed{db = -\frac{(R + R_H)}{2}\sin\frac{\theta_s}{2}d\theta}. \tag{17.6}$$

The negative sign in equation (17.6) implies that as the scattering angle (θ_s) decreases the impact parameter (b) increases.

Now we have to ask the following two questions:

Q1. How do we find the number of particles that are scattered into a certain patch or solid angle ($d\Omega$)?

Q2. What must be the impact parameter (b) for incident particles to scatter into the particular patch ($d\Omega$) after interaction?

Let us begin by answering Q1. First, we have to define what we mean by a patch. A patch can be thought to lie on a sphere of radius $\mathfrak{R}$ centered at the middle of the incident particle at the instant of collision, as seen in figure 17.4. The sides of patch will be ($\mathfrak{R}d\theta_s$) and ($\mathfrak{R}\sin\theta_s d\phi$), where ϕ is the azimuthal angle and θ_s is the polar angle from point C. The patch gives us a certain set of directions (virtual lines) that start at point C and pass through the patch. To characterize the bundle of directions, we have to specify the angle (θ_s) and the azimuthal angle (ϕ). Then, mathematically,

$$d\Omega = \frac{\text{patch area}}{\mathfrak{R}^2}$$
$$= \frac{(\mathfrak{R}d\theta_s)(\mathfrak{R}\sin\theta_s\ d\phi)}{\mathfrak{R}^2}$$
$$\Longrightarrow d\Omega = \sin\theta_s\ d\theta_s\ d\phi.$$

Recall that for radian measure of angles this is called the *infinitesimal solid angle* of the same name ($d\Omega$).

Next, let us turn to answering Q2. What is the impact parameter (b) for incident particles to scatter into the patch ($d\Omega$) after interaction? To answer this question, we

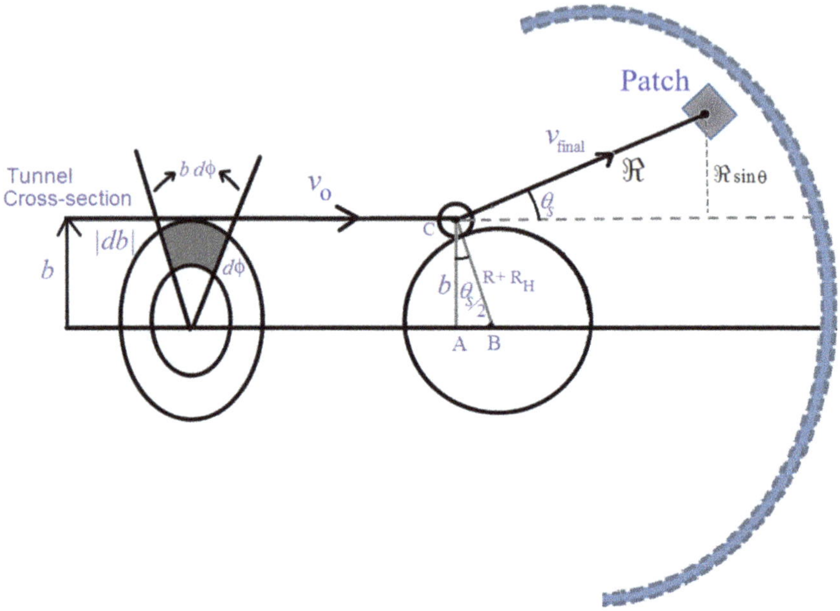

Figure 17.4. Defining patch and tunnel area.

will do the same thing as we did to solve Q1. Define a patch that has the area that is held in the path of the incident particles. All the particles that go through this patch will scatter into solid angle ($d\Omega$). Suppose we call this a *tunnel area*[1] and mathematically express it as ($d\sigma$). Then from figure 17.4 we can say $d\sigma$, formally called the cross-section for scattering, will have sides (db) and ($bd\phi$) and will mathematically be given as

$$\begin{aligned} d\sigma &= (db)(bd\phi) \\ &= b\ db\ d\phi \\ &= \left((R + R_H)\cos\frac{\theta_s}{2}\right)\left(\frac{-(R + R_H)}{2}\sin\frac{\theta_s}{2}d\theta_s\right)d\phi \\ \Longrightarrow d\sigma &= \left(\frac{R + R_H}{2}\right)^2(-\sin\theta_s)d\theta_s\ d\phi. \end{aligned}$$

Here we have used the trigonometric identity $2\cos\frac{\theta}{2}\sin\frac{\theta}{2} = \sin\theta$.

17.3.1 Differential cross-section

From the answers to Q1 and Q2, we can obtain the differential scattered cross-section simply by dividing $d\sigma$ (the tunnel area) by ($d\Omega$ (the patch area) as

[1] Tunnel area terminology first named by Annie Loelinger, Newtonian Mechanics class, 26 October 2017.

$$\left|\frac{d\sigma}{d\Omega}\right| = \frac{\text{cross} - \text{section for scattering into } d\Omega}{\text{size of } d\Omega}$$

$$= \frac{\left(\frac{R+R_H}{2}\right)^2 (\sin\theta_s) d\theta_s\, d\phi}{\sin\theta_s d\theta_s\, d\phi}$$

$$\boxed{\left|\frac{d\sigma}{d\Omega}\right| = \left(\frac{R+R_H}{2}\right)^2}. \quad (17.7)$$

Equation (17.8) shows that the differential cross-section is completely independent of the angle of scatter (θ_s). Keep in mind that this is the idealized case and is not absolutely true in all circumstances. There will be always be some dependence on θ_s.

For hard spheres, the cross-section per unit solid angle is the same for all scattered directions, which means that the scattered particles are distributed uniformly in all directions. Hence the scattering is *isotropic*.

17.3.2 Total cross-section

To find the total cross-section, we simply integrate the differential cross-section:

$$\sigma = \int_{\text{sphere}} \frac{d\sigma}{d\Omega} d\Omega$$

$$= \int_0^{2\pi} \int_0^{\pi} \frac{1}{4}(R + R_H)^2 \sin\theta_s d\theta_s\, d\phi$$

$$= \frac{1}{4}(R + R_H)^2 \int_0^{\pi} \sin\theta d\theta_s \int_0^{2\pi} d\phi$$

$$= \frac{1}{4}(R + R_H)^2 (2)(2\pi)$$

$$\boxed{\sigma = \pi(R + R_H)^2}. \quad (17.8)$$

This is exactly what we obtained before, $(R + R_H) = b$, which is the impact parameter.

17.4 Capture cross-section

When the incident and target particles are oppositely charged, or if either the incident or the target is neutral and the other has a charge, the incident particle will be attracted to the target particle upon approach and they will 'fall' towards the target and may be held at the surface of the target. If the incident particle is held by the target particle, the incident particle can be considered as *captured*.

We can also have capture in certain incidences called *molecule formation*. Going back to the example from section 17.1.1, suppose the particle X is charged and has a charge $(-Ze)$, where Z is a positive integer and (e) is the charge of an electron. When

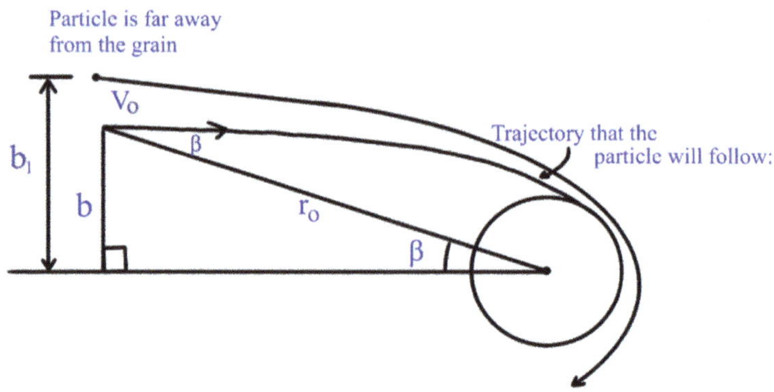

Figure 17.5. Capture cross-section.

the ionized atoms in the incident beam come close to particle X, they will be attracted to particle X. When the incident ion strikes the surface of particle X it will be held there and can be considered to be captured.

The next question is, what will be the cross-section for such a capture? Clearly, the attractive force between the incident ion and particle X bends the initial trajectory towards particle X. The impact parameter (b) shown in figure 17.5 is the decisive quantity.

Next question: how small would the impact parameter b have to be for the incident ion to be captured, assuming all collisions result in a capture? To answer this question, let us start by finding the total energy at any point in the path. The total energy will be a sum of the kinetic (T) and potential (V) energies, as

$$\begin{aligned} E &= \frac{p^2}{2m} + V(r) \\ &= \frac{(mv)^2}{2m} + V(r) \\ &= \frac{L^2}{2mr^2} + V(r) \\ E &= \frac{L^2}{2mr^2} + V(r), \end{aligned} \tag{17.9}$$

where L is the angular momentum ($L = mvr$), and r is the incident ion's position relative to the center of particle X. When the mass of particle X is much larger than the mass of the incident ion, we can consider particle X at rest and take the reduced mass to be just the mass of the incident ion. At minimum separation along the trajectory, the angular momentum $L = mv_0b$ and equation (17.9) becomes

$$E = \frac{(mv_0b)^2}{2mr_{\rm min}^2} + V(r_{\rm min}). \tag{17.10}$$

The critical value of the impact parameter occurs when $r_c = r_{\rm min}$, let us call that value b_c. Then equation (17.10) may be rewritten as

$$E = \frac{(mv_0 b_c)^2}{2mr_c^2} + V(r_c)$$

$$\frac{1}{2}mv_0^2 = \frac{m^2 v_0^2 b_c^2}{2mr_c^2} + V(r_c)$$

$$\frac{1}{2}mv_0^2 - V(r_c) = \frac{mv_0^2 b_c^2}{2r_c^2},$$

which can be simplified by dividing both sides by $\frac{1}{2}mv_0^2$ to give the expression

$$\boxed{b_c^2 = \left(1 - \frac{V(r_c)}{\frac{1}{2}mv_0^2}\right) r_c^2}, \tag{17.11}$$

which means:

- All ions with $b > b_c$ will miss particle X.
- All ions with $b \leqslant b_c$ will strike particle X and will be captured.

This implies that the capture cross-section will be given as

$$\sigma_{\text{capture}} = \pi b_c^2 = \pi \left(1 - \frac{V(r_c)}{\frac{1}{2}mv_0^2}\right) r_c^2, \tag{17.12}$$

keeping in mind that $V(r_c) = \frac{kZZ_i e^2}{r_c}$.

Cross-section dependence on energy

Another thing to keep in mind is that the cross-section depends on the energy of the ions:

- If the energy is much greater than the absolute value of potential energy at the grain surface then

$$\sigma_{\text{capture}} \simeq \pi R^2, \tag{17.13}$$

  which is the geometric cross-section. This makes sense, since incoming particles with very large energy will be moving very fast ($E = \frac{1}{2}mv_0^2$). This implies that this ion must be headed for a direct collision with particle X in order to be captured by particle X.
- If the ion moves slowly as it comes in from a distance, the electrical attraction can pull it in to be captured even if impact parameter b is very large.

From these statements we can deduce that cross-section increases as energy decreases.

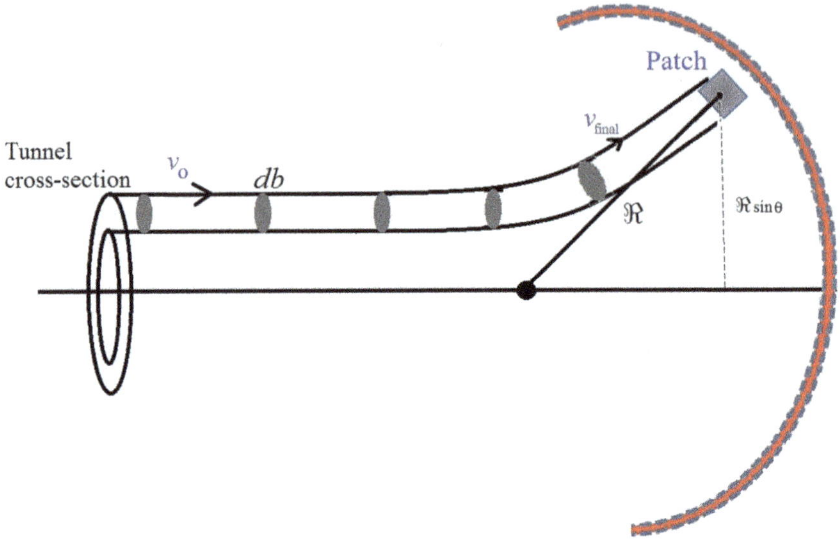

Figure 17.6. Repulsive scattering.

17.5 Repulsive cross-section

Like charges repel each other when they are brought close to each other. Hence, in the case where both the incident and the scattered particle have the same type of charge we call the interaction *repulsive scattering*, which corresponds to a *repulsive cross-section* (figure 17.6).

From section 17.3 we know that the number of scattered particles that pass through the *patch* cross-section ($d\Omega$) per unit time is equal to the number of incident particles per unit time coming through the *tunnel* cross-section (db). This means we can say the flux of particles passing through tunnel cross-section (db) is equal to the number of particles scattered into the patch solid angle ($d\Omega$). This is mathematically expressed as

$$d\Omega = \sin\theta_s d\theta_s d\phi. \tag{17.14}$$

Then the total incident flux will be equal to the number of particles scattered into total solid angle (Ω_T):

$$\Omega_T = \int d\Omega = \sin\theta_s d\theta_s \int_0^{2\pi} d\phi = 2\pi \sin\theta_s d\theta_s. \tag{17.15}$$

Let us call this the 'ring of fire'[2]. This means that all the incident particles coming from the tunnel ring will hit the 'ring of fire', as seen in figure 17.7.

[2] The term 'ring of fire' was coined during a lecture by Hannah Sage, 3 October 2017.

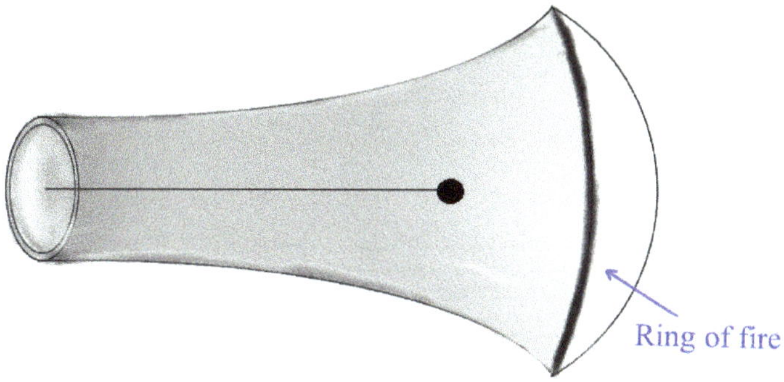

Figure 17.7. The ring of fire.

Differential cross-section in terms of incident particles
If the total number of particles coming from the tunnel cross-section is N then the total incident flux will be

$$\text{incident flux} = N(2\pi \sin \theta_s d\theta_s).$$

We can also define the differential cross-section in terms of incident particles as

$$\frac{d\sigma}{d\Omega} = \frac{N}{\text{incident flux}}. \tag{17.16}$$

Cross-section in terms of impact parameter (b)
Suppose there are N particles incident on a target. Then we can say that the number of particles approaching the target in the (db) tunnel strike the patch ($d\Omega$) and the particles approaching in the db ring tunnel strike the ring of fire $d\Omega_T$.

The area of the ring tunnel is given as ($2\pi bdb$) and the total number of such incoming particles is equal to the total number of scattered particles through a given angle ($d\theta_s$):

$$N(\text{area of ring tunnel}) = N(d\Omega_T)(\Omega(\theta_s))$$
$$N(2\pi bdb) = N(2\pi \sin \theta_s d\theta_s)(\Omega(\theta_s))$$
$$bdb = \sin \theta_s d\theta_s \Omega(\theta_s).$$

Hence the total cross-section in terms of impact parameter can be written as

$$\boxed{\Omega(\theta_s) = \frac{b}{\sin \theta_s}\left|\frac{db}{d\theta_s}\right|}. \tag{17.17}$$

I have added the absolute value in this equation, which is necessary because the derivative is negative, which implies as (b) reduces (θ_s) increases, as seen in figure 17.8.

In order to find the path the incident particle will follow we have to solve the energy equation of orbit, as done below in section 17.5.1.

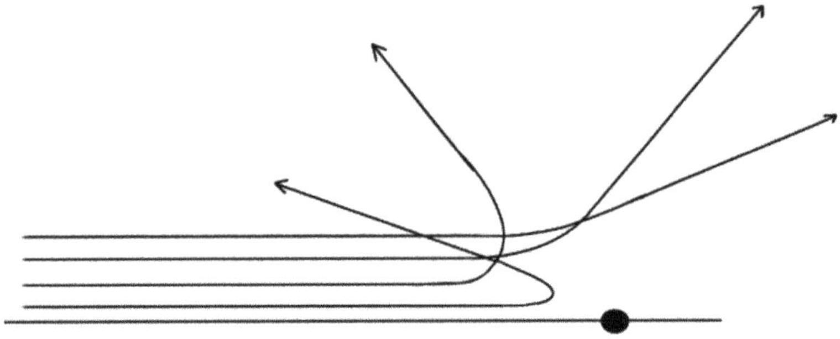

Figure 17.8. Scatter angle (θ_s) and impact parameter (b).

17.5.1 General energy equation of orbit

Consider an incident particle of charge (q) and mass (m) passing near a heavy particle of charge Q. Assume that particle Q is located at the origin, O. If we consider the case of a repulsive force, the incident particle will be repelled with a force given by Coulomb's law:

$$F(r) = \frac{Qq}{r^2}. \tag{17.18}$$

We know that the total energy at any instant of time in the path is given as

$$E_{\text{total}} = T + V, \tag{17.19}$$

where T and V are kinetic and potential energies, respectively. In an inverse-square field, we can express the energy in polar coordinates as

$$E_{\text{total}} = \frac{1}{2}m(\dot{r}^2 + r^2\dot{\theta}^2) + V(r). \tag{17.20}$$

To obtain the energy of the orbit, we need to change variables for mathematical expediency and put $r = \frac{1}{u}$, differentiating and knowing the angular momentum ($l = r^2\dot{\theta}$ = constant),

$$dr = -\frac{1}{u^2}du = -\frac{1}{u^2}du\frac{d\theta}{d\theta} = -r^2\dot{\theta}\frac{du}{d\theta} = -l\frac{du}{d\theta}.$$

Hence we can write the above equation as

$$E_{\text{total}} = \frac{1}{2}ml^2\left[\left(\frac{du}{d\theta}\right)^2 + u^2\right] + V(u^{-1}), \tag{17.21}$$

which is called the energy equation of orbit. In the given case we know that the potential energy is proportional to r^{-1}. Hence,

$$V(u^{-1}) = -ku \tag{17.22}$$

and the energy equation of orbit becomes

$$E_{\text{total}} = \frac{1}{2}ml^2\left[\left(\frac{du}{d\theta}\right)^2 + u^2\right] + (-ku), \tag{17.23}$$

which by a bit of algebra can be shown to become

$$d\theta = \frac{1}{\sqrt{\frac{2E_{\text{total}}}{ml^2} + \frac{2ku}{ml^2} - u^2}}du. \tag{17.24}$$

Integrating this equation, using $\int \frac{du}{\sqrt{au^2 + bu + c}} = \frac{1}{\sqrt{-a}} \cos^{-1}\left(-\frac{b + 2au}{\sqrt{b^2 - 4ac}}\right)$, it can be shown that

$$\frac{k}{ml^2}\left[\sqrt{1 + \frac{2E_{\text{total}}ml^2}{k^2}} \cos(\theta - \theta_o) + 1\right] = u, \tag{17.25}$$

which, knowing ($r = 1/u$), is written as

$$\boxed{r = \frac{\frac{ml^2}{k}}{1 + \sqrt{1 + \frac{2E_{\text{total}}ml^2}{k^2}} \cos(\theta - \theta_o)}}, \tag{17.26}$$

which is the equation of a hyperbola, hence the incident particle will follow a hyperbolic path. This may also be seen from the physical fact that the energy E is always greater than zero in a repulsive field of force. (In our case $E = T + V = \frac{1}{2}mv^2 + k\frac{Qq}{r}$.) Hence, the eccentricity ϵ, the coefficient of $\cos(\theta - \theta_o)$, is greater than one, which means that the orbit must be hyperbolic.

We can also write the equation of the orbit, knowing ($k = -Qq$) in the form:

$$r = \frac{1}{\left[\sqrt{\left(\frac{Qq}{ml^2}\right)^2 + \frac{2E_{total}}{ml^2}} \cos(\theta - \theta_o) - \frac{Qq}{ml^2}\right]}. \tag{17.27}$$

A special case: at closest approach

At $r = \infty$.

Let us first consider that the incoming particle is far away. We can consider this point where $r = \infty$; at this point the denominator in equation (17.27) must equal zero:

$$1 + \sqrt{1 + \frac{2E_{\text{total}}ml^2}{(Qq)^2}} \cos(\theta - \theta_o) = 0, \tag{17.28}$$

which can be simplified to

$$\cos(\theta - \theta_o) = \frac{1}{\sqrt{1 + \frac{2Eml^2}{(Qq)^2}}}. \tag{17.29}$$

The incident particle approaches along one asymptote and recedes along the other, as shown in figure 17.9.

At $r = r_{\min}$.
We know that the scattering depends on the impact parameter (b). The minimum distance during a hyperbolic orbit is called the closest approach ($r_{\min}$). At closest approach, as can be seen in figure 17.9:

$$\pi = 180^\circ = 2\theta_o + \theta_s$$
$$\frac{\pi - \theta_s}{2} = \theta_o$$
$$\theta_s = \pi - 2\theta_o.$$

Using these equations we can derive the following:

$$\begin{aligned}\cot\frac{\theta_s}{2} &= \frac{\cos\left(\frac{\pi - 2\theta_o}{2}\right)}{\sin\left(\frac{\pi - 2\theta_o}{2}\right)}\\ &= \frac{\cos\frac{\pi}{2}\cos\theta_o + \sin\frac{\pi}{2}\sin\theta_o}{\sin\frac{\pi}{2}\cos\theta_o + \cos\frac{\pi}{2}\sin\theta_o}\\ &= \frac{\sin\theta_o}{\cos\theta_o}\\ &= \tan\theta_o.\end{aligned}$$

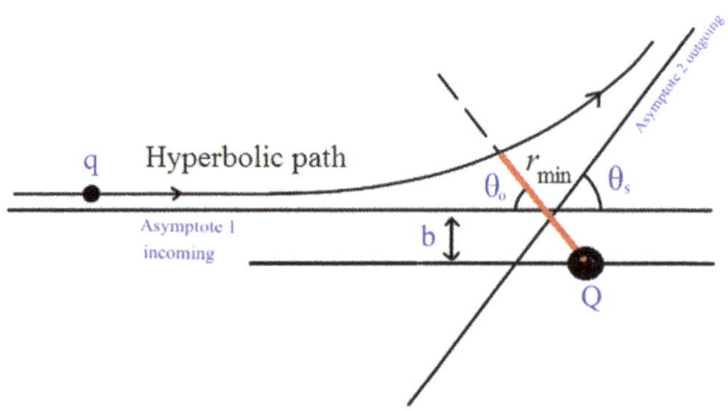

Figure 17.9. Closest approach.

Then the closest approach as a function of impact parameter (b) is given as

$$r_{\min} = \frac{b\cos\left(\frac{\theta_s}{2}\right)}{1 - \sin\left(\frac{\theta_s}{2}\right)}. \tag{17.30}$$

This has a limit for back-scattering which is just determined by setting the initial kinetic energy equal to the final potential energy. This will correspond to a stopping potential and send it backwards, so at its closest approach all the energy is electric potential energy. For example, if we take the charge of the target to be (Ze) and the charge of the incoming particle to be (e) then $r_{\min}$ will be given as

$$r_{\min} = \frac{Zke^2}{T}, \tag{17.31}$$

where the kinetic energy (T) is $\frac{1}{2}mv^2$.

Impact parameter

The impact parameter is the perpendicular distance from the origin (scattering center) to the initial line of motion of the particle. Starting from equation (17.29), we can use values from equation (17.5.1) and take $\theta = 0$:

$$\cos(\theta - \theta_o) = \frac{1}{\sqrt{1 + \frac{2Eml^2}{(Qq)^2}}}$$

$$\cos\theta_o = \frac{1}{\sqrt{1 + \frac{2Eml^2}{(Qq)^2}}}$$

$$\cos\left(\frac{\pi - \theta_s}{2}\right) = \frac{1}{\sqrt{1 + \frac{2Eml^2}{(Qq)^2}}}.$$

For small angles $\sin \theta/2 \cong \tan \theta/2$, and ignoring 1 w.r.t $\frac{2Eml^2}{(Qq)^2}$. Using the trigonometric expansion for cosine $\cos\frac{\pi - \theta_s}{2} = \sin\frac{\theta_s}{2}$, we will obtain

$$\tan\frac{\theta_s}{2} = \frac{1}{\sqrt{\frac{2Eml^2}{(Qq)^2}}}. \tag{17.32}$$

Rearranging we can write

$$\cot\frac{\theta_s}{2} = \sqrt{\frac{2Eml^2}{(Qq)^2}} \tag{17.33}$$

and combining the above results we can write

$$\tan\theta_o = \sqrt{\frac{2Eml^2}{Q^2q^2}} = \cot\frac{\theta_s}{2}. \tag{17.34}$$

If we write angular momentum as $l = b^2v_o^2$ and energy as $E = \frac{1}{2}mv^2$, then the above equation becomes

$$\begin{aligned}\tan\theta_o &= \sqrt{\frac{2\left(\frac{1}{2}mv^2\right)mb^2v_o^2}{Q^2q^2}} \\ &= \frac{mbv_o^2}{Qq},\end{aligned}$$

which can be rewritten as

$$\boxed{\tan\theta_o = \cot\frac{\theta_s}{2} = \frac{mbv_o^2}{Qq} = \frac{2bE}{Qq}}. \tag{17.35}$$

17.6 Scattering of alpha particles

In the early 1900s the atom was thought to be a distributed blob of positive charge within which the negatively charged electrons were embedded randomly. Think of it as a chocolate chip cookie in 3D, with the cookie as the atom and chocolate chips as electrons. This model was called the 'Thomson atom' after J J Thomson (1856–1940 CE).

In 1907, Hans Geiger (1882–1945 CE) was in the process of testing the validity of the Thompson atomic model. Things behaved more or less as expected, except there were many more large-angle scatterings than could be accounted for by the Thomson model. Around 1911, Ernest Rutherford (1871–1937 CE) joined in the effort of figuring out the problem of the large scattering angles of alpha particles by thin metal foils. It was as though an on-rushing freight train had been hurled backward upon striking a toy car sitting in the middle of the track.

What model could explain that? The closest thing Rutherford could think of was a comet swinging around the Sun and coming back out again. This suggested a hyperbolic orbit for a positively charged alpha particle attracted by a negatively charged nucleus. Rutherford realized that this was the familiar inverse-square nature of the force law which inevitably leads to conic sections as solutions for the orbit. It would be the same whether the force was attractive or repulsive.

Rutherford used conservation of angular momentum and energy along with a theorem about conics from geometry that related the eccentricity of the hyperbola to the angle between its asymptotes to obtain a complete solution to the alpha particle scattering problem. This gave us the currently accepted model of the nuclear atom in only a very broad sense, since we know that the electrons do not orbit the nucleus in ellipses.

17.6.1 The Rutherford scattering formula

To derive the Rutherford scattering formula we start from equation (17.35):

$$\cot\frac{\theta_s}{2} = \frac{2bE}{Qq}.$$

Differentiating w.r.t. θ_s,

$$\cot\frac{\theta_s}{2} = \frac{2bE}{Qq}$$

$$\frac{1}{\sin^2\frac{\theta_s}{2}}\frac{1}{2} = \frac{2E}{Qq}\left|\frac{db}{d\theta_s}\right|.$$

We know the differential cross-section in terms of impact parameter is given by equation (17.17), so we can write

$$\sigma\frac{\sin\theta_s}{b} = \left|\frac{db}{d\theta_s}\right| \tag{17.36}$$

and we can put this value in the above equation:

$$\frac{1}{2\sin^2\frac{\theta_s}{2}} = \frac{2E}{Qq}\left|\sigma\frac{\sin\theta_s}{b}\right|$$

$$\frac{Qq}{4E\sin^2\frac{\theta_s}{2}}\frac{b}{\sin\theta_s} = \sigma.$$

Using trigonometric identity $\sin\theta = 2\sin\frac{\theta}{2}\cos\frac{\theta}{2}$, and using equation (17.35), $\cot\frac{\theta_s}{2} = \frac{mbv_o^2}{Qq}$, we can rewrite the above equation as

$$\frac{Qq}{8E\sin^3\frac{\theta_s}{2}\cos\frac{\theta_s}{2}}\left(\frac{Qq}{mv_o^2}\right)\cos\frac{\theta_s}{2}\sin\frac{\theta_s}{2} = \sigma$$

$$\frac{Q^2q^2}{16E^2\sin^4\frac{\theta_s}{2}} = \sigma.$$

This is the famous Rutherford scattering formula and it shows that the differential cross-section is proportional to the fourth power of sine of $\frac{\theta_s}{2}$:

$$\boxed{\sigma \propto \frac{1}{\left(\sin\frac{\theta_s}{2}\right)^4}}. \tag{17.37}$$

This was verified in the first part of the twentieth century and was considered an early milestone of nuclear physics.

Example 17.1

A beam of 5 MeV alpha particles is incident on a gold nucleus. Most of the particles pass through unchanged but two are deflected, one at an angle of 90° and another at an angle of 45°. Calculate the value of the impact parameter for both particles.

Hint: alpha particles have a charge $2e$ and a gold particle has charge $79e$.

Solution

Using equation (17.35) for each case, given the scatter angle (θ_s) and knowing $1\ \text{eV} = 1.6 \times 10^{-19}$ J, we obtain

$$\cot\frac{\theta_s}{2} = \frac{2bE}{Qq}$$

$$b = \frac{Qq}{2E}\cot\frac{\theta_s}{2}.$$

For a scattering angle of 90°: $b = 2.5 \times 10^{-18}$ m.
For scattering angle of 45°: $b = 6.1 \times 10^{-18}$ m.

Concept questions 17.1

1. Statistically, we see that passengers in a heavier full-sized car are more likely to survive a car crash than occupants of a lighter compact car. Do you expect this to be true? Please explain in your own words.
2. If you throw a basketball at an approaching car, the ball will bounce back at you with an increased speed. Explain in your own words.
3. Why did the observation of large scattering angles in the alpha particle scattering experiment seem so surprising at the time?

Exercises 17.2

1. An alpha particle of incident energy 7.7 MeV is incident on a gold nucleus. For what value of impact parameter will we obtain a deflection of 30°?

2. Compare the probabilities for deflection by 1° with the probability for deflection by angles of 30° for the alpha particle given in question 1 incident on a gold atom.

Further reading

Arya A 1990 *Introduction to Classical Mechanics* 2nd edn (Upper Saddle River, NJ: Prentice-Hall)
Baierlein R 1983 *Newtonian Dynamics* (New York: McGraw-Hill)
Fowles G 1986 *Analytical Mechanics* 4th edn (New York: CBS College Publishing)
Fowles G and Cassiday G 2005 *Analytical Mechanics* 7th edn (Belmont, CA: Thomson Learning)
Goldstein H 1980 *Classical Mechanics* 2nd edn (Reading, MA: Addison-Wesley)
Landau L D and Lifshitz E M 1960 *Mechanics* (Reading, MA: Addison-Wesley)
Ohanian H 1985 *Physics* (New York: WW Norton)
Resnick R, Halliday D and Krane K 2002 *Physics* vol 1, 5th edn (New York: Wiley)
Stephenson R 1961 *Mechanics and Properties of Matter* 2nd edn (New York: Wiley)
Symon K 1971 *Mechanics* 3rd edn (Reading, MA: Addison-Wesley)

Appendix A

Unit conversion

Length

1 mile = 5280 feet = 1.61 km
1 inch = 25.4 mm = 2.54 cm
1 foot = 12 inches = 30.48 cm
1 meter = 1.094 yd =
1 yd = 3 feet = 36 inches

Mass

1 pound = 16 ounces
1 gallon = 4 quarts
1 quart = 2 pints
1 pound = 454 grams
1 liter = 1.06 quarts
1 ton = 908 kg

Time

1 year = 12 months = 365 days = 3.15×10^7 s
1 day = 24 h = 1440 min = 86 400 s

Capacity

1 fl oz = 29.57 ml
1 quart (qt.) = 2 pints (pt.) = 0.946 l
1 pt = 0.473 l
1 gallon (gal) = 4 qt = 3.785 l

doi:10.1088/2053-2563/aae1b9ch18

Appendix B

Velocity and acceleration in various coordinates

B.1. Velocity and acceleration in polar coordinates

In polar coordinates, the position of a particle is found by knowing the radial distance (r) from a user defined origin (O) and the angle (ϕ) (figure B.1).

The position vector of a particle

The trajectory of a particle is found by evaluating r and ϕ as a function of t. The position vector of a particle is mathematically given as

$$\vec{r} = |\vec{r}|\hat{r}. \tag{B.1}$$

Velocity and acceleration of a particle

To find velocity and acceleration, let us begin with unit vectors $\hat{r}$ and $\hat{\phi}$ in terms of their Cartesian components along $\hat{i}$ and $\hat{j}$. The unit vectors expressed in Cartesian coordinates are

$$\begin{cases} \hat{r} &= \cos\phi\hat{i} + \sin\phi\hat{j} \\ \hat{\phi} &= -\sin\phi\hat{i} + \cos\phi\hat{j}\,. \end{cases} \tag{B.2}$$

Differentiating we obtain

$$\begin{cases} \dfrac{d\hat{r}}{dr} &= 0 \\ \dfrac{d\hat{r}}{d\phi} &= -\sin\phi\hat{i} + \cos\phi\hat{j} \equiv \hat{\phi} \end{cases} \tag{B.3}$$

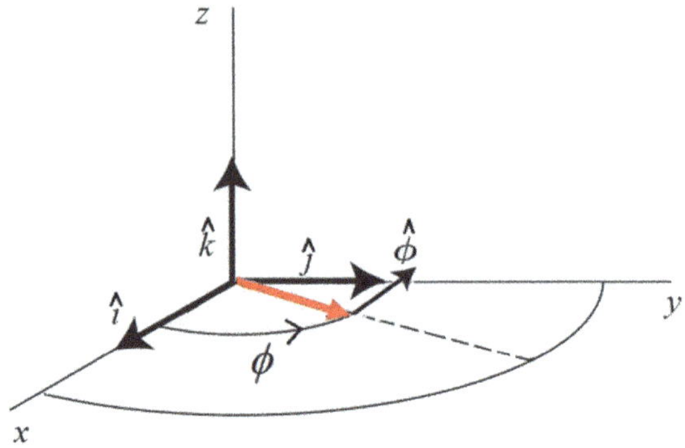

Figure B.1. Polar coordinates.

$$\begin{cases} \dfrac{d\hat{\phi}}{dr} &= 0 \\ \dfrac{d\hat{\phi}}{d\phi} &= -\cos\phi\hat{i} - \sin\phi\hat{j} \equiv -\hat{r}. \end{cases} \tag{B.4}$$

Velocity vector
We can now derive equation (B.1) with respect to time and write

$$\vec{v} = \dot{r}\hat{r} + r\dot{\hat{r}} = \dot{r}\hat{r} + r\dot{\phi}\hat{\phi}. \tag{B.5}$$

Here, $\dot{r} \equiv v_r$ is called the *radial* velocity component and $r\dot{\phi} \equiv v_\phi$ is called the *transverse* or the *circumferential* velocity component. Note that, $\sqrt{v_r^2 + v_\phi^2} \equiv v$.

Acceleration vector
Differentiating equation (B.5) with respect to time, we obtain the acceleration as

$$\vec{a} = \ddot{r}\hat{r} + \dot{r}\dot{\hat{r}} + \dot{r}\dot{\phi}\hat{\phi} + r\ddot{\phi}\hat{\phi} + r\dot{\phi}\dot{\hat{\phi}}. \tag{B.6}$$

In this equation we can rewrite the second term as

$$\dot{r}\dot{\hat{r}} = \dot{r}\frac{d\hat{r}}{dt}\frac{d\phi}{d\phi} = \dot{r}\frac{d\hat{r}}{d\phi}\frac{d\phi}{dt} = \dot{r}\hat{\phi}\dot{\phi}$$

and the last term can be rewritten as

$$r\dot{\phi}\dot{\hat{\phi}} = r\dot{\phi}\frac{d\hat{\phi}}{dt}\frac{d\phi}{d\phi} = r\dot{\phi}\frac{d\hat{\phi}}{d\phi}\frac{d\phi}{dt} = r\dot{\phi}(-\hat{r})\dot{\phi} = -r\dot{\phi}^2\hat{r}.$$

Putting these values in equation (B.6) and rearranging we obtain

$$\vec{a} = (\ddot{r} - r\dot{\phi}^2)\hat{r} + (r\ddot{\phi} + 2\dot{r}\dot{\phi})\hat{\phi}, \tag{B.7}$$

where $a_r \equiv (\ddot{r} - r\dot{\phi}^2)$ is called the *radial* acceleration component, and $a_\phi = (r\ddot{\phi} + 2\dot{r}\dot{\phi})$ is called the *transverse* or the *circumferential* acceleration component. Please note that $\sqrt{a_r^2 + a_\phi^2} \equiv a$.

B.2. Velocity and acceleration in cylindrical coordinates

In the case of 3D motion the position vector of a particle in cylindrical coordinators is given as (figure B.2):

$$\vec{r} = r\hat{r} + z\hat{z}. \tag{B.8}$$

We know from equation (B.2) the unit vectors $\hat{r}$ and $\hat{\phi}$, which are the same for cylindrical coordinates. Additionally we will have $\hat{z} = \hat{k}$ as the third coordinate. Their derivatives will be given as

$$\begin{cases} \dot{\hat{r}} &= \dot{\phi}\hat{\phi} \\ \dot{\hat{\phi}} &= -\dot{\phi}\hat{r} \\ \dot{\hat{k}} &= 0. \end{cases} \tag{B.9}$$

The kinematic vectors can now be expressed relative to the unit vectors. Thus, the velocity vector will be given as

$$\vec{v} = \dot{r}\hat{r} + r\dot{\phi}\hat{\phi} + \dot{z}\hat{k}. \tag{B.10}$$

Taking the derivative w.r.t. time again, we obtain acceleration vector as

$$\vec{a} = (\ddot{r} - r\dot{\phi}^2)\hat{r} + (r\ddot{\phi} + 2\dot{r}\dot{\phi})\hat{\phi} + \ddot{z}\hat{k}. \tag{B.11}$$

Note the similarities between equations (B.5) and (B.10) for the velocity vectors and between equations (B.7) and (B.11) for acceleration vectors.

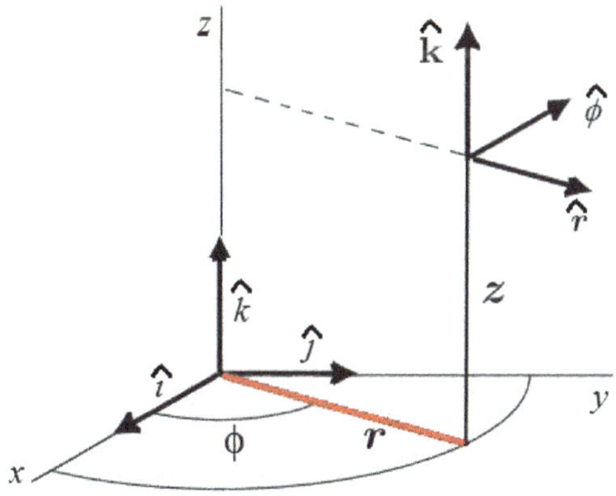

Figure B.2. Cylindrical coordinates.

B.3. Velocity and acceleration in spherical coordinates

In spherical coordinates, we use a distance (r) and two angles (θ, ϕ) to specify the position of a particle as shown in figure B.3.

The unit vectors written in Cartesian coordinates are

$$\begin{cases} \hat{r} &= \sin\theta\cos\phi\hat{i} + \sin\theta\sin\phi\hat{j} + \cos\theta\hat{k} \\ \hat{\theta} &= \cos\theta\cos\phi\hat{i} + \cos\theta\sin\phi\hat{j} - \sin\theta\hat{k} \\ \hat{\phi} &= -\sin\phi\hat{i} + \cos\phi\hat{j}\,. \end{cases} \tag{B.12}$$

It can easily be seen that the derivatives of these unit vectors may be simplified using equation (B.12) as

$$\begin{cases} \dfrac{d\hat{r}}{dt} &= \dot{\phi}\sin\theta\hat{\phi} + \dot{\theta}\hat{\theta} \\ \dfrac{d\hat{\theta}}{dt} &= -\dot{\theta}\hat{r} + \dot{\phi}\cos\theta\hat{\phi} \\ \dfrac{d\hat{\phi}}{dt} &= -\dot{\phi}\sin\theta\hat{r} - \dot{\phi}\cos\theta\hat{\phi}. \end{cases} \tag{B.13}$$

The kinematic vectors can now be expressed relative to the unit vectors. Thus, the position vector will be given as

$$\vec{r} = r\hat{r}. \tag{B.14}$$

The velocity vector will be given as

$$\vec{v} = \dot{r}\hat{r} + r\dot{\phi}\sin\theta\hat{\phi} + r\dot{\theta}\hat{\theta}. \tag{B.15}$$

Taking derivative w.r.t. time again we obtain the acceleration vector as

$$\begin{aligned} \vec{a} = &(\ddot{r} - r\dot{\phi}^2\sin^2\theta - r\dot{\theta}^2)\hat{r} + (r\ddot{\theta} + 2\dot{r}\dot{\theta} - r\dot{\phi}^2\sin\theta\cos\theta)\hat{\theta} \\ &+ (r\ddot{\phi}\sin\theta + 2\dot{r}\dot{\phi}\sin\theta + 2r\dot{\theta}\dot{\phi}\cos\theta)\hat{\phi}. \end{aligned} \tag{B.16}$$

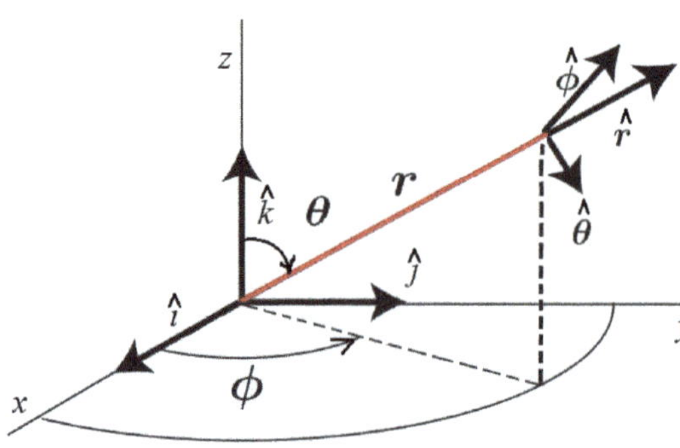

Figure B.3. Spherical coordinates.

Appendix C

Noether's theorem

Noether's theorem was presented in 1918 CE by Emmy Noether and is one of the most beautiful theorems in physics. Noether's theorem is very useful in providing insights into general theories in physics. Noether's theorem allows us to relate symmetries of a theory with its laws of conservation. For example:

- If a system is invariant (does not change) under spatial translation we obtain the law of conservation of linear momentum.
- If a system is invariant under rotation we obtain the law of conservation of angular momentum.
- If a system does not change over time we obtain the law of conservation of energy.

Why do we even expect conservation laws to be related to symmetry?
At the outset, it is not at all clear what the relation between conservation laws and symmetries is. We resort to a simple example. Consider Newton's second law: the rate of change of linear momentum equals the net external force acting on a body, i.e.

$$\vec{F} = \frac{dp}{dt}. \tag{C.1}$$

Suppose that the force field $\vec{F}$ is derived from an external potential, i.e. $\vec{F} = -\Delta V$, and the problem in 1D is then

$$\frac{dp}{dt} = -\frac{\partial V}{\partial x} = \lim_{\varepsilon \to 0} \frac{V(x+\varepsilon) - U(x)}{\varepsilon}. \tag{C.2}$$

Note that when potential (V) is a constant then $V(x+\varepsilon) = V(x)$. This means that $\vec{F} = 0$ and hence from Newton's law, we can write $\frac{dp}{dt} = 0$ with linear momentum $(p) =$ constant.

doi:10.1088/2053-2563/aae1b9ch20

Next, when we say $V(x) = V(y)$ it is the same as saying that potential V is invariant under translations $x \to x'$, i.e. that translation in space is a symmetry of the function V. Hence we have found that the translational symmetry of V implies conservation of linear momentum.

Noether's theorem establishes such relationships between symmetries of the action and quantities conserved along the trajectory.

IOP Publishing

Techniques of Classical Mechanics
From Lagrangian to Newtonian mechanics
Samya Zain

Appendix D

Configuration space

Configuration space is also called *C-space*. In classical mechanics, the parameters that define the configuration of a system are called generalized coordinates and the vector space defined by these coordinates is called the *configuration space* of the physical system. Configuration space is a way to visualize the state of an entire system as a single point in a higher dimensional space:

- A single particle's position in a 3D Euclidean 3-space is

$$\vec{r} = (x, y, z), \tag{D.1}$$

  then its C-space = R^3.

  If a particle is constrained in some way, for example, if it can only move along a wire, then the C-space is the subset of R^3 that defines point on the wire, i.e. S^1.

  If the particle can only move in 2D, for example when we play air-hockey, the puck can only move on the table-top. Then the C-space is given as S^2.

  Similarly if the particle moves on a sphere, the C-space is also given as S^2.
- For a system made up of n-particles, the C-space will be given by R^{3n}.

www.ingramcontent.com/pod-product-compliance
Ingram Content Group UK Ltd.
Pitfield, Milton Keynes, MK11 3LW, UK
UKHW051906100926
8556IPUK00005B/786

9 780750 32074C